AMONG OUR BOOKS
IN THE ENGLISH LANGUAGE

- Advances in Organic Geochemistry, 1973.
- Drilling Data Handbook.
- Energy Conversion in Refining and Petrochemistry.
- International Symposium. Geodynamics in South-West Pacific. Noumea (New-Caledonia). August 27 to September 2, 1976.
- International Symposium. Structural History of the Mediterranean Basins. Split (Yugoslavia). October 25-29, 1976.
- Jamin B.- Efficiency versus Number of Transfer Units for Pure Counter-Flow and (1-n) Exchangers.
- Madrelle A.- Strength Calculations for Oil Country Tubular Goods.
- Mercier C.- Petrochemical Industry and the Possibilities of its Establishment in the Developing Countries.

Production Course
- Enhanced Recovery, by M. Latil.
- Properties of Reservoir Rocks. Core Analysis, by R. Monicard.
- Recent Technological and Economical Developments in the Petrochemical Industry.
- Reyre D. and Mohafez S.- A first Contribution of the NIOC-ERAP Agreements to the Knowledge of Iranian Geology.
- Safety Recommendations for Marine Operations.
- Safety Recommendations. Operating Land Oil Well Drilling Rigs.

Schilling A.-
- Tome 1. Motors Oil and Engine.
- Tome 2. Automobile Engine Lubrication.
- Seismic Filtering.
- Symposium : Sedimentology and the Oil Industry. Fifth World Petroleum Congress, New York, 1959.
- Valais M. and Hiegel H.- The Gas Industry in the World.
- Viscosity and Density of Light Paraffins, Nitrogen and Carbone Dioxide.
- Zambrano E., Vasquez E., Duval B., Latreille M. and Coffinieres B.- Paloegeographic and Petroleum Synthesis of Western Venezuela.

INSTITUT FRANÇAIS DU PÉTROLE PUBLICATIONS

PIERRE LE TIRANT

Head Marine Geotechnics Department
Institut Français du Pétrole

Professor of Marine Geotechnics
École Nationale Supérieure
des Techniques Avancées

Seabed Reconnaissance and Offshore Soil Mechanics

for the Installation of Petroleum Structures

Preface by

JACQUES DELACOUR

Director of the Division
of Drilling and Production Techniques
of the Institut Français du Pétrole

English Translation by
JOHN CHILTON-WARD

1979

ÉDITIONS TECHNIP 27, RUE GINOUX 75737 PARIS CEDEX 15 technip

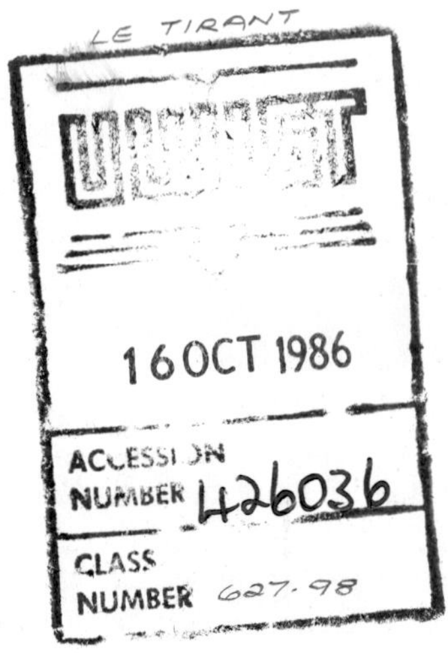

ISBN 2-7108-0333-X

Printed in France
by Imprimerie Louis-Jean, 05002 Gap

PREFACE

When we consider the dimensions attained by the most recent structures implanted on the continental shelf for the needs of offshore petroleum drilling and production, we cannot help being struck by the daring of the builders who thus openly defy the bad old reputation given to those who "build on sand".

To comprehend the scope of this challenge, we must also understand that these constructions exposed to the onslaught of the elements – and, in the case of the North Sea, its violence is well known – in water depths that are constantly increasing – reaching 300 m for the most recent fields – are designed to be a completely reliable support, during at least a quarter of a century, for equipment whose cost and risk density is unequalled in any other industrial complex existing in the world except perhaps nuclear power plants which, however, are located on dry land.

Of course, the extraordinary advances stemming from the development of computers and mathematical models have resulted in the implementation of computing methods making it possible to consider all cases of static and dynamic loads for such structures when they are subjected to the action of the elements in situations which can also be simulated almost without limits by mathematical modeling.

Therefore, we can be confident in the actual mechanical design of such structures, whether made of steel or concrete, according to configurations that are as diverse as they are original.

At the same time, it is no less true that they are sitting on the seabed where they must find a permanent, stable and reliable foundation so as to guarantee their longevity.

We can thus comprehend the absolutely major importance of obtaining as exact an understanding as possible of the properties of the subsea soils on which such gigantic structures are to be implanted with their mass and size roughly convering a foundation volume having a diameter of approximately 150 m.

The present book had to be written first of all to demonstrate that this essential phase of offshore exploitation had not been neglected as compared with the designing of spectacular structures.

Indeed, over the last ten years a great many reconnaissance methods – either indirect ones using geophysical processes or direct ones using mechanical coring and *in situ* measurements – have gradually been developed and improved so as to provide builders with the required data both for choosing the type of structure as a function of the soil properties and then of designing its foundations.

At the same time, for engineers in charge of exploiting subsea hydrocarbon fields, this book is an essential guide so that they can assume the role of an industrial architect by choosing the structures that will make up the production complex, i.e. platforms on pilings, gravity structures, anchored floating structures, pipelines, etc.

With the help of specialists from French oil and service companies, the author has done a fine job of presenting the current state of the means available and the state of the art in the field under consideration.

In so doing, he has also marked out the boundaries of current knowledge and consequently has opened up the path toward future developments which have already reached the point where speedy preparations have been engaged for producing hydrocarbons from well beyond the continental shelf in deep offshore areas where almost nothing is now known about the subsea soils.

JACQUES DELACOUR
Director of the Division
of Drilling and Production Techniques
of the Institut Français du Pétrole

FOREWORD

This book was first published in French in 1976 and has now been updated and translated into English to include recent developments in offshore technology. It is the result of several years of collective work carried out at the request of the leading French oil companies so as to provide practicing engineers with indispensable information on the reconnaissance of subsea soils and on making use of such results for installing petroleum structures. Therefore, it is not just another treatise on soil mechanics but rather an attempt to synthesize current techniques and methods used in offshore geotechnics.

• Part One begins by reviewing the elementary definitions and concepts which are indispensable in soil mechanics, and then it looks at aspects of seismic reflection which are useful for understanding geophysical reconnaissance methods used for determining seafloors and subsea soils.

• Part Two gives the principle, description, implementation, performances and application limits of the leading soil reconnaissance techniques capable of being used on the continental shelf and beyond. These techniques are classified in four groups depending on the goals aimed at :

– reconnaissance of the bathymetry and morphology of seabeds by geophysical and visual methods ;
– reconnaissance of superficial layers by means of high-resolution seismic methods ;
– geotechnical reconnaissance of soils by different coring methods for the purpose of identifying and determining mechanical properties ;
– geotechnical reconnaissance of soils by *in situ* measurements and logging.

For geophysical methods and *in situ* measurements, an effort is made to give the indications required for analyzing and interpreting the recordings obtained as a function of different types of equipment and oceanographic conditions. For coring, the text emphasizes the representativity of samples taken depending on the conditions governing the implementation of equipment and techniques and the nature of the formations encountered.

The problem of offshore positioning is not taken up in this book.

• Part Three describes the different geotechnical problems involved in the offshore installation of the principal structures used for petroleum exploration and production :

– steel platforms (jackets) attached to the seabed by pilings ;
– gravity structures, in particular concrete ones, i.e. platforms and/or storage tanks ;
– jack-up platforms ;
– anchorages for floating structures (using anchors, deadweights, pilings) ;
– pipelines.

The problems taken up for the different types of structures include :

– their installation, stability and, eventually, their removal ;
– reconnaissance of seabeds and soils prior to their installations ;
– how to calculate their stability under the effect of fixed loads and cyclic, horizontal or vertical stresses.

Forces resulting from the action of the elements are assumed to be known, i.e. wind, waves, currents, tides, etc.

Despite the careful organization of this volume, the dual desire to avoid too frequent cross references and to make it readily consultable by engineers on the job has resulted in various necessary repetitions, in particular concerning the description of the conditions governing the implementation of equipment and techniques for geotechnical reconnaissance.

The units used in the book have been deliberately chosen with a regard toward practical applications, while at the same time conforming to the recommendations of the International System of Units (SI). For example, stresses in the soil are expressed in t/m^2, and fluid pressures in pipes are given in kg/cm^2 with the mention in parentheses of the values in kPa. When Anglo-American units are habitually used as a matter of practice, their values are also given in metric units in parentheses.

A detailled Table of Contents at the beginning of each chapter will enable readers to quickly find the information they are seeking.

This book is also the textbook used as a basis for :

– the Marine Geotechnics Course given as part of the oceanology major at the *École Nationale Supérieure des Techniques Avancées (ENSTA)* in Paris ;
– talks and papers on the reconnaissance of offshore soils presented at various seminars (in particular the *École Nationale Supérieure du Pétrole et des Moteurs*).

I would like to thank the Management of the Division of Drilling and Production Techniques of the *Institut Français du Pétrole* which has greatly encouraged the writing of this book and then its translation into English.

But the preparation of this book could not have been carried out successfully without the collaboration of a great many scientists in the IFP Group and in French companies. In particular, I would like to thank :

– MM. BECUE, CASTEL, MONTARGES, NAUROY, PUECH and DES VALLIERES of the Division of Drilling and Production Techniques of the *Institut Français du Pétrole*, for their collaboration in the preparation of some chapters, the updating of the English edition and the proofreading of the manuscript and proofs ;

– MM. Cassand and Dubois of the Geophysics Division of the *Institut Français du Pétrole* and MM. Amar, Kuhn and Lapierre of the *Bureau d'Etudes Industrielles et de Coopération de l'Institut Français du Pétrole (BEICIP)* for their contribution to the chapters on geophysical reconnaissance methods for the seabed and subsea soils ;
– MM. Cochard, Gabaix-Hialle, Mizikos and Sagot of *Société Nationale Elf-Aquitaine, Production (SNEA(P))* and MM. Bares, Geoffrois, Kervadec, Kyvellos and Parrot of *Compagnie Française des Pétroles (CFP),* thanks to whom fruitful exchanges were developed between the *Institut Français du Pétrole* and French oil companies in the field of offshore geotechnics.

I would also like to thank MM. Courtois and Hai for producing most of the drawings which illustrate this book.

Pierre Le Tirant
July 1978

CONTENTS

(A detailed Summary has been inserted at the beginning of each Chapter)

1 // Preliminary Concepts

CHAPTER I

Concepts of Soil Mechanics

CONTENTS

INTRODUCTION

The subject of this book is reconnaissance of soils and siting of structures at sea. It is not a treatise on soil mechanics, for which the reader can refer to the numerous works on soil mechanics at present available [1] [2] [3] [4].

In this preliminary chapter, it nonetheless appears necessary to recall certain concepts and definitions relevant particularly to the identification, shear strength and consolidation of soils (1) [5].

1. Owing to its granular structure, soil is a material which is discontinuous on a microscopic scale (i.e. that of the grain). However, from the engineer's standpoint, the concept of continuity generally applies on the macroscopic scale.

The porous volume between the grains of soils is filled with a fluid, namely water and/or gas. Because of this, the distortion of the skeleton is linked to the appearance of a variation of pressure in the fluid and hence to hydrodynamic movements.

Under the effect of external loads, stresses between the grains and non-linear deformations occur. Increased stresses result in slip between grains, thus destroying the initial structure of the soil: the limit strength is attained.

2. Marine soils are not significantly different from terrestrial soils. However, some particularities characterize them, namely:

– the diagenesis of the soil is often less ancient at sea than on land. Nonetheless, many exceptions to this observation are encountered, depending notably on local geology (for example the north part of the North Sea),

– the proportions of the various elements constituting the soils are sometimes fairly different: frequent calcareous sand at sea, a clay-to-sand ratio which is often very high, particularly in great depths of water, etc.,

– the total saturation of the marine soils generally simplifies problems of soil mechanics. However, owing to the decompression of core samples, a considerable quantity of dissolved gases (particularly in deltaic zones) is sometimes released,

– erosion phenomena also bring about differences which are often considerable in the properties of surface soils.

---

(1) These concepts are very clearly summarized in the excellent work entitled "Précis de géotechnique" by P. Habib [5].

3. Marine soil mechanics therefore do not essentially differ from terrestrial soil mechanics. However, the conditions of environment give rise to many difficulties that should be borne in mind when siting marine structures:

– marine soil reconnaissance techniques generally only differ from terrestrial techniques (in situ measurements and core drilling) in their method of application. However, owing to the difficulties encountered in carrying out these techniques and the high cost of the operations involved, reconnaissance of marine soils is invariably less comprehensive than that for terrestrial soils and can often even be inadequate,

– the frequent uncertain knowledge of the mechanical properties of soils, added to the often very restricted experience on some sites, often leads to overdimensioning the foundations of the structures, in conjunction with safety factors which are well above those applied ashore,

– the considerable dimensions and conditions of installation of the foundations for marine structures are often uncommon on land: insertion of piles down to about 100 m, gravity bases with a diameter of 100 m and over, laid on a more or less irregular bottom, etc.,

– the lateral forces to which marine structures are subjected are generally very considerable and in most cases completely incomparable with those to which terrestrial structures are subjected.

4. The laboratory analysis of soils sampled by various techniques (percussion, rotation, vibration, pressure, etc.) has three objectives which are essential to the knowledge of the properties of soils:

– identification of the soils,

– measurement of the shear strength of the various strata, in order to predict the structure under static and dynamic conditions,

– determination of the deformation characteristics of the soils, in order to estimate the foreseeable settlements beneath the structure.

The results of laboratory tests made on land are open to much criticism owing to:

– the poor conservation of core samples,

– the considerable handling resulting from transport, etc.,

to such an extent that most of the identification tests and simple measurements of undrained shear strength are now more and more made at once on the reconnaissance vessel.

I.1. IDENTIFICATION OF SOILS

Identification of a soil consists in characterizing the soil with enough precision to enable it to be compared to other soils. All perfectly standardized identification tests are of purely conventional nature: the scrupulous compliance with the operating modes enables relatively comparable results to be obtained.

Here, we shall confine ourselves to the principle of the tests and common measurements without entering into the operating details, which are covered by specialized publications [6].

I.1.1 Water content and densities

The water content and wet density of the soil are determined on the vessel, as soon as the sample has been taken.

I.1.1.1 Water content

By definition, the water content $w\%$ is the ratio of the weight of water to the weight of the dry soil:

$$w\% = \frac{P_w}{P_{\text{dry soil}}}$$

where:

P_w is by convention the loss of weight of the sample after remaining in drying oven at a temperature of 105° C for 24 h.

The water contents of the soils vary considerably with their consolidation. For instance:

- normally consolidated clays: $w \approx 30$ to 60%,
- very rigid clays: $w \approx 10$ to 15%,
- overconsolidated clays in the North Sea: $w \approx 20$ -- 25%,
- soft, highly colloidal clays: $w \approx 200\%$,
- sands: $w \approx 15$ to 35%.

I.1.1.2 Densities, porosity, void ratio and relative density

In soil mechanics, the concepts of "density" and "weight per unit volume" are considered in practice as the same. This (inaccurate) definition which is today universally accepted will therefore be used.

The various densities considered in soil mechanics are the following:

- wet density γ: weight per unit volume of soil, including its water.
- dry density γ_d : weight per unit volume of soil after remaining 24 h in an exsiccator at 105° C,
- submerged density γ': weight per unit volume of submerged soil,
- specific weight G: weight per unit volume of solid grain.

The **porosity** n is the ratio of the volume of pore spaces to the total volume.
The **void ratio** e is the ratio of the volume of pore spaces to the volume of solids. It follows that:

$$n = \frac{e}{1 + e}$$

The degree of saturation $s\%$ of a soil (i.e. the ratio of the volume occupied by the water to the total volume of pore spaces) is of no interest for marine soils, where $s = 100\ \%$ (saturation water content w_s).

On the basis of the various definitions above, one can immediately deduce the following simple equations, assuming the density of the water γ_w to be 1 (the density of sea water is in fact 1.02 to 1.03):

$$\gamma = \gamma_d (1 + w)$$

$$n = 1 - \frac{\gamma_d}{G}$$

$$w_s = \frac{1}{\gamma_d} - \frac{1}{G} \qquad \text{or} \qquad \begin{cases} \gamma_d = \dfrac{1}{w_s + \dfrac{1}{G}} \\[2ex] \gamma = \dfrac{1 + w_s}{w_s + \dfrac{1}{G}} \end{cases}$$

$$\gamma = (1 - n) G + n$$

$$\gamma_d = (1 - n) G$$

$$\gamma' = \gamma - 1$$

For most common minerals in soil, G can be assumed to be approximately 2.7.

The determination of the water content of marine soils (invariably saturated) directly leads to a very good estimate of the wet and dry density (γ and γ_d).

Table I.1.1 gives some values of w, γ, γ_d for common water contents of marine soils.

TABLE I.1.1

Water content $w\%$	**Wet density** γ **(t/m³)**	**Dry density** γ_d **(t/m³)**	**Remarks**
15	2.21	1.92	Overconsolidated clays of the North Sea
20	2.10	1.75	
30	1.94	1.49	
40	1.82	1.30	
50	1.72	1.15	Soft clays: deltaic zones, abyssal plain of the Sea of Iroise
75	1.56	0.89	
100	1.46	0.73	

In practice, the most frequent values are the following:

$\gamma \doteq$ from 1.7 to 2.2

$\gamma_d = \begin{cases} \text{clays: } 1.3 \text{ to } 1.7 \\ \text{sands: } 1.6 \text{ to } 1.7 \end{cases}$

$\gamma' =$ from 0.7 to 1.2 (γ' is often assumed to be approximately 1 for soils of average consolidation)

In the case of sands, the concept of **relative density** D_r is frequently used (for the purposes of comparison):

$$D_r = \frac{e_m - e}{e_m - e_M} = \frac{\gamma_{dM}}{\gamma_d} \cdot \frac{\gamma_d - \gamma_{dm}}{\gamma_{dM} - \gamma_{dm}}$$

where indices m and M correspond respectively to minimum and maximum compacting.

The difficulty lies in determining γ_{dm} and γ_{dM}, which are highly sensitive to the operating mode.

This concept nonetheless enables sands to be classified as very loose, loose, average, compact and highly compact.

For example, the relative density of sands in place varies from 20 or 30% up to 100% and over (for highly dense North Sea sands).

I.1.2 Size distribution and Atterberg limits

The size distribution and Atterberg limits are determined in the laboratory (ashore).

I.1.2.1 Determination of size distribution

This is carried out:

– by screening for grains above 100 μ in size,
– by sedimentation in a liquid for the finer grains.

Both for screening and settling, the operating conditions are of vital importance in comparing different results and particularly:

– the method of separation of the grains,
– the method of maintaining separation (deflocculating agents, etc.), which completely govern the results.

For screening, the AFNOR standard has adopted screens defined by mesh openings of the Renard $\sqrt[10]{10}$ series.

For sedimentation, Stoke's law is applied, assuming the particles to be spherical.

By convention, the following designations are generally adopted (2):

– gravel = 20 to 2 mm,
– sand = 2 mm to 50 μ,
– silt = 50 to 5 μ,
– clay (3) = $< 5\ \mu$.

As an example, sands encountered in the North Sea at various sites near the surface typically display:

– a fairly uniform size distribution with an average value of $D_{50} \approx 100$ to 200 μ (Fig. I.1.1),
– a frequently high density such that $\gamma \approx 2$ to 2.1 and $D_r \approx 100\ \%$.

---

(2) The size distribution ranges may vary with the standards accepted.

(3) This definition is purely granulometric and makes no assumptions as to the mineralogical nature of the particles.

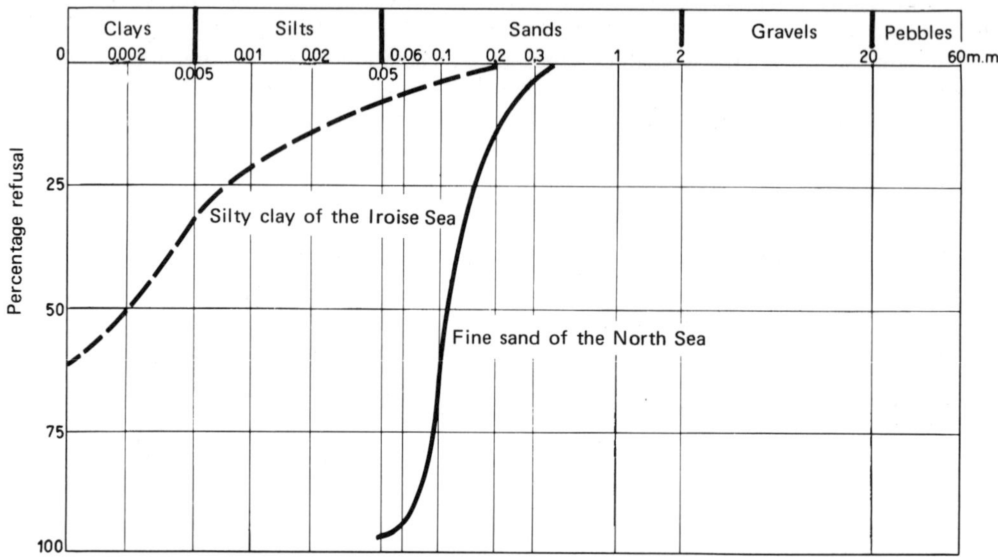

FIG. I.1.1. – Conventional grain size distribution limits.

I.1.2.2 Atterberg limits

The Atterberg limits are conventional water contents which characterize the state of the soil. They are determined on the fraction of soils < 0.4 mm.

By definition:

– **the limit of liquidity** W_L is the water content above which the soil behaves as a semi-liquid and flows under its own weight,

– the **limit of plasticity** W_P is the water content below which the soil loses its plasticity and becomes friable.

Conventional tests to determine the Atterberg limits using well defined operating modes are perfectly reproductible. The values of W_L and W_P obtained correspond to distinct changes in the mechanical behaviour of the soil.

The **index of plasticity** I_P is defined as:

$$I_P = W_L - W_P$$

And is greater, the more "clayey" the soil:

– if $I_P \approx 10$, the soil is only slightly clayey,
– if $I_P > 30$, the soil is highly clayey.

Overconsolidated clayey soils encountered on several sites in the North Sea have:

– an index of plasticity $I_P \approx 20$ to 30%,
– a water content of 20 to 25%, generally very close to the limit of plasticity W_P.

A highly useful classification of soils is rendered possible by means of the Casagrande relationship between the index of plasticity I_P and the liquidity limit W_L (Fig. I.1.2):

$$I_P = 0.73(W_L - 20)$$

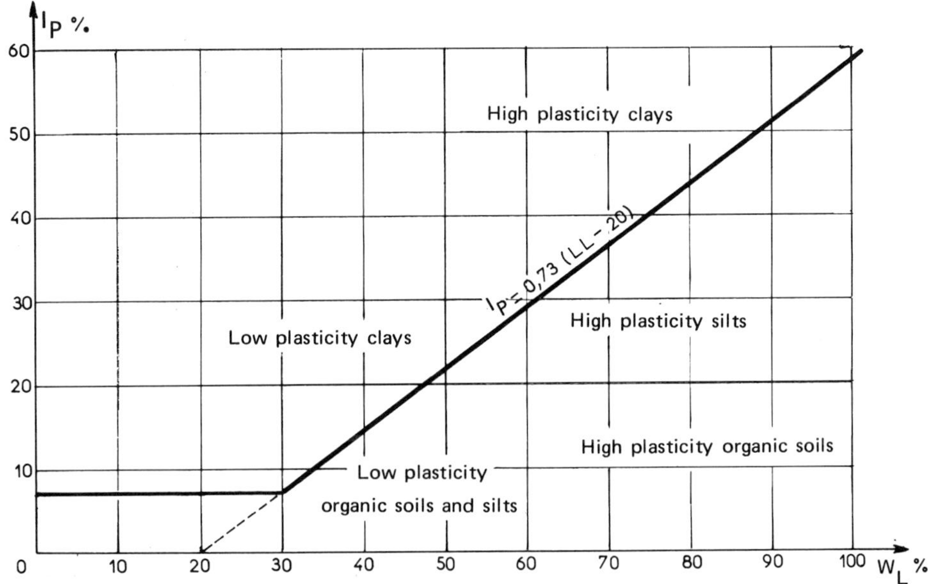

FIG. I.1.2. – Casagrande soil classification diagram.

I.1.3 Permeability of soils

I.1.3.1 Darcy's law

Darcy's law expressing the flow of water in soil is as follows:

$$V = Ki$$

where:

V = percolation velocity,
K = coefficient of permeability,
i = hydraulic gradient.

In this expression, K has the dimensions of velocity (and implicitly contains the viscosity of the water).

In the case of water, where the viscosity $\eta \approx 1$ cP, the equivalence between the permeabilities in cm/s and Darcys is expressed by the approximate relationship:

$$10^{-3} \text{ cm/s} \longleftrightarrow 1 \text{ Darcy}$$

I.1.3.2 Some values of the permeability of soils

The permeability of soils can only be validly determined by means of injections tests or in situ pumping.

For sands, an approximation can be given by using Hazen's formula:

$$K \text{ (cm/s)} = 100\, D_{10}^2 \text{ (cm)}$$

where D_{10} represents the diameter of the grain for which 10% of the elements in the soil are smaller in size.

The permeability varies considerably with the nature of the soil and, for the same type of soil, from one site to another. As an example, the following values for K can be given, expressed in centimetres per second:

– sands = 10 to 10^{-4} cm/s (the dense sands of the North Sea, with a relative density ≈ 100%, have a permeability $K \approx 10^{-3}$ cm/s),
– silts = 10^{-4} to 10^{-7} cm/s,
– clays = 10^{-7} to 10^{-10} cm/s.

I.2 THE SHEAR STRENGTH OF SOILS

If in a mass of soil the shear stress exceeds a certain value, plastic deformations appear and slip occurs : the soil breaks by shearing.

I.2.1 Shear strength and intrinsic curve of a soil

A distinction will be drawn between sands and clays.

I.2.1.1 Sands

On all facets situated along a line or a rupture surface of a sand, proportionality exists between the tangential stress τ and the normal stress σ: the shear stress is linked solely to the friction between grains. At the onset of slip, one can write:

$$\tau = \sigma \operatorname{tg} \varphi$$

where φ is the internal angle of friction of the sand (essentially a function of its compaction). This is Coulomb's law.

In the Mohr representation (σ, τ), the relationship $\tau = \sigma \operatorname{tg} \varphi$ expresses the location of the points where the limit equilibrium of the sand is reached: this is the **intrinsic curve.**

For sands, the following equation can be verified by experiments:

$$e \operatorname{tg} \varphi = \mathrm{C}$$

where $0.45 < C < 0.55$. In actual fact, C depends on the shape, nature, size distribution, etc., of the grains.

The critical density γ_c is the initial density of a sand corresponding to zero variation in volume during shear (γ_c is clearly a function of the normal stress σ).

When one approaches γ_c, the angle of friction of the sands is about 32°. In the case of highly dense sands ($D_r \approx 100\%$), the angle of friction attains 40 to 45°. In the case of loose sands ($\gamma < \gamma_c$) the angle of friction is below 30°.

I.2.1.2 Clayey soils

In clays and silts, the shear strength is related both to the contact actions (friction) and physical links between particles. The limit equilibrium equation is then expressed as:

$$\tau = c + \sigma \operatorname{tg} \varphi$$

where c designates the cohesion (or the shear strength with zero normal stress).

Since clay or silt is saturated, the normal stress σ is partially balanced by an interstitial pressure u. The above equation can then be written:

$$\tau = c + (\sigma - u) \operatorname{tg} \varphi$$

where $\sigma - u$ designates the normal stress to which the solid skeleton is actually subjected. This is Hvorslev's relationship.

In general it is accepted that:

$$\sigma' = \sigma - u$$

where σ' designates the effective normal stress.

Just as for sands, in the Mohr diagram (τ , σ), the relation $\tau(\sigma)$, being the locus of points corresponding to the limit equilibrium of soil, is known as the intrinsic curve.

Remark. Under dynamic stresses (repeated loading under wave action, pile driving, vibrations, etc.), both sands and clays may liquify, thus losing all shear strength.

The density γ of the soil then becomes (see Section I.1.1):

$$\gamma = \frac{(1 + w_l)\, G}{1 + G w_l}$$

where w_l is the water content on liquefaction.

I.2.2 Determination of the shear strength of soils

Two types of tests: direct shear and the triaxial test enable the shear strength of soils to be determined.

I.2.2.1 Direct shear test

The direct shear test is performed using the direct shearing device or shearing box (the most familiar being the Casagrande device) (see Fig. I.2.1a).

The sample of soil is enclosed between two semi-boxes capable of sliding with relation to each other.

The following are applied to the samples :

- a constant normal stress σ,
- a shear stress τ, increasing to rupture.

The locus of points representing pairs of values (σ , τ) on rupture constitutes the intrinsic curve (Fig. I.2.1a).

In practice, the direct shear device has several defects, the most important being:

- the variation in the surface of the plane of slip during the test,
- the non-homogeneous distribution of the shear stresses in the plane of slip,
- the variation in the water content.

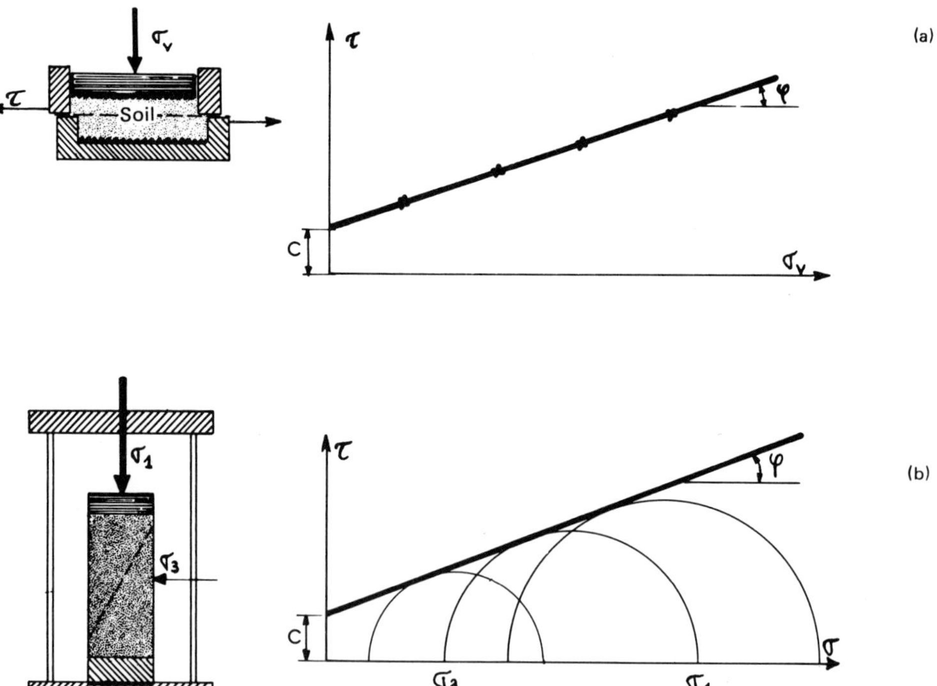

FIG. I.2.1. – Determination of the shear strength of soils:
a. Principle of direct shear test to determine the intrinsic curve.
b. Principle of triaxial test to determine the intrisic curve.

I.2.2.2 Triaxial test

In the triaxial test, the sample of soil enclosed in a sealed sheath (Fig. I.2.1b) and placed in a cell is subjected:

– laterally to a stress σ_3 $(= \sigma_2)$ by means of a fluid,
– vertically to a stress σ_1 by a piston.

While maintaining σ_3 constant, σ_1 is gradually increased to rupture. The "axial stress $\sigma_1 - \sigma_3$/axial deformation ε" curve is recorded. Figure I.2.2 represents a number of types of "stress/deformation" curves for sands and clays.

Rupture takes the form:

– either of a slip plane (in rigid soils),
– or excessive deformation (particularly in soft clays).

The test determines the limit equilibrium Mohr circle. The envelope on Mohr circles forms the intrinsic curve which is practically linear in the case of soils subjected to ordinary stresses of a few tens of t/m² (a few hundred of kPa) or more (Fig. I.2.1b).

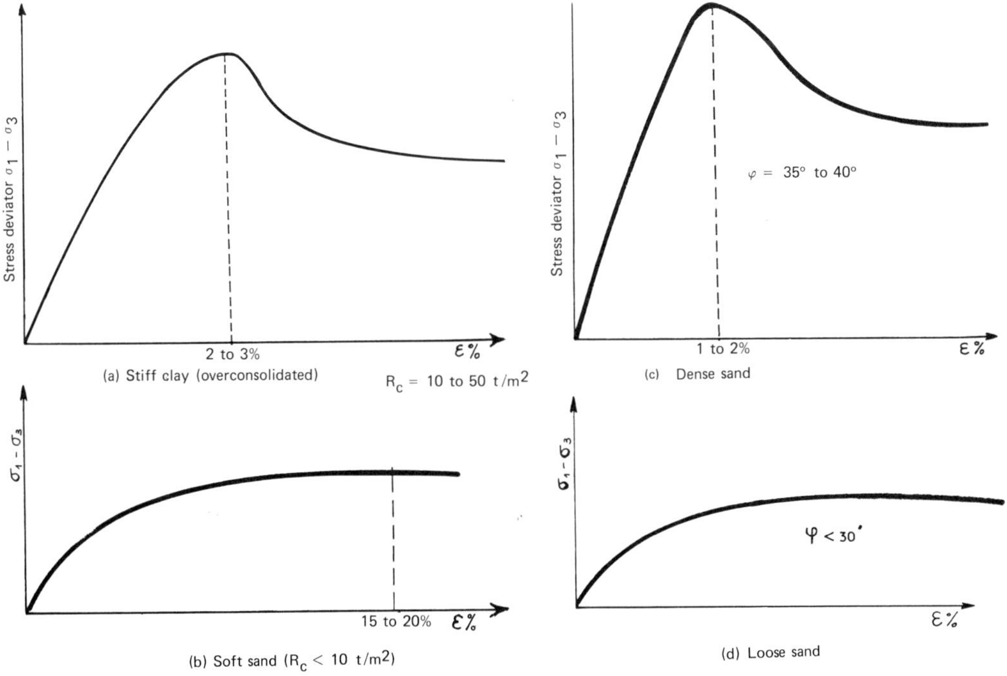

FIG. I.2.2. – "Stress-strain" curves of various types of soil.

In the case of **clayey soils**, depending on the speed with which the triaxial test is made, the interstitial pressure u induced by the increase in normal stresses may dissipate or not.

This being so, one must distinguish between (see Section I.2.3):

– the "unconsolidated – undrained" or "unconsolidated – fast" test; the interstitial pressure cannot be dissipated,

– the "consolidated – undrained" or "consolidated – fast" test; the interstitial pressure induced during shear cannot be dissipated,

– the "consolidated – drained" or "consolidated – slow" test with dissipation of the interstitial pressure.

In the case of **sands**, the loading rate is generally not so important as with clayey soils. However, if the sand is subjected to fast loading, for instance under the effect of waves (with a cycle of about 10 s), drainage does not occur with each loading cycle, and the interstitial pressure therefore does not dissipate.

The friction angle of sand under undrained conditions (fast test) corresponding to the maximum of the "stress – deformation" curve is less than the friction angle under drained conditions (slow test):

– in the undrained (fast) test, the friction is only partially brought to bear. Deformation remains very low,

– in the drained (slow) test, friction intervenes completely. Deformation attains 2 to 3%.

As an example, for dense sands ($D_r \approx 100\%$) in the North Sea, one observes:

$$\varphi_{\text{drained}} \approx 43^\circ$$
$$\varphi_{\text{undrained}} \approx 34^\circ$$

I.2.3 Undrained and effective cohesions of clays

Various tests enable the shear strength of soils to be determined in the laboratory.

I.2.3.1 Shear with neither consolidation nor draining

The rate of axial deformation commonly accepted varies from 0.5 to 2% per minute.

If tests are made starting from various states of stress $\sigma_1 = \sigma_3$, the envelope (4) of the Mohr rupture circles corresponding to the total stresses is a straight line parallel to the axis of the normal stresses, the equation being:

$$\tau = c_u$$

where:

c_u designates the **undrained cohesion** of the soil (Fig. I.2.3).

With no consolidation, all the Mohr circles express as effective stresses (σ_1', σ_3') are theoretically coincident (the increase in σ_3 equalling the increase in interstitial pressure).

---

(4) This envelope of Mohr circles, defined both by the intergranular stresses and the interstitial water pressure, cannot be called an "intrinsic curve", which characterizes a given solid.

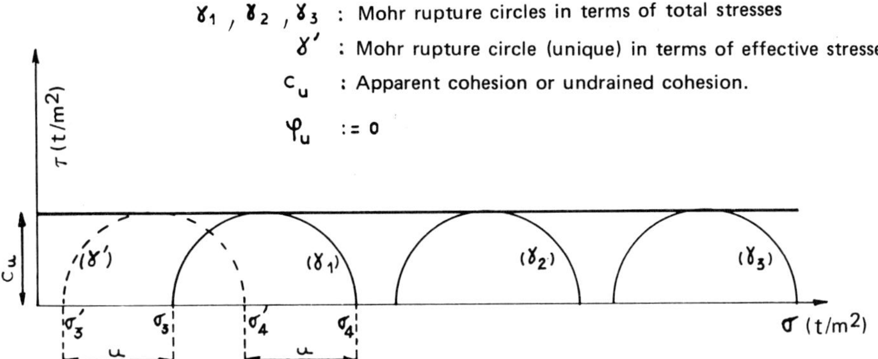

FIG. I.2.3. – Results of unconsolidated and undrained triaxial tests.

It appears that the values of undrained cohesion c_u determined by these tests are generally pessimistic:

– on the one hand, owing to the rehandling of the core samples taken,
– on the other, owing to the lack of consolidation of the sample used.

I.2.3.2 Shear after consolidation, with or without draining

The duration of consolidation, which varies with the permeability of clayey soil, is generally several days.

For each state of consolidation, there is a corresponding value of the shear strength known as the apparent or **undrained cohesion** c_u (Fig. I.2.4). For clays, it is observed that c_u varies linearly with the consolidation stress, i.e. in actual fact with depth.

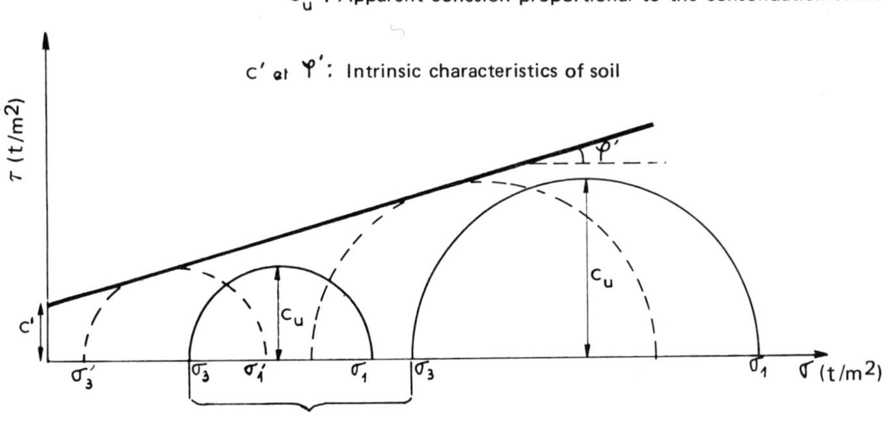

FIG. I.2.4. – Results of consolidated and undrained triaxial tests.

The envelope of effective stress Mohr circles forms the intrinsic curve of the equation:

$$\tau = c' + \sigma' \operatorname{tg} \varphi'$$

where:

c' and φ' respectively designate the **effective cohesion** and the **effective angle of friction** of the soil.

Intrinsic characteristics of the soil c' and φ', deduced from the tests **with** and **without** draining, and theoretically the same, may differ slightly owing to the experimental conditions linked to measurement of the interstitial pressure.

I.2.3.3 Simple methods for estimating undrained cohesion

Various quick methods that can be used immediately aboard a vessel provide an estimate of the undrained cohesion c_u (or undrained shear strength s_u) of clayey soils:

– unconfined compression test (the particular case of the triaxial test where $\sigma_3 = 0$). The cohesion $c_u \approx \frac{R_c}{2}$ where R_c designates the unconfined compression strength,

– penetration of the "fall-cone" or pocket penetrometer generally yields in addition a value of c_u,

– the laboratory vane is used, especially if $c_u < 20$ to 30 t/m² (200 to 300 kPa).

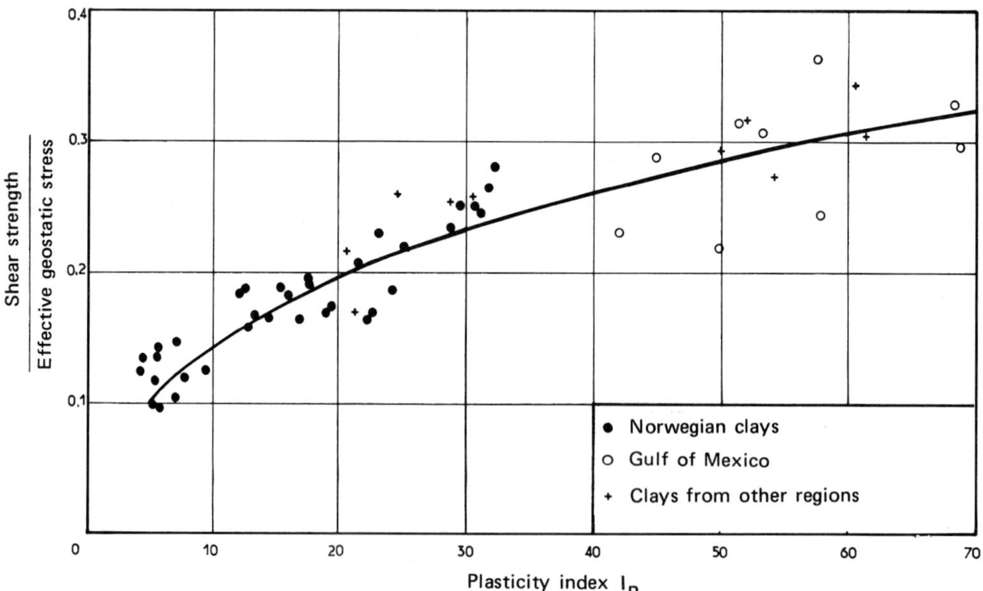

FIG. I.2.5. – Variation of the "cohesion-effective geostatic stress" ratio with the plasticity index (according to Bjerrum).

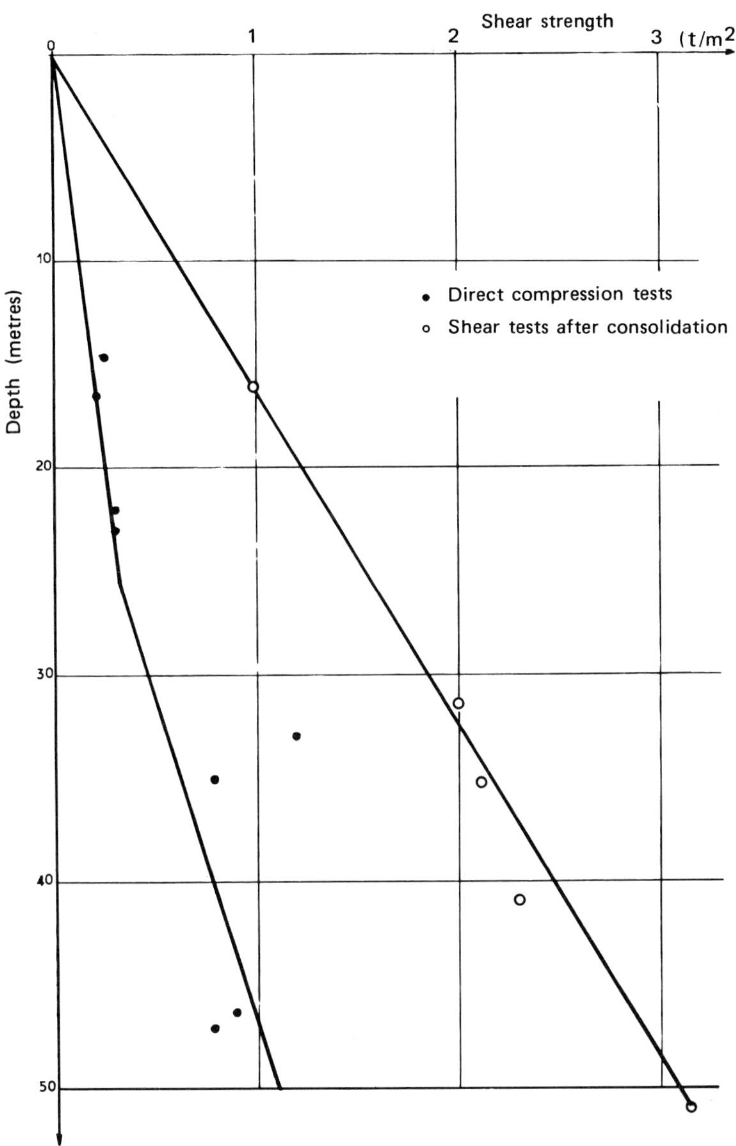

FIG. I.2.6. – Shear strength of a highly sensitive soft clay.

Many empirical relationships have been proposed to estimate the shear strength of clayey soils on a basis of the density and Atterberg limits. We shall merely mention here:

–Skempton's equation:

$$\frac{s_u}{p} = 0.11 + 0.0037\, I_P$$

where p is the submerged geostatic stress ($p = \gamma' h$ where $\gamma' = \gamma - 1$),

– the experimental results of Bjerrum (see Fig. I.2.5).

Remark. The development of the shear strength of the soil with its state of consolidation under the effect of repeated stressing (during a storm) requires:

– either triaxial tests with repeated loading,
– or linear shear tests, alternated and repeated.

I.2.3.4 Examples of the shear strength of clayey marine soils

Figure I.2.6 represents the undrained shear strength of a soft Mediterranean clay measured by unconfined compression test and by triaxial test after consolidation. It can be remarked that the shear strength after consolidation is distinctly greater than that obtained under unconfined compression. The difference is probably all the greater, the more sensitive the clay.

Figures I.2.7 and I.2.8 reproduce the values of the undrained shear strength of overconsolidated clays encountered in the North Sea, measured by various laboratory techniques : unconfined compression test, triaxial test, vane test, "fall-cone", pocket penetrometer, etc.

The results obtained:

– display very considerable scatter to a ratio of up to 1 to 3 or more (Fig. I.2.7) One may wonder to what extend this scatter is inherent in the core drilling and/or measuring technique. A considerable dispersion in the implementation of the results obviously follows from this scatter,

– indicate a minimum shear resistance at a depth of 20-30 m (Fig. I.2.8). This anomaly, which is confirmed by penetrometer tests, may find an explanation in the geology and in the influence of the waves.

Remark. Various in situ measuring devices : vane test, penetrometer, pressuremeter used to reconnoitre soils at sea, enable a value of the undrained shear strength to be measured or deduced (see Chapter VI, Seabed exploration by in situ measurements). The values of the undrained cohesion c_u measured in this way nonetheless vary considerably from one device to another.

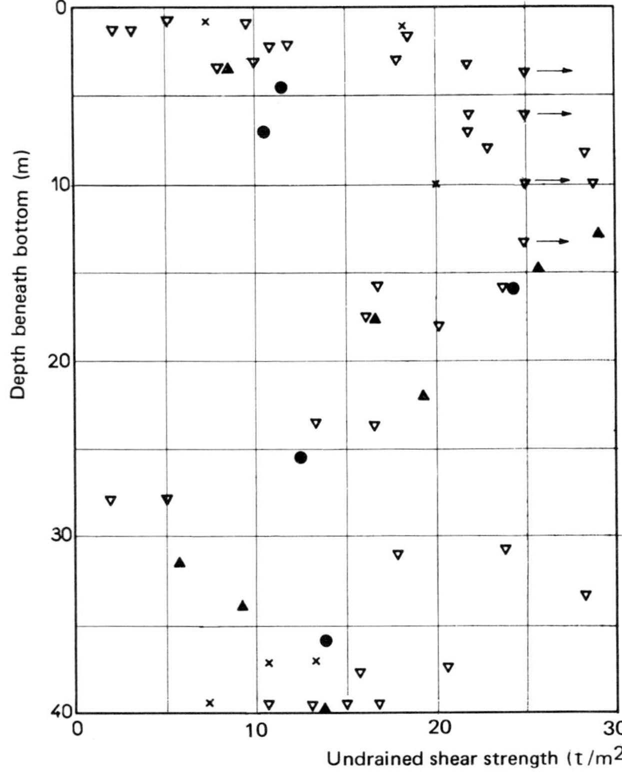

FIG. I.2.7. – Example of core profile in the North Sea (according to Ove Eide).

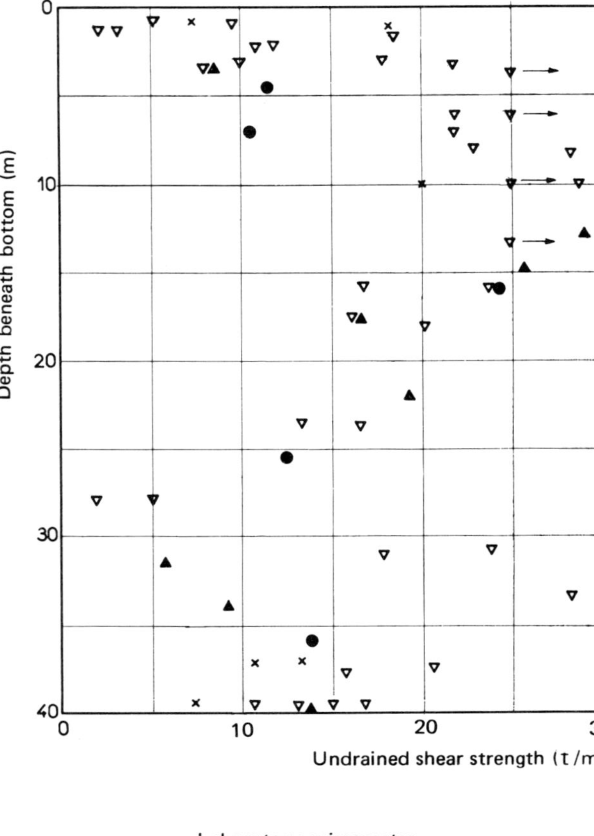

FIG. I.2.8. – Example of reduction of shear strength between 10 and 20 m depth in the North Sea (according to Ove Eide).

I.3 THE COMPRESSIBILITY OF SOILS

The granular skeleton of a soil subjected to a given state of stress undergoes gradual compression. In the case of sands, the settlement is generally small and does not call for protracted tests. On the other hand, study of the consolidation of clays implies long duration laboratory tests.

I.3.1 Consolidation of soils

I.3.1.1 Consolidation phases of clayey soils

The consolidation of a clay comprises three distinct phases corresponding to three different physical phenomena (Fig. I.3.1).

The instantaneous settlement of the soil, before expulsion of the water, corresponds to the undrained deformation of the soil.

The primary settlement (or hydrodynamic consolidation corresponds to the dissipation of the interstitial pressure u (Fig. I.3.1). The duration of this phase (for $u = 0$) can be estimated, provided the configuration of the layer (thickness, draining layers), compressibility and permeability of the soil be known.

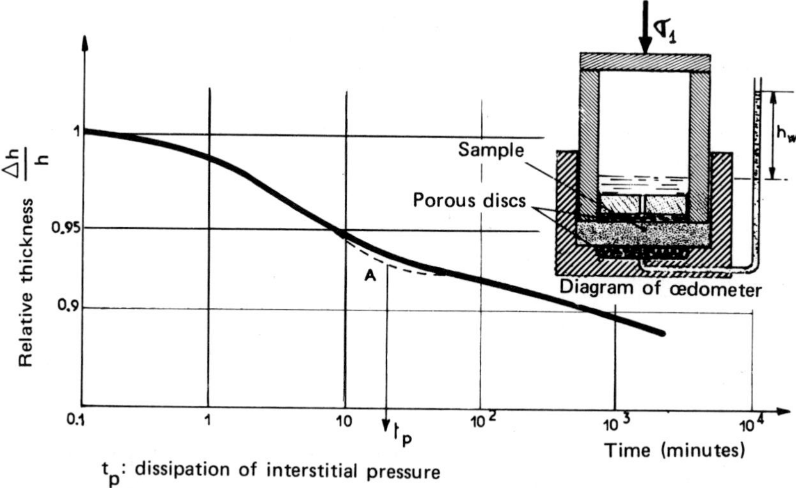

FIG. I.3.1. – Example of consolidation curve obtained with the œdometer at constant pressure.

The secondary settlement, which is the consequence of creep, obeys a linear law in logarithm of time. This additional settlement, which is practically negligible in the case of consolidated or overconsolidated clays, is of considerable importance, on the contrary, for muds or soft clays.

I.3.1.2 Determination of the settlement of soils. Interpretation of the oedometric diagram

The determination of the settlement of soils is carried out in the laboratory:

– either using the conventional Terzaghi oedometer (this being the conventional method) – the test is conducted on a thin core (1 to 2 cm) to reduce the duration of the primary consolidation (which is proportional to the square of the thickness),

– or using the triaxial, by means of a long duration drained test.

On the linear part of the "void ratio e – vertical stress applied log σ_1" oedometric diagram, the different characteristics of compressibility of the soil are defined (Fig. I.3.2), namely:

– the compressibility constant C, such that:

$$\epsilon_1 = \frac{\Delta h}{h} = \frac{\Delta e}{1 + e} = -\frac{2{,}3}{C} \log \left(1 + \frac{\Delta \sigma_1}{\sigma_1}\right)$$

– the compression index C_c, by the equation:

$$\Delta e = - C_c \log \left(1 + \frac{\Delta \sigma_1}{\sigma_1}\right)$$

– the oedometric modulus E_0, such that:

$$E_0 = \frac{2{,}3\, \sigma_1 \,(1 + e)}{C_c}$$

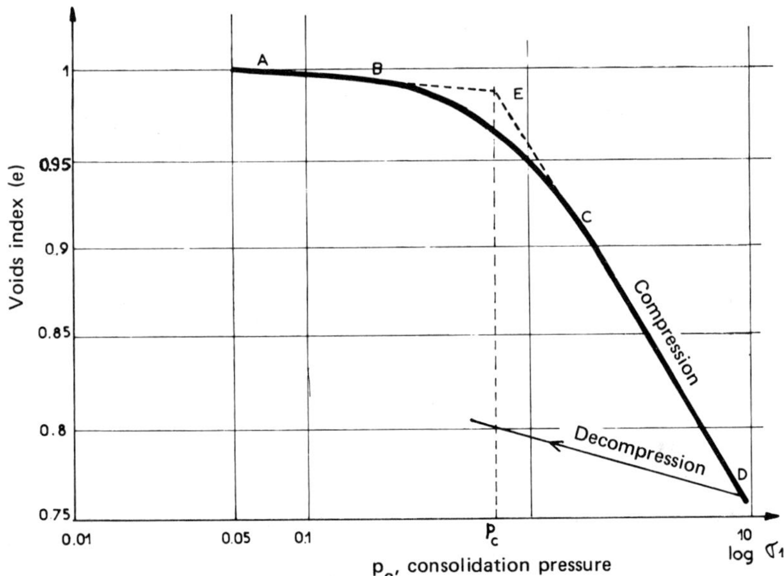

FIG. I.3.2. – Example of œdometric curve of a normally consolidated clay.

In these relationships, Δe designates the reduction in the voids index corresponding to the increase $\Delta\sigma_1$, in vertical stress σ_1.

Table I.3.1 gives a number of order of magnitudes of the compression index C_c and the oedometric modulus E_0 for different types of soils.

TABLE I.3.1

	Sands	Stiff or fairly stiff clays ($c_u > 5$ t/m^2)	Soft clays
Compression index C_c	0.02 – 01	0.1 – 0.25	$\geqslant 2$
Oedometric modules E_0 (t/m^2)	Several hundred	2,000 to 500	200 to 50

I.3.1.3 State of consolidation of clayey soils

From the "$e - \log \sigma_1$" oedometric diagram, the geotechnical signification of the state of consolidation of clayey soils is defined.

In principle, an oedometric diagram takes the form of two branches *AB* and *CD*, intersecting at *E*, the abscissa of which represents the consolidation pressure p_c of the soil (Fig. I.3.3):

- if $p_c = \gamma' h$, the clay is normally consolidated (Fig. I.3.3b),
- if $p_c > \gamma' h$, the clay is overconsolidated and the coefficient of overconsolidation is defined by the ratio $p_c/\gamma_d h$ (Fig. I.3.3c),
- if $p_c < \gamma' h$, the clay is underconsolidated, consolidation of the recent mud sediments not being complete (Fig. I.3.3a).

When the sample has undergone considerable handling as a result of core drilling, diagram *ABCD* is replaced by *A'D'*, which tends to match *CD* for high stresses σ_1 (Fig. I.3.3b).

The overconsolidation ratio varies considerably with the geological history of the sediment:

- from 2 to 4 in the case of certain clays encountered in the north part of the North Sea,
- up to 10 and more on certain sites.

I.3.2. Estimation of settlement on the basis of oedometric and pressuremetric tests

I.3.2.1 Validity of laboratory tests

The validity of laboratory tests for estimating settlement is highly variable, depending on the nature of the soils.

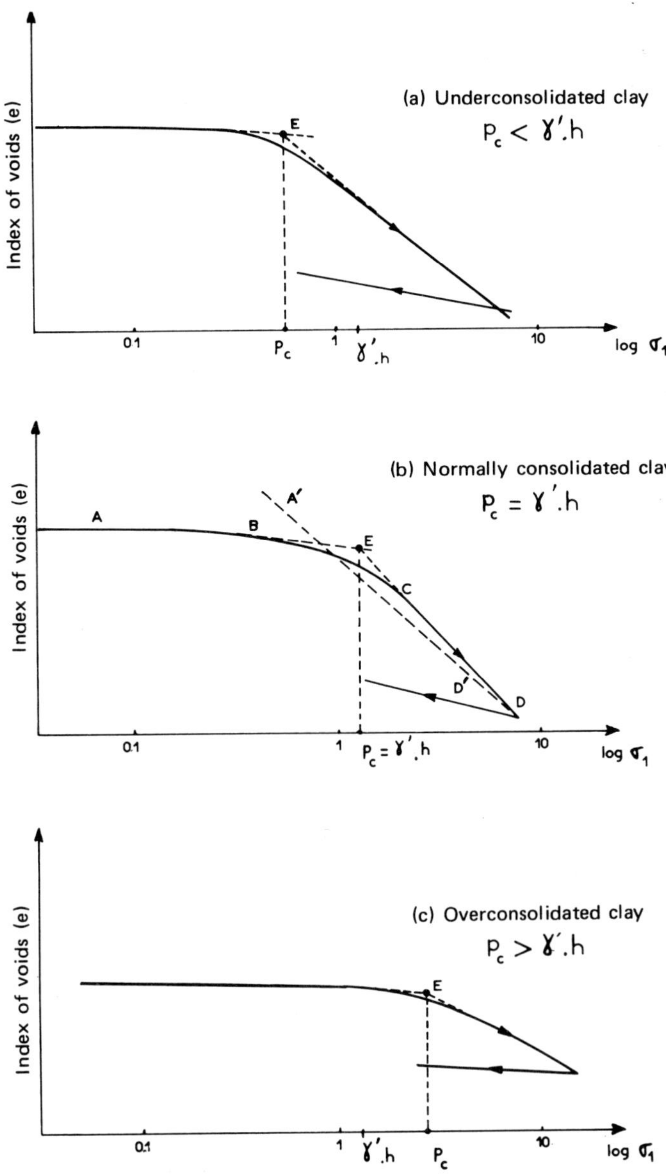

FIG. I.3.3. – State of consolidation of clays (according to œdometric diagram).

In the case of homogeneous clays, normally consolidated and of little sensitivity, the results of laboratory tests appear representative of the properties of the soils in place. The settlement can be calculated to an error of less than 10%.

In the case of thin heterogeneous layers, many tests may be needed to determine the different characteristics of compressibility and their mean value. In addition, the true distribution of the stresses becomes highly complex or even impossible, owing to:

– layers of very different consistency situated immediately beneath the foundation,
– the "arch effect" caused by a resistant layer overlying a more compressible layer,
– isolated slices of stronger soils or soils with a non-uniform surface.

In these cases, the error in the settlement may exceed 50% or even 100% in highly overconsolidated or sensitive soils. It should be noted that **in almost all cases, the calculated settlement is higher than the real life settlement.**

In any case, the difficulty lies in the estimation of the consolidation duration.

I.3.2.2 Validity of pressuremetric tests

According to Menard, settlement W can be estimated from the pressuremetric modulus E by the formula:

$$W = \frac{1.33}{3\,E} pR_0 \left[\lambda_1 \frac{R}{R_0}\right]^{\alpha} + \frac{\alpha}{4.5\,E} pR\lambda_2$$

This formula, deduced from the theory of elasticity, is corrected empirically by a coefficient α which is a function of the nature and state of the soil:

– the first term represents the deformation under the effect of the deviational stress tensor and allows for the influence of the dimensions of the foundations,

– the second term is an attempt to approach the consolidation phenomenon (spherical stress tensor).

In this equation:

p designates the mean stress on the soil,
R_0 is a reference dimension,
λ_1 and λ_2 are coefficients of shape (also used by other authors),
α is a coefficient varying from 1/3 to 1, depending on the particular terrain.

It would appear from observation of the settlement of various types of foundation (on land), that forecasts of settlement by the pressuremetric method are generally fairly satisfactory.

One should however remark that the absence of draining and handling of the soil near the probe have opposing effects on the evaluation of the pressuremetric modulus, which explains to a certain extent the good relative agreement observed between the settlements predicted and those measured.

In the case of large size foundations, it appears that evaluation of the settlement on the basis of the pressuremetric test would be correct for:

– consolidated or overconsolidated clays,
– highly compact sands,

where the pressuremetric modulus lies between about 500 and 2,500 t/m^2 (5 and 25 MPa). For soft soils or loose sands, it is essential to use the oedometric method (at least as a check).

Remark. Certain attempts have also been made to link the cone resistance measured on the penetrometer with the compressibility of the soil.

BIBLIOGRAPHY

[1] CAQUOT (A.) and KERISEL (J.). – *Traité de mécanique des sols.* Gauthier-Villars, Paris, 1966.

[2] TERZAGHI (K.) and PECK (R.B.). – *Soils Mechanics in Engineering Practice.* John Wiley and Sons, New-York.

[3] LAMBE (T.W.) and WHITMAN (R.V.). – *Soil Mechanics.* Wiley and Sons, Inc., New York, 1969.

[4] SUKLJE (L.). – *Rheological Aspects of Soil Mechanics.* Wiley – Interscience, Londres, 1969.

[5] HABIB (P.). – *Précis de géotechnique.* Dunod, Paris, 1973.

[6] *Modes opératoires des essais de sols des Laboratoires des Ponts-et-Chaussées.* Dunod, Paris.

CHAPTER II

Aspects of seismic Reflection

CONTENTS

INTRODUCTION

The description of the techniques of the geophysical prospecting of sea bottoms and marine soils by echo sounders, sediment sounders, boomers, sparkers, etc., first necessitates a review of the basic concepts of seismic reflection [1] [2].

1. The fundamental principle of seismic reflection consists in measuring the time between the emission of an acoustic signal in the ground and its return in echo form.

Owing to the behaviour of the elastic waves in the subsoil, each echo corresponds to an acoustic discontinuity generally amenable to interpretation from the geological standpoint (boundary of a layer, lithological variation, etc.).

2. The method implies the use of three instruments (Fig. II.1.1):

– an acoustic transmitter E (located at A),
– an acoustic receiver R (located at B),
– a recorder D, to note travel times.

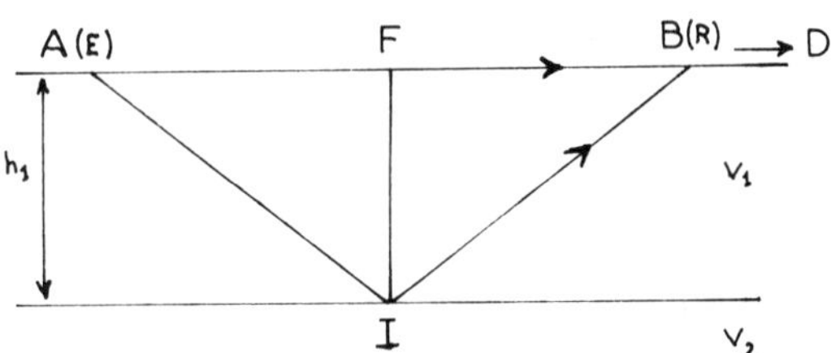

FIG. II.1.1. – Principle of seismic reflection.

3. The fact that a time t is associated with a reflector does not directly yield the position of this reflector, since the time t depends on the nature of the path followed AIB.

To eliminate this ambiguity, the seismic technician refers to a travel time t' corresponding to a vertical path IF (with F midway between A and B), such that:

$$t' = \frac{FI + IF}{V_1} = \frac{2h_1}{V_1}$$

The simplest solution which leads to this result consists in combining E and R at F, or ensuring that distance AB is small compared to IF. In this case $t \# t'$.

The reflector has a travel time which corresponds to twice the depth of the reflector. This is termed the two-way time (*twt*).

4. We shall confine ourselves in the present chapter:

– to the principle and application of seismic reflection to continuous seismic sounding,
– to generalities concerning seismic techniques: transmitters (sources), receivers and recorders.

The inherent characteristics of each type of transmitter: echo sounder, sediment sounder, boomer, sparker, will be set forth in the chapters concerning "Seabed exploration: bathymetry and topography" and "Seabed exploration by high resolution seismic prospecting".

The detailed theory of receivers and the technology of receivers and recorders goes beyond the scope of this book.

II.1 PRINCIPLE AND APPLICATION OF SEISMIC REFLECTION

Seismic methods use either reflection or refraction of waves to detect discontinuities in the subsoil. At present, practically only seismic reflection is used for the reconnaissance of marine soils.

II.1.1 Basic theory of seismic reflection

Seismic reflection consists in recording the echoes of an acoustic pulse reflected by the discontinuities of the subsoil. The phenomena of propagation and reflection of the sound wave can be explained by the elementary laws of optics.

II.1.1.1 Propagation in a continuum

If a disturbance is provoked in the soil, owing to the elastic nature of the environment, this disturbance is observed to propagate in all directions.

In the case of a homogeneous and isotropic medium, the "envelope" of disturbance is a sphere known as the **wave front** (Σ), centred on the origin of the disturbance (O). This sphere propagates at a constant speed V_1 which is characteristic of the medium (Fig. II.1.2).

Rather than follow the disturbance envelope, it is sufficient to study the development of a normal to the envelope, or the radius OR. When the speed is constant, the radius is a straight line.

II.1.1.2 Propagation in a discontinuum

The soil invariably consists of layers of different densities d_i and propagation speeds V_i.

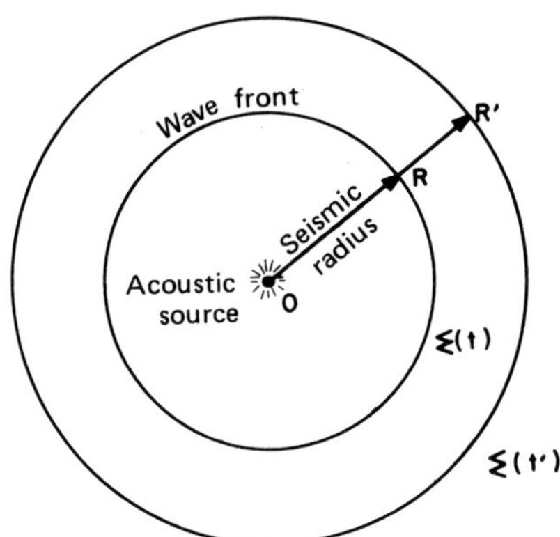

FIG. II.1.2. – Propagation of wave front through an isotropic medium.

Each medium is characterized by its acoustic **impedance**, being the product of the density of the medium multiplied by the velocity of the sound waves through it (Fig. II.1.3):

$$Z_1 = V_1 d_1$$

$$Z_2 = V_2 d_2$$

The surface of separation π between the two media is known as the seismic **mirror** or **reflector**.

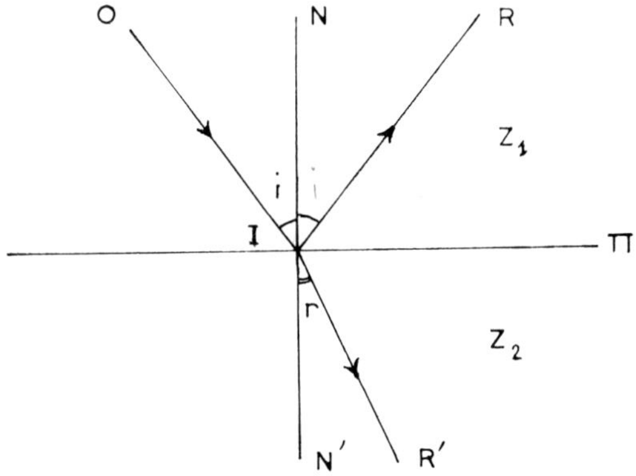

FIG. II.1.3. – Path of rays through a discontinuous medium.

When the disturbance originating from O arrives at the discontinuity, the phenomena of reflection and refraction in the wave occur, following well-known optical laws (the laws of Descartes):

– the incident ray (OI), reflected ray (IR) and refracted ray (IR') stay in the same plane (the incident plane), perpendicular to the reflective surface,

– the angle of incidence (OIN) equals the angle of reflection (NIR),

– the angles of refraction ($N'IR'$) and incidence (OIN) are related by the equation:

$$\frac{\sin i}{V_1} = \frac{\sin r}{V_2}$$

In addition, the phenomena which occur on the surface π can be characterized by the distribution of the energy between the reflected ray IR and the ray transmitted into the second medium IR'. For simplification, assuming the case where OI is perpendicular to π (incident angle normal), these equations become relatively simple.

If E is the energy of the incident ray OI:

– the energy of the reflected ray IR is given by:

$$E_1 = r \cdot E$$

– the energy transmitted into the second medium (ray IR') is given by:

$$E_1' = t \cdot E = (1 - r)E$$

In these equations:

r, the **coefficient of reflection** on the surface π is defined by:

$$r = \frac{Z_2 - Z_1}{Z_2 + Z_1} = \frac{d_2 V_2 - d_1 V_1}{d_2 V_2 + d_1 V_1}$$

t, the **coefficient of transmission,** is given by:

$$t = \frac{2Z_1}{Z_2 + Z_1} = \frac{2d_1 V_1}{d_2 V_2 + d_1 V_1}$$

II.1.1.3 Diffraction.

When a reflector comprises accidents (flaws) or local heterogeneities (erratic blocks, irregularities caused by erosion surfaces, etc.), parasite reflections can be observed, linked to diffraction: an angular point is scattering the energy in all directions.

Where a diffractive point P lies on the bottom (isolated block) at depth h (Fig. II.1.4), the equation giving the echo return time t in terms of the distance d of the vessel vertically above the point of diffraction (O) is:

$$t^2 = \frac{d^2 + h^2}{V^2}$$

This equation represents a hyperbole. Since time t is assigned to the vertical of point E, as one approaches O, one inscribes a hyperbole (P', P''. . .) with its apex at P.

It should be noted that a similar phenomenon can occur in the presence of a sharp relief off the course of the vessel, resulting in what is known as a lateral echo.

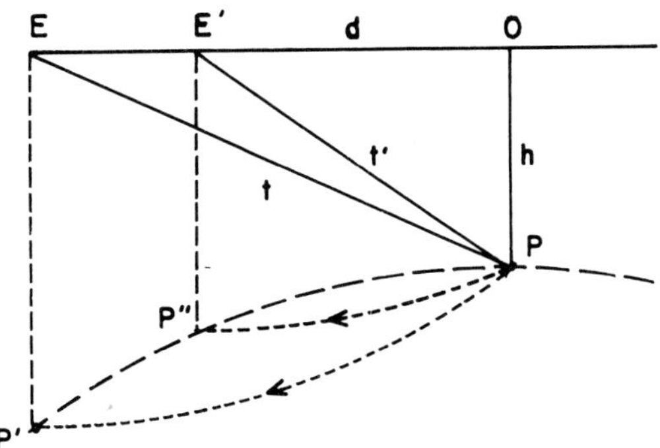

FIG. II.1.4. – Principle of diffraction.

II.1.2 Seismic reflection resolution

II.1.2.1 Definition of resolution

Resolution is the minimum time (or distance) (e) separating two reflectors (strata) that can be separately distinguished on the recordings. Several cases have to be considered.

First, it is assumed that the soil beneath the sea floor consists of a series of reflectors. In addition, it is assumed (Fig. II.1.5) that:

- the reflectors lie close to one another (equidistance h),
- the signal $s(t)$ emitted by the source of sound S is infinitely short,
- the paths of the waves are approximately vertical.

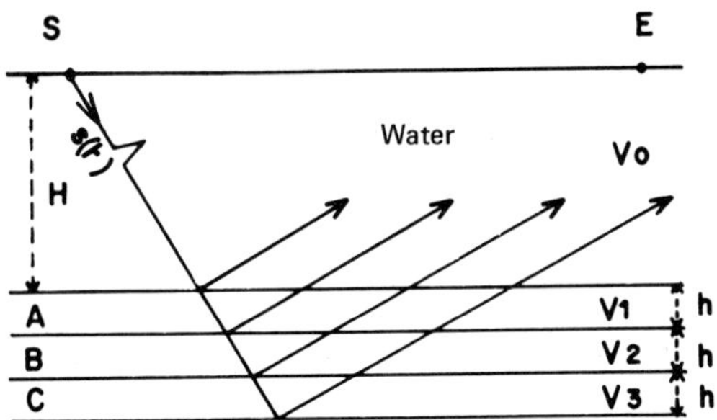

FIG. II.1.5. – Response of a series of adjacent reflectors.

The signal reflected from the first reflective surface (the bottom) arrives at the recorder after an interval $t_1 = \frac{2H}{V_0}$.

The signal from the second reflective surface will arrive t_2 $\left(= \frac{2h}{V_1}\right)$ seconds later.

The signal from the third surface will arrive t_3 $\left(= \frac{2h}{V_2}\right)$ after the previous one, and so on.

The recording obtained is of the type shown in Fig. II.1.6.

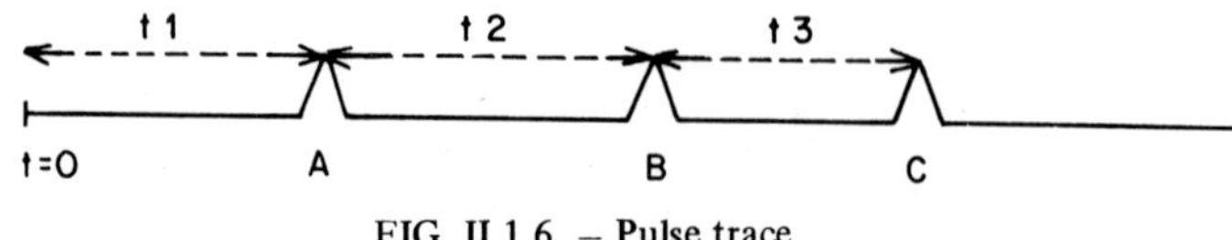

FIG. II.1.6. – Pulse trace.

Assuming now that the reflectors invariably lie close to one another, but that the signal $s(t)$ is no longer infinitely short. Its duration is $T > t_2, t_3, \ldots$

Let this signal be represented by $A(t)$. The plot ([1]) obtained $R(t)$ is shown on Fig. II.1.7.

It can be seen that it is no longer possible on plot $R(t)$ to discern separately the echo from reflectors A, B, C.

The resolving power is therefore linked to the duration of the signal emitted by the acoustic source.

If a geological strata has a thickness e and an acoustic velocity V, it will only appear on the plot provided the duration of the signal t is such that:

$$t \leqslant \frac{2e}{V}$$

For example, for a resolution of 1 m in sediments with an acoustic velocity of 2,000 m/s, the signal must have a duration less than or equal to: $t = 2/2{,}000 = 1$ ms.

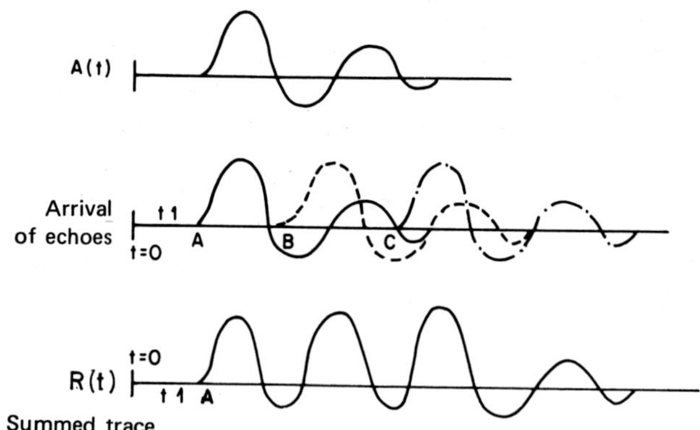

FIG. II.1.7. – Production of the true trace.

(1) A **plot** is the curve of response of the hydrophones recorded at a given point as a function of time.

II.1.2.2 Frequency spectrum and resolution

Aside from the assumption concerning the velocity and equidistance of the reflectors, the ideal resolving power is attained for a signal (known as the **Dirac signal**) which is infinitely short (Fig. II.1.8a). This signal has a flat spectrum, or in other words it contains all frequencies equally (Fig. II.1.8b).

The signals emitted by an acoustic source are never "Dirac" type signals, but have a certain oscillation period and hence are **frequency spectrum**. Consequently, true sources have characteristics which limit the resolving power.

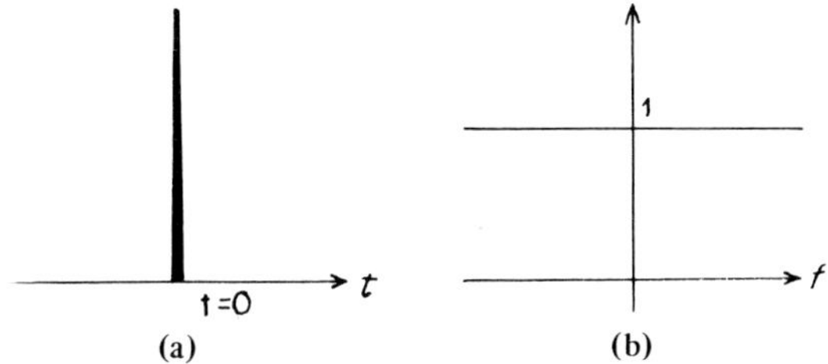

FIG. II.1.8. – Representation of the Dirac signal.

II.1.3 Limitation of the fineness of seismic reflection recordings

In addition to the frequency spectrum, the fineness of the seismic recordings is limited in particular by the filtering effect of the soil, parasite reflections, bubble effect and cavitation.

II.1.3.1 Filtering by the soil

From the standpoint of the signal, the soil is not neutral in behaviour. It absorbs frequencies selectively. It therefore acts as a filter with respect to the signal. One can say that as an approximation it absorbs the high frequencies preferentially. This means that the penetration (2) of the signals is all the more limited, the higher the emission frequencies from the sources. For instance, echo sounders, which are sources of high frequencies, have practically zero penetration and give only the image of the bottom.

It should be noted that even were one to be able to emit a Dirac signal, the selective absorption of the high frequencies would from the point of view of the recording give the same result as the emission of a signal with a certain duration, by deforming the flat Dirac spectrum (Fig. II.1.9).

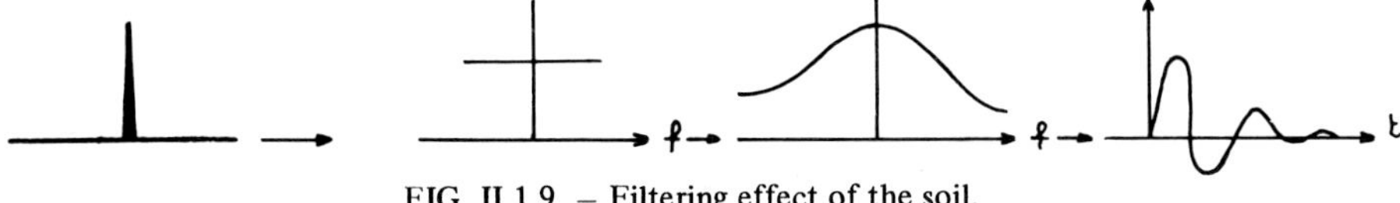

FIG. II.1.9. – Filtering effect of the soil.

(2) **Penetration** infers the two-way time (or depth) characterizing the deepest reflections discernable on the sections.

II.1.3.2 Parasite reflections

When the coefficient of reflection of certain surfaces has very high energy, resonance phenomena or trapping of the energy of the signal may occur, resulting in **multiple reflections** (Fig. II.1.10).

Several cases may occur, as can be seen on Fig. II.1.11:

– "pedalling" (Fig. II.1.11a): the energy is trapped between the bottom and the surface,
– resonance in a geological stratum ("pedleggs") (Fig. II.1.11b),
– a single internal "zig-zag" (Fig. II.1.11c).

Another factor, which is particularly disturbing to the recordings, is the **shadow**, where the energy reflected by the air-water interface quickly follows the signal sent to the bottom and doubles up all the reflections on the recording with a delay of $t = \frac{2h}{V}$ (Fig. II.1.11d).

From the practical standpoint, the shadow can be eliminated by placing the emitter very close to the surface. Another method would be to submerge the emitter to a sufficient depth so that the shadow arrives after the useful reflections.

II.1.3.3 Bubble effect

An acoustic source, like the sparker, will generate a certain amount of gas resulting from decomposition of the water after it has emitted the shock wave.

This gas rises to the surface in the form of a bubble which, allowing for the initial kinetic energy of the gas, will oscillate around ambient pressure, without ever reaching equilibrium. Each reversal of motion of the walls of the bubble results in a secondary pulse thus extending the duration of the initial signal.

Figure II.1.12 shows an example of how bubble effect influences a seismic plot.

II.1.3.4 Cavitation

If at a point A in a liquid under a hydrostatic pressure P_0, a disturbance is created, the total pressure at this point P_t representing the sum of the hydrostatic pressure and the acoustic pressure P, which is of vibratory nature, may become negative when $P > P_0$. Changeover to negative pressures of the medium results, in accordance with the law of dissolution of gases, in the appearance of bubbles of the gases which were originally dissolved in the liquid or, if the liquid is perfectly pure, its evaporation.

When the pressure P_t again becomes positive, the compression of the bubbles which arose in the negative pressure zone results in a veritable implosion, accompanied by the appearance of noises (cavitation noises), a considerable rise in temperature (up to 10,000° C) and a considerable loss of the energy released as a result of absorption:

– the cavitation noise modifies the signal,

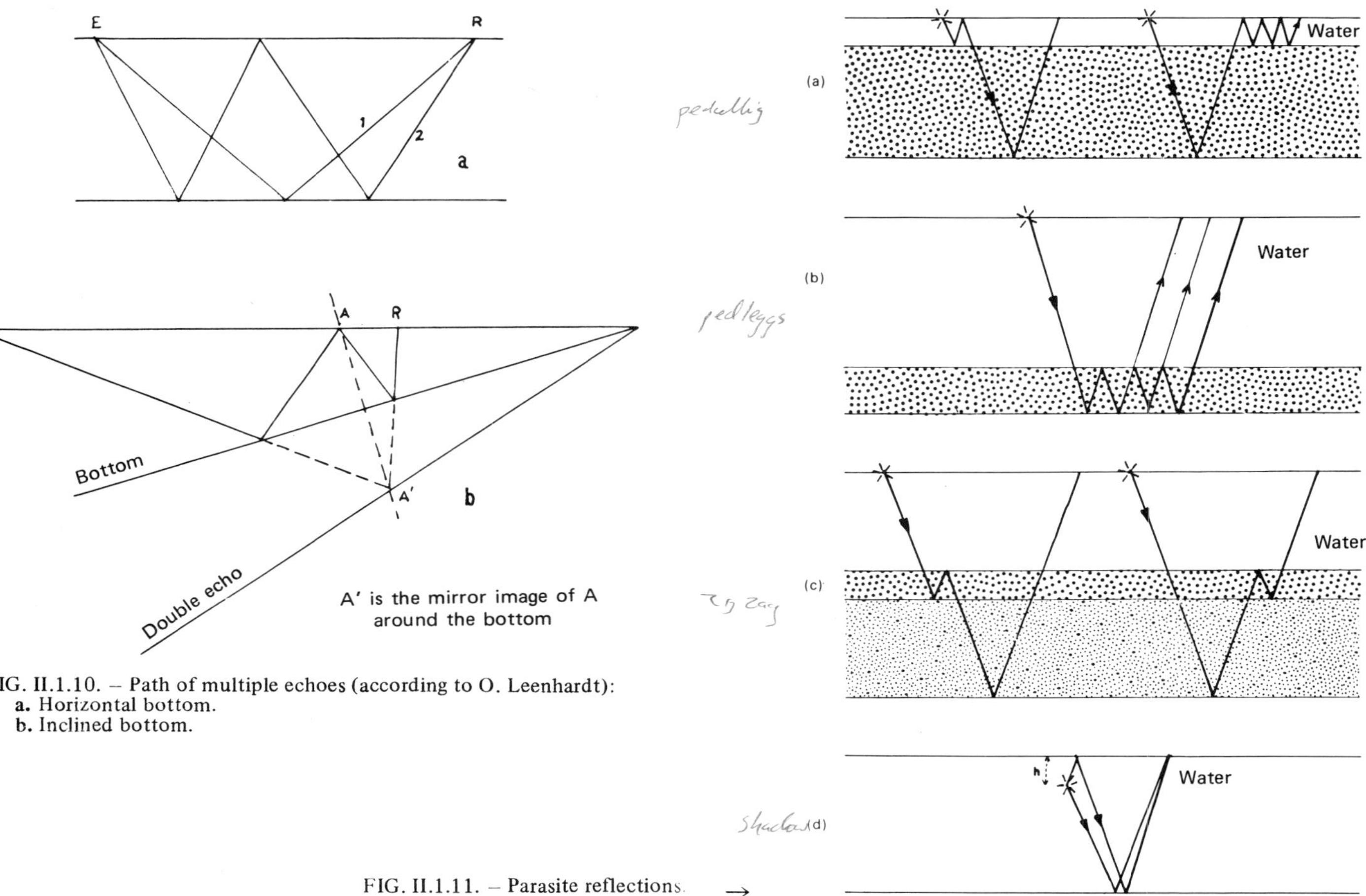

FIG. II.1.10. – Path of multiple echoes (according to O. Leenhardt):
a. Horizontal bottom.
b. Inclined bottom.

FIG. II.1.11. – Parasite reflections. →

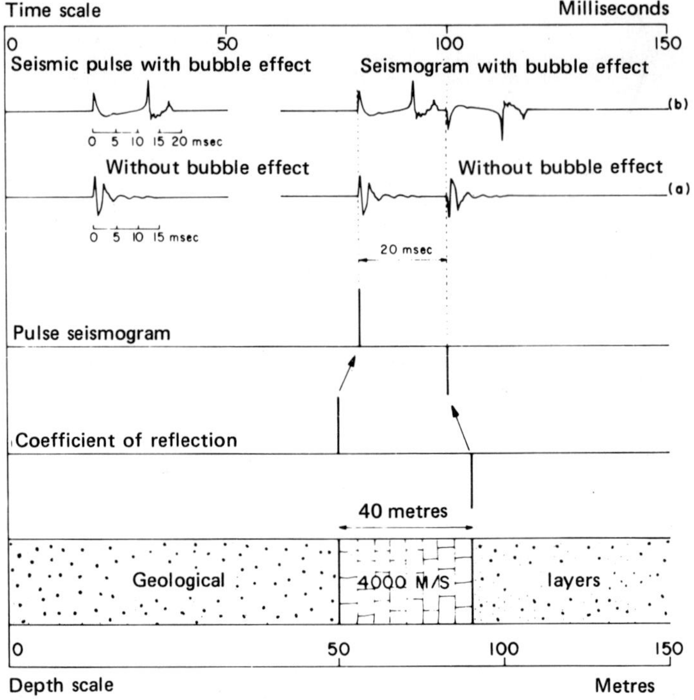

FIG. II.1.12. – Influence of bubble effect.

– the temperature-rise and (partly) noise deteriorate the material used to emit the signal (erosion of the plates),

– the absorption limits the energy radiated from the source.

The cavitation phenomenon therefore has very important consequences on the acoustic emission. It should be recalled that in practice cavitation appears when the emission threshold reaches a level of 125 dB.

II.1.4 Application of seismic reflection : continuous seismic sounding

Continuous seismic sounding (or continuous high resolution seismic reflection) covers the techniques using the seismic reflection method in order to obtain as accurate a section as possible of about the first 100 m of soil beneath the bottom.

II.1.4.1 Continuous seismic sounding procedure

As the vessel progresses (Fig. II.1.13):

- a source emits short acoustic pulses towards the bottom at a very high rate,
- a receiver receives the echos reflected from the bottom and the underlying strata,
- a graphic recorder records the echos received (after first amplifying and filtering them).

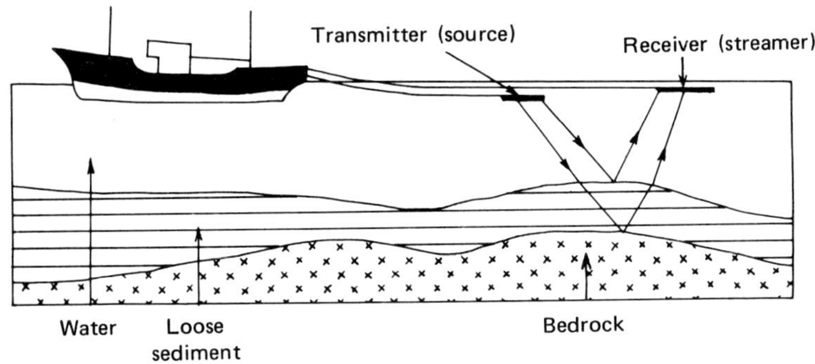

FIG. II.1.13. – Diagrammatic representation of continuous seismic sounding.

With the vessel advancing at about 5 knots and at a rate of emission of several pulses per second, a time section is obtained providing an all but continuous representation of the various reflective layers of the subsoil vertically beneath the path of the vessel.

Continuous seismic sounding is a function of the perfection of the interface between the transmitter and the water.

Provided in addition that the distance between the transmitter and the receiver (streamer) is sufficiently small compared to the depth of water (the first reflecting surface being the bottom), the acoustic paths will be practically vertical so that the echos will provide the position of the reflective surfaces in terms of the two-way-time.

II.1.4.2 Quality of recordings with continuous seismic sounding

The phenomena indicated above (see II.1.3) – filtering by the soil, parasite reflections, bubble effect and cavitation – all result in extension of the duration of the signal, or in other words deviate from the ideal pulse plot.

Certain processing methods enable the pulse to be contracted, thus improving the resolving power. This technique, known as pulse compression, is not very widely applied in conventional high resolution seismics; one then has to seek for optimum efficiency by appropriate selection of the equipment and correct implementation.

The resolving power of the recorder is itself limited:

– in particular, the graphic recorder, which is generally used in high resolution seismics, has a stylus the thickness of which cannot be neglected, and which may make it impossible to discriminate between two signals very close to each other,

– in addition, the dynamics of the receiver and recording equipment (logarithm of the ratio of the strongest to the weakest recordable signal) is limited compared with the dynamics of the echos from the soil.

II.2 GENERAL INFORMATION CONCERNING SEISMIC TECHNIQUES

As pointed out above, seismic methods call for the use of three types of device:

- transmitters or sources,
- receivers,
- recorders.

In the present section, we propose to describe the general characteristics of use of these three types of device.

II.2.1 Transmitters (or sources)

The choice and use of a source for a given purpose depends on the following:

- its physical characteristics,
- the application parameters,
- the utilization security,
- the cost of the transmitter.

These various types of data more or less relate one to the other and only exact knowledge of the influence of each parameter can guarantee satisfactory results.

II.2.1.1 The signature and spectrum of a source

The **signature** of a source is the representation against time of the amplitude of the signal emitted by this source. This signature is characteristic of the source. The duration of the signal, which is particularly important for the resolving power, can display oscillations linked to the bubble phenomenon (see Section II.1.3.3) or mechanical "bounce" of the source itself. For instance, the reply (b) of the initial pulse (a) caused by the oscillation of the bubble can be seen on Fig. II.2.1.

Only the initial peak of the pulse (first echo to the receiver) is of interest to the seismic expert. If the distribution of the amplitude (energy) is not highly concentrated on this peak, usable energy is lost.

This concept can be specified by considering the **spectrum** of the signature. This spectrum shows the energy distribution with the frequency band considered. Owing to the selective filtering of the high frequencies by the soil, it is vital that most of the energy be distributed in the usable frequency band, which is determined by the depth of the target.

II.2.1.2 Directivity of a source

The directivity of a source characterizes its property of spatial distribution of the energy in a given manner.

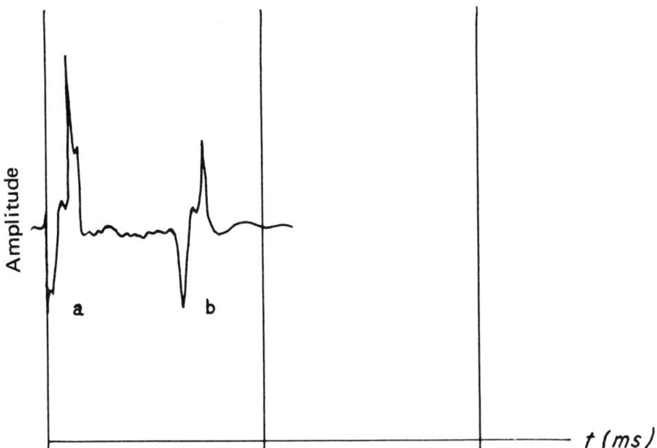

FIG. II.2.1. – Example of the signature of a source comprising bubble effect.

An effort is made to send most of the energy vertically so as to excite only a limited surface of the sea floor and hence reduce lateral diffractions and reverberations.

II.2.1.3 Transmission rate and fidelity

The transmission rate is the number of shots per unit time. It is vital that this transmission rate be sufficiently high so as to reconstitute the structure of the subsoil with the greatest possible fidelity. Among other things, the continuity of the reflections can only be guaranteed for a high shooting rate, all the more so the more complex the geology. The shooting rate depends on the power of the transmitter.

On the basis of this shooting rate and the speed of the vessel, the restitution accuracy can be appraised. For instance, at a speed of 5 knots and a rate of 2 shots/second (power = 300 J) a reflection point is formed every 1.25 m.

The transmission fidelity, or in other words the capacity of the source to reproduce a signal identical to itself, is an important characteristic of use.

II.2.1.4 Characteristics of application of sources

The arrangement of the source –attached to the hull of the vessel, towed, etc,– varies with the type of device and will be examined for each particular case echo sounders, sediment sounders, boomers, sparkers (see Chapters III and IV).

The depth of submersion of the source has a major effect on the shape of the signal and the bubble effect period, as can be seen from Fig. II.2.2.
As before, this depth is adapted to suit the type of device.

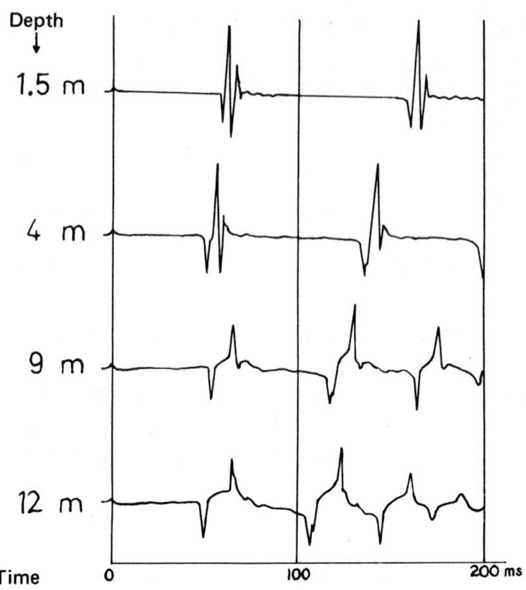

FIG. II.2.2. – Influence of depth of submersion on signal.

II.2.2 Receivers

The function of an acoustic receiver is to transform the acoustic energy into electrical energy in the form of a variable voltage.

Receivers may be:

– **point receivers**, i.e. consisting of a single element,
– or **linear receivers**, i.e. consisting of a group of receivers forming a **streamer**.

The basic receiver unit (or hydrophone) generally consists of piezoelectric ceramics.

II.2.2.1 Electrical coupling

Sensitivity is the prime quality required of a hydrophone: it is the ability of the receiver to reveal small variations in pressure.

Sensitivity must be independent of the depth of immersion: it is generally in the range of 90 dB/V/μbar.

The **frequency response curve** of the receiver must also be as flat as possible in the usable frequency band lest distortion of the signal be introduced.

Hydrophones can be considered as voltage generators, the impedance of which is essentially capacitive: a hydrophone can be defined by its capacitance C and voltage V (in response to a unit variation in pressure).

II.2.2.2 Filtering caused by the submersion of the receiver

Owing to the submersion of the receiver, the rising signal is (see Fig. II.2.3):

- initially recorded upwards,
- then re-recorded following reflection at the air-water interface with:
  - a delay $\tau = \dfrac{2h}{V}$,
  - reverse polarity from that of the initial signal (the water-air reflection coefficient $r \# -1$).

This therefore results in elongation of the signal in turn having a filtering effect.

The submersion of the streamer is favourable to the low frequencies.

Raising of the streamer is favourable to the high frequencies and improves the resolving power by reducing the time between the direct signal from the bottom and the shadow signal.

As an example, when submerged to a depth of 0.25 m, frequencies of about 1,500 Hz are favoured.

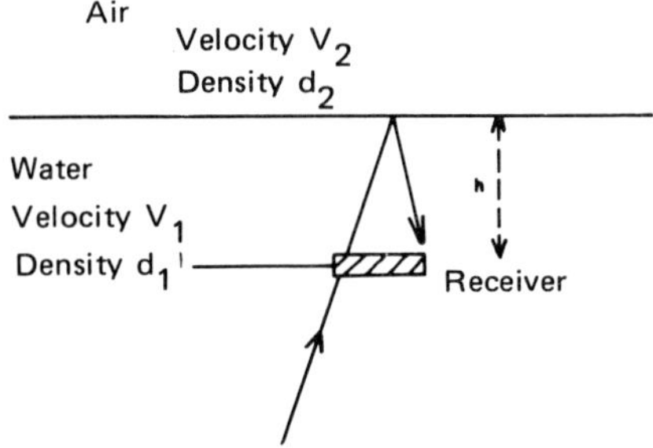

FIG. II.2.3. – Reflection of the signal beneath the surface.

II.2.2.3 Spatial filtering and the directivity of a receiver

The quality of a receiver can be defined by its signal/noise ratio. Positioning of n receivers improves this ratio by $\sqrt{n}$.

In addition, grouping of several receivers together results in **spatial filtering** depending on the distance between the receivers and the apparent wave length along the device. For instance, the interval between hydrophones favours certain wavelengths and hence certain frequencies.

The length of the streamer, which is generally a few metres, is a compromise between filtering and resolving power.

For a single transducer, the **directivity** diagrams on transmission and reception are identical. For a group of receivers, they differ.

II.2.2.4 Implementation characteristics of receivers (streamers)

The aim in implementation must be to reduce the various types of noise:

– **ambient noise** (towing vessel, sea, traffic, etc.). A linear receiver configuration reduces the noise from the vessel,

– **firing noise**: in shallow water, the direct wave may mask the reflections. This is corrected by adjusting the gap between transmitter and receiver,

– **induced noises** of various kinds:

  – the turbulence of the layer of water in contact with the streamer; this is attenuated by increasing the gap between receivers,
  – the changes in the speed of the vessel; these are compensated by a dampener at the head of the streamer,
  – vibrations owing to poor balancing of the streamer; the constructor corrects these by submerging the receivers in silicone oil.

Figure II.2.4 is a diagrammatic representation of a streamer satisfying the various stability requirements. The stabilization tail at the end of the streamer minimizes effects resulting from the state of the sea in particular.

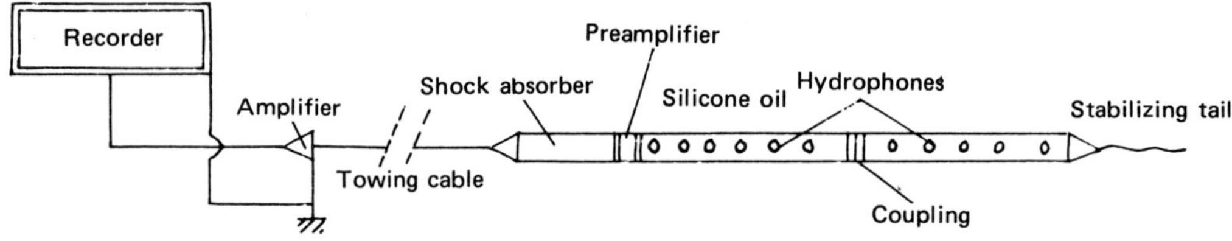

FIG. II.2.4. – Diagrammatic representation of a streamer.

The depth of submersion must if possible be identical to that of the transmitter. In practice, this factor is difficult to control.

This depth decreases with speed. In high resolution seismics, the streamer is generally towed on the surface, which implies a relatively satisfactory state of the sea (force $\leqslant 5$).

II.2.3 Recorders

The recorder amplifies, filters and retranscribes in **analog** or **digital** form the signals originating from the receiver.

These signals occurs in the form of a variable voltage of very low level (1 μV at 100 mV).

II.2.3.1 Functions of the recorder

The signal undergoes double amplification on leaving the receiver:

– the preamplifier of the streamer raises the signal to a higher level than that of the original electronic noise,

– the amplifier of the recorder adapts the characteristics of the signal to those of the recorder.

Adaptation of the signal can be defined in two ways:

– the signal can be recorded on electrosensitive paper or a magnetic tape. In either case, the recorder is sensitive in only one amplitude band. The signals therefore have to be adapted to this band,

– the dynamics of the seismic signal is about 100 dB. The dynamics (3) of a paper recorder is 25 dB and that of a digital recorder 84 dB. The dynamics of the signal therefore has to be set to that of the recorder.

The purpose of filtering is twofold:

– first, to adapt the spectrum of the signal (which is generally a very wide band signal) to the desired spectrum,

– second, to eliminate noise which may disturb the recording (motors).

II.2.3.2 Analog recording on paper

The analog recording uses the principle of belinography. On the signal, which is represented in the form of a variable voltage at the receiver output:

– first, the negative amplitudes are suppressed and unduly large positive amplitudes are clipped (Fig. II.2.5a),

– then, an electrosensitive paper is inscribed in proportion to the amplitude of the clipped signal (Fig. II.2.5b).

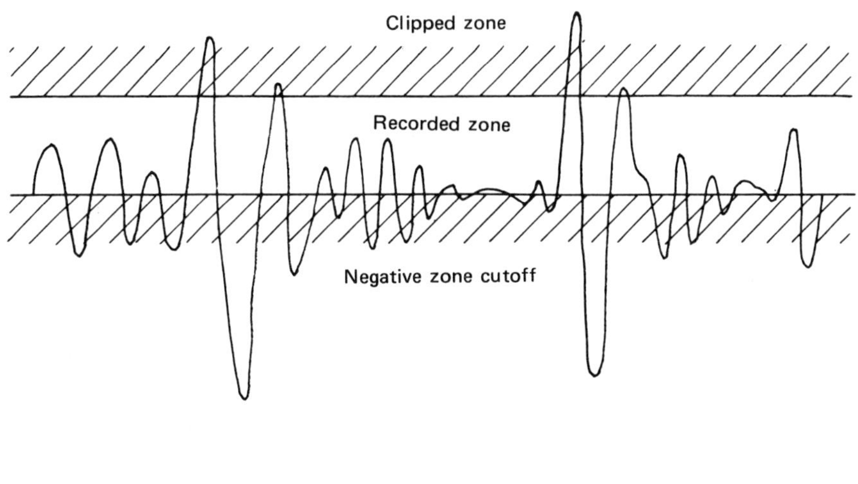

FIG. II.2.5. – Graphic analog recording.

(3) The dynamics of a paper is defined as the logarithm of the ratio of the amplitude of the signal which marks the paper to that of the signal which inscribes the lowest visible trace.

The paper is then run between two electrodes, one of which consists of a stylus (or screw) raised to a voltage which is proportional to the amplitude of the signal.

The stylus (or screw) makes a constant-speed **vertical sweep** during a period corresponding to the two-way propagation time of the signal from the surface down to the deepest reflector:

– adjusting this speed provides several recording scales,

– offsetting the origin prevents the unnecessary recording ot the travel time through the water in great depths.

The rate of advance of the paper (giving the horizontal scale) must be modifiable so as to obtain a recording displaying the least possible exaggeration of scale.

The sweeps must be close enough to provide good optical correlation.

Most recorders comprise **time lines** or depths calibrated in milliseconds, seconds or metres (assuming a value of about 1,500 m/s for the velocity of sound in water).

Paper-type digital recorders can be classified:

– either by their recording mechanism:
  – screw recorders,
  – stylus or friction pad recorders,

– or by the type of paper used:
  – Simrad, Mufax, Alfax wet paper,
  – Timefax, NDK, Teledeltos dry paper.

There is a practical relevance between the two classifications, screw recorders generally using wet paper and stylus or friction pad recorders using dry paper.

II.2.3.3 Digital recording

This technique is beginning to appear in high-resolution seismics.

On leaving the amplifier, the analog form signal is sampled, i.e. periodically measured, the measuring interval being termed the **sampling step.** This sampling step is defined by the characteristics of the spectrum of the signal. It must be at least 0.2 ms in high resolution seismics.

The samples are then recorded on magnetic tape in binary form. There are several advantages to this recording technique:

– storage of the data is denser than that with analog recording,

– the data are reproducible as desired and do not deteriorate,

– the binary representation of the data enables them to be processed on a computer.

Three types of digital recorder can be mentioned:

– the first renders simple visual digital control of an instantaneous value of the analog system possible at the time it is recorded (digital indicator),

– the second is a veritable digital recorder, i.e. all the data are digitized and stored on tape,

– the third, which is fully digital, also enables the data on the plot recorded to be summed.

BIBLIOGRAPHY

[1] LEENHARDT (O.). –*Le sondage sismique continu.* Masson, Paris, 1972.

[2] INGHAM (A.E.). – *Sea surveying* (Vol. I and II). John Wiley and Sons, London-New-York, 1975.

2 // Seabed Reconnaissance

CHAPTER III

Seabed Exploration : Bathymetry and Topography

CONTENTS

INTRODUCTION

1. The purpose of geophysical surveying of sea bottoms and soils is to obtain the following information:

– first, on the surface of the soil, bathymetry, topography, wrecks and obstacles, by **acoustic** methods (echo sounder, side-scan sonar) and if necessary by the magnetometric method,

– second, information concerning the geometry, structure and configuration of approximately the first hundred metres of subsoil, using **seismic reflection** methods or very occasionally seismic refraction.

2. The various geophysical surveying techniques are generally carried out simultaneously and the results obtained are entirely complementary.

In the present chapter, only surveying techniques for sea bottoms will be described (excluding seismic techniques for the reconnaissance of the subsoil, which will be described in the following chapter):

– **bathymetry** using the echo sounder, if necessary confirmed in greater detail using the pressure sensor applied from a submarine,

– **topography** (morphology) using the side-scan sonar, often investigated locally in greater detail by visual observation (television camera or submarine).

3. One should stress two elementary though fundamental points for using the results of geophysical surveying:

– first, particular care is required in making the measurements, particularly in using the instruments and equipment,

– second, the quality of the programme – locational accuracy, choosing the best equipment, density of the mesh for the profiles – governs the accuracy of the documents prepared.

III.1 BATHYMETRY TECHNIQUES : ECHO SOUNDER AND PRESSURE SENSOR

Bathymetric observations:

– are made in all cases with an echo sounder,

– may also be further specified over a limited area by means of a pressure sensor used from a submersible vessel (submarine or undersea robot).

III.1.1 Principle and equipment of the echo sounder

III.1.1.1 Principle of the echo sounder

The echo sounder puts out a brief ultrasonic pulse which is reflected from the sea bottom. The return echo is amplified and then continuously recorded.

Let V be the speed of sound in water and t the time interval between the emitted and return echo, the depth H is given by:

$$H = \frac{Vt}{2}$$

III.1.1.2 Equipment of the echo sounder

Transmission and reception are ensured by a common electro-acoustic transformer or **transducer** which converts the mechanical vibrations into electrical vibrations of the same frequency [1].

Coupled to an electric pulse generator, the transducer converts the electrical energy into acoustic energy on transmission, and conversely the reflected acoustic signal is converted into an electrical signal.

The most widely used transducers are based on the piezoelectric properties of certain ceramics (barium titanate, zirconate). They vibrate at a certain resonance frequency. These vibrations, transmitted to the water, act as sound pulses.

The optimum frequency range, which depends on the dephts of water and nature of the bottom, extends from about 15 to 200 kHz, depending on the type of device. The higher the frequency, the more efficient the absorption.

At the recording end, the propagation times measured are converted into depth, depending on the speed of sound in water (from 1,460 to 1,560 m/s in sea water). For a given speed, the rate of the stylus, which inscribes along a strip of paper, determines the scale of the soundings, namely the number of metres of water represented on the width of the recording paper.

III.1.2 Characteristics of transducers and resolving power of echo sounders

III.1.2.1 Characteristics of transducers

Transducers are characterized by their nominal frequency, directivity and level of energy.

The nominal frequency of a transducer designates its transmission frequency under permanent excitation (i.e. resonance).

The directivity of the transducer corresponds to the angle of its lobe (see Fig. III.1.1).

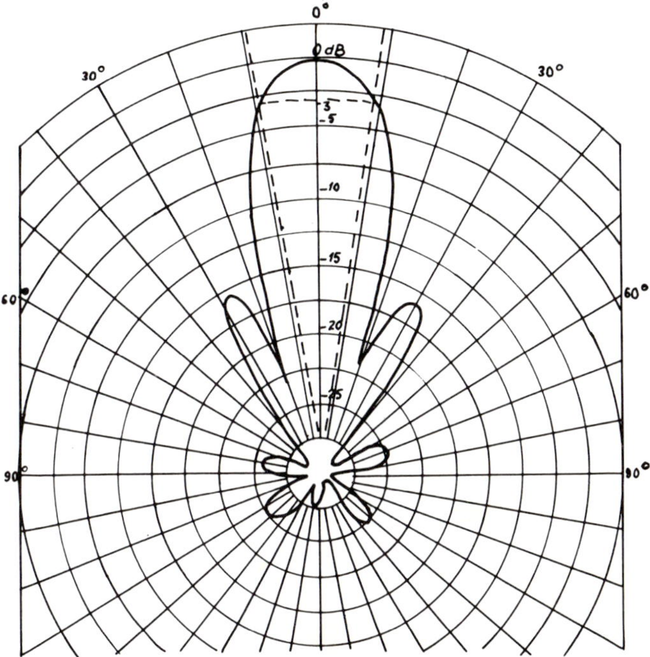

FIG. III.1.1. – Example of directional diagram of transducer.

For precision echo sounders, used for bathymetry, the sound beam is relatively narrow. The following are typical orders of magnitude:

– for common echo sounders:

10-20° at 50-30 kHz

– for large diameter echo sounders with very narrow beams, used at great water depths:

3-6° at 30-15 kHz

The transmission level of a transducer is a measure of the energy transmitted along the axis of the transducer, measured one metre away. A high transmission for the same electric power is the sign of better efficiency.

III.1.2.2 Resolving power of an echo sounder

Resolving power of an echo sounder essentially depends on the duration of the pulse, the angle of the ultrasonic beam, the depth of the water and the topography of the bottom.

A resolving power is limited by the fact that it is impossible to transmit an extremely brief signal.

If Δt is the shortest discernible time interval between two echoes, then the depth resolution is:

$$\Delta H = \frac{V}{2} \, . \, \Delta t$$

where: V is the speed of sound in water.

Figure III.1.2 shows the maximum resolution that can be expected from a bathymetic system in terms of pulse duration.

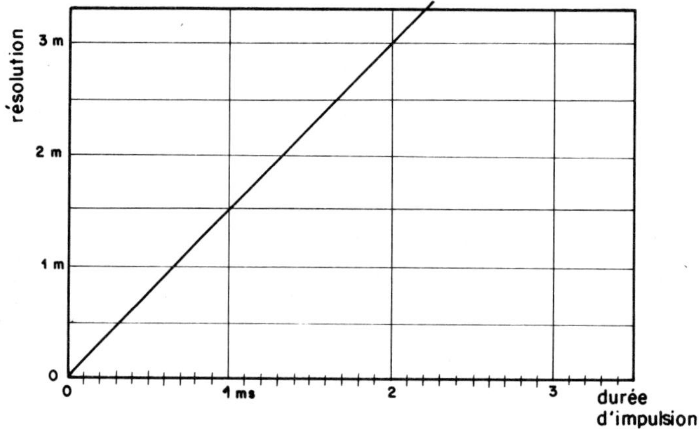

FIG. III.1.2. - Resolution in terms of pulse duration (for $V = 1{,}500$ m/s.

In order to improve this parameter, a recent technique known as pulse compression makes it possible by digital processing of the recording to convert the signal observed on the recording into a shorter arbitrary theoretical signal (correlation between the transmitted and received signals).

The angle of the ultrasonic beam results in space in a cone shape. The surface S swept (circle or ellipse) depends on the angle of the cone, the depth of the water and the topography of the bottom.

The following formula is for a flat bottom (see Fig. III.1.3):

$$L = 2d \operatorname{tg} \frac{\theta}{2}$$

where: d is the distance between the transducer and the sea bottom.

The first echo from the bottom results from the interaction between the wave front and the bottom (plane tangential to a sphere): the echo therefore orginates from the top most point of the surface swept and not necessarily from the point vertically beneath the transmitter.

The duration of the reflected signal depends on the time needed to develop the surface Σ on the bottom (see Fig. III.1.4) and the duration of the initial pulse. In practice, allowing for the complex shape of the bottom and the distortion of the wave front (anisotropy of the speed of sound in water), it is all but impossible to predict the duration of the signal.

As a result, without the pulse compression technique, the diagram of Fig. III.1.2 represents an optimum which is never actually reached.

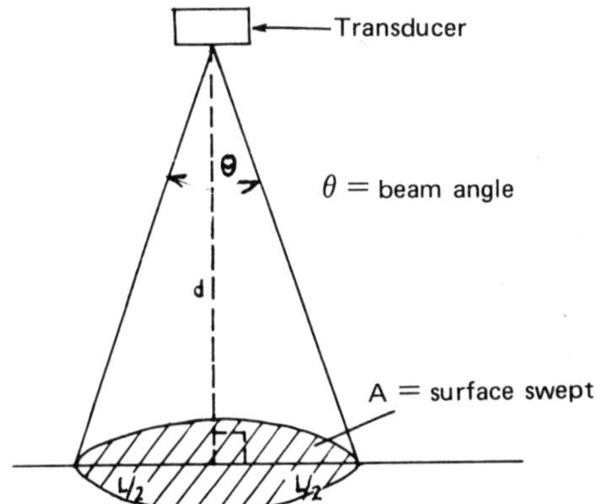

FIG. III.1.3. – Geometry of sound beam.

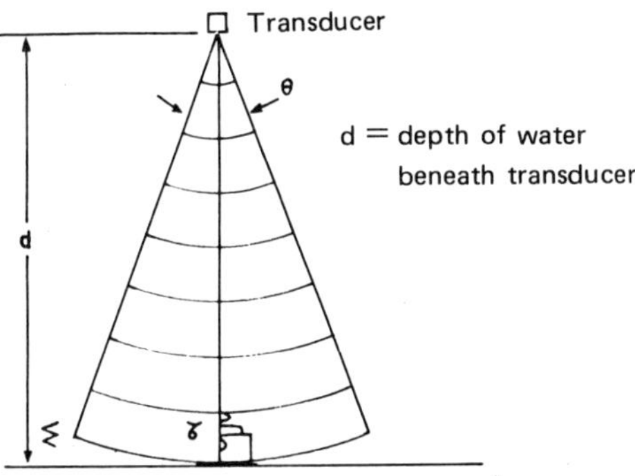

FIG. III.1.4. – Influence of shape of wave front.

III.1.3 Implementation of echo sounders and existing equipment

III.1.3.1 Implementation of echo sounders

Transducers ("bases") are generally installed on the hull of the vessel, with the beam aimed vertically downwards.

They may be:

– either permanently secured to the ship (which is invariably the case for narrow beam transducers installed on certain oceanographic vessels),

– or installed temporarily on the side of the ship, at the end of an outboard boom.

The effect of turbulence is avoided by placing the transducer ahead of the vessel and submerging it to a slight depth (1 to 5 m).

The effects of the motion of the ship (roll and pitch) are minimized or even completely counteracted by using the transducer in the central well or by stabilizing its base by means of a gyroscope system.

Depending on the accuracy needed, the optimum speeds of the vessel are from 3 to 6 knots, depending on the depth of water and the relative complexity of the relief of the bottom.

The amplitude of the waves must not be higher than 1.5 m, nor must be wind speed exceed 20 knots (force 5).

Utilization of the results implies:

– first, very frequent calibrating measurements,
– second and if necessary, recording of the roll, pitch and heave motions.

III.1.3.2 Main precision echo sounders

There can be no question of making a catalogue of all precision echo sounders, particularly all the existing types of various constructors.

Table III.1.1 summarizes the essential characteristics of the main current units (generally with beam angles of 10 to 20°) used in maximum water depths of a few hundred metres [1].

Large diameter (about 1 m) transducer echo sounders with narrow beams (between about 3 and 6°) are permanently mounted on certain oceanographic research vessels. For instance:

– the "d'Entrecasteaux" of the *French Navy* is equipped with a Edo Western unit with a beam angle of 6°, 4.7° or 2.8° (at frequencies of 16.25 or 34 kHz) used in the "Famous 73" operation at a depth of 2,000-3,000 m of water,

– the "Valdivia" of *Preussag* is equipped with an Elac Schelfrandlot unit with a beam angle of ±1.4° (30 kHz), used for surveying at depths of 2,000 m in the Red Sea.

The "Charcot" of *CNEXO* (France) is equipped with a "Seabeam" unit (sixteen beams) for surveying at large depths (2000-3000 m of water).

III.1.4 Improved bathymetry techniques and their implementation

One has to consider:

– first, possible improvements in the implementation of echo sounders from the surface,
– second, the carrying-out of bathymetry techniques from a submarine.

III.1.4.1 Echo sounders secured to the hull of a vessel

There are various possibilities of improving the quality of bathymetry observations made from a surface support vessel.

TABLE III.1.1
GENERAL CHARACTERISTICS OF A NUMBER OF ECHO SOUNDERS

Maker and type	Kelvin Hugues MS 26 types A and F	Krupp Atlas-Deso 10	Elac Laz 17 Type CAST OR	Elac Laz 71 R	Raytheon and other
Frequency Energy	30 kHz and 32 kHz	Double frequency among : 30, 33, 80, 100, 210 kHz 300, 150, 150, 150, 150 W	50 kHz	12, 15, 20, 30/190, 50, 200 kHz depending on model + option	Depending on trans- ducer model (16 to 208 kHz)
Beam angle	15° (30 kHz), 13° (32 kHz)	15/20° (30 kHz), 19 and 10° (80 kHz), 15° (100 kHz), 8° (210 kHz)	10-15°/15-20°	10-15°/15-20°	Variable: 8° at 208 kHz Possibility of narrow beam (6.5° at 16 kHz–3° at 200 kHz)
Pinging rate	267 to 553 pulses/min	0-20 m = 10 pulses/s 0-40 m = 5 pulses/s 0-80 m = 2,5 pulses/s	Up to 540 pulses/min	20 to 150 pulses/min	(Type DE-7,198-208 kHz) 534 pulses/min for range of 120 m
Range	80 and 160 m	Maximum range: 1,400 m	Maximum range: 320 m	Maximum range: 2,000 m	
Propagating speed calibration	By adjustment of stylus speed	1,360–1,480 m/s 1,420-1,540 m/s (standard) 1,480-1,600 m/s		Variable from 1,400 to 1,600 m/s	Variable (1,300-1,700 m/s)
Accuracy		Scale 0-20: ± 5 cm 0-40: ± 10 cm 0-80: ± 20 cm		± 0,5%	± 0,25%
Recorder – Paper type width – Paper advance rate – Dimensions – Weight	Circular – angle of stylus 60° Dry 15.5 cm Standard: 2.5 cm/mn 59.5 x 34.5 x 24 cm 41 kg	Dry 23 cm 1.25-2.5-5-10 cm/min 44 x 44 x 19.5 cm 21 kg	Dry 20 cm Adjustable between 0 and 48 cm/min	Dry 20 cm from 0.4 to 6 cm/min 50 x 51 x 63.5 cm	Model UGR 196 C 48 cm 87 x 51 x 29 cm 55 kg
Observations	Circular recorder not very practical A MS 45 model exists	Can be coupled to a digital indicator (Atlas-Edig 10) Combined version (with Atlas- Digigraph digital indicator)		Can be coupled to an electronic scale expansion device enabling an expanded scale to be recorded. Can be coupled to a digital recorder (DAZ 6 or 8)	Can be coupled to a pulse correlater (CESP II) and/or a digital indicator (POD-200)

By reducing the angle of the ultrasonic beam, the irregularities of the bottom can be studied in greater detail. Use of a narrow beam echo sounder (a few degrees) is absolutely indispensable for water depths of more than a few hundred metres.

The stabilization of the transducer base, by adding a gyroscopic device, eliminates the effect of the vertical motion with relation to sea level.

The measuring quality is improved by digital recording of the depth (use of all-electronic time bases, elimination of range change errors, electronic processing of the echo signal, etc.). However, analog recording (which conveys richer information) must be retained and used at the same time.

Automation of the reading of the positions and depths facilitates the use of the results.

Improved knowledge of the tide on continental shelves enables a considerable number of depth correction errors to be avoided.

Simultaneous use of an echo sounder and a side-scan sonar (which displays the bottom between two echo sounder profiles) would appear highly desirable. Fig. III.1.5 shows that it is possible to "miss" a relief using an echo sounder alone. When used simultaneously, the side-scan sonar shows the need to record an additional echo sounder profile at the relief thus detected.

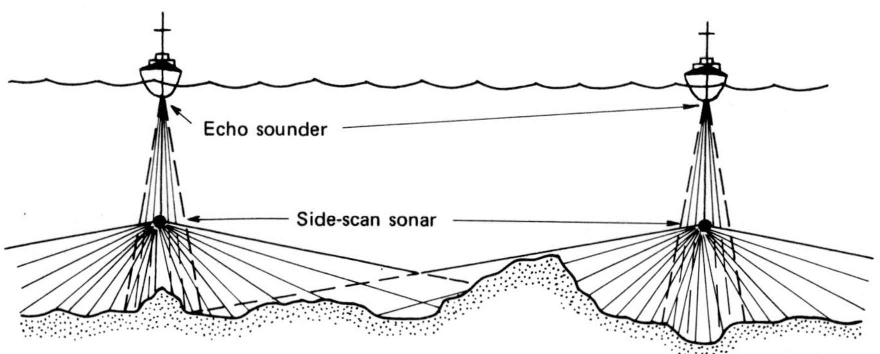

FIG. III.1.5. – Bathymetry through the combined use of an echo sounder and a side-scan sonar.

III.1.4.2 Use of an echo sounder or pressure sensor from a submarine

Two techniques can be considered:

– first, use of an echo sounder near the bottom,
– second, use of a pressure sensor from a submarine.

Better accuracy in great water depths may be obtained by reducing the distance between the sea bottom and the echo sounder:

– either by towing equipment near the bottom,
– or by installing them on submarines working near the bottom.

The search for great **relative** accuracy (compared to a reference point), for siting gravity structures, for instance, is satisfied thanks to a sensor installed on a submarine (resting on the bottom) and connected:

- on the one hand to a constant pressure cell equivalent to the hydrostatic pressure of the reference point,
– on the other, to a device enabling the pressure variations resulting from the relief and differences in level of the bottom to be followed.

The procedure requires various corrections: tide, drift with time (particularly as a function of temperature), waves, atmospheric pressure, etc.

The relative accuracy obtained is about 10 cm.

Taking bathymetry readings from a submarine calls for location by means of pingers moored to the bottom.

The location and very appreciable increase in the cost of these new bathymetry techniques limits their application to relatively small areas previously reconnoitred by conventional echo sounder methods used from a surface support.

III.2 ANALYSIS AND INTERPRETATION OF ECHO SOUNDER RECORDINGS

The search for accurate bathymetry (to the nearest 10 cm for siting gravity structures, for instance):

– sets considerably more stringent requirements than for obtaining other geophysical data,
– and calls for very rigorous analysis and interpretation of the recordings made.

III.2.1 Plotting and description of echographs

III.2.1.1 Plotting of echographs

The depth is obtained by continuous analog recording. Because of this, the recording does not give the depth at a single point vertically below the vessel, but the minimum distance to the bottom within the cone generated by the beam angle (see Paragraph III.1.2.2).

On the Continental Shelf, the error in measurement of the depths vertically beneath the sounder is generally very small when precision echo sounders with a beam angle of 10 to 20° are used.

In great depths, only a narrow vertical beam sounder with an angle of just a few degrees, often used together with a conventional sounder, can provide a bathymetric recording which is exact in its details.

III.2.1.2 Description of recordings

Echo sounder recordings take the form of depth sections (for a given speed of sound in water generally defined as 1,500 m/s). They show:

– a thick black line at the top corresponding to the transmitted pulse,
– below, the profile of the sea bottom.

Figure III.2.1 shows an example of an echograph obtained from an echo sounder.

The horizontal and vertical scales are generally different, with a vertical exaggeration of a few tens, with typical recording conditions (speed of vessel 4 to 5 knots, paper speed 1 to 2 cm/min, paper about 20 cm wide).

III.2.2 Analysis of recordings

The true depth at each point is given by the following relation:

true depth = depth measured ± calibration correction – tide correction.

III.2.2.1 Measurement of depth

The accuracy in depth is first defined by the operational quality of the power supply (voltage regulation) and the equipment (adjustment of the stylus drive motor).

To attain the exact depth, one must first of all "trim" the measurement from the effect of vertical motions of the transducer with relation to sea level:

– in the case of a uniform bottom, hand smoothing of the analog recording, enables the profile of the bottom to be obtained with sufficient accuracy,
– in the case of an irregular (rocky) bottom or one with a microrelief (ridges), the amplitude of which may approach that of the vertical movements of the ship, all the recording details will be masked by smoothing. Simultaneous recording of heave and pitch and roll will therefore facilitate the task of obtaining the exact depth.

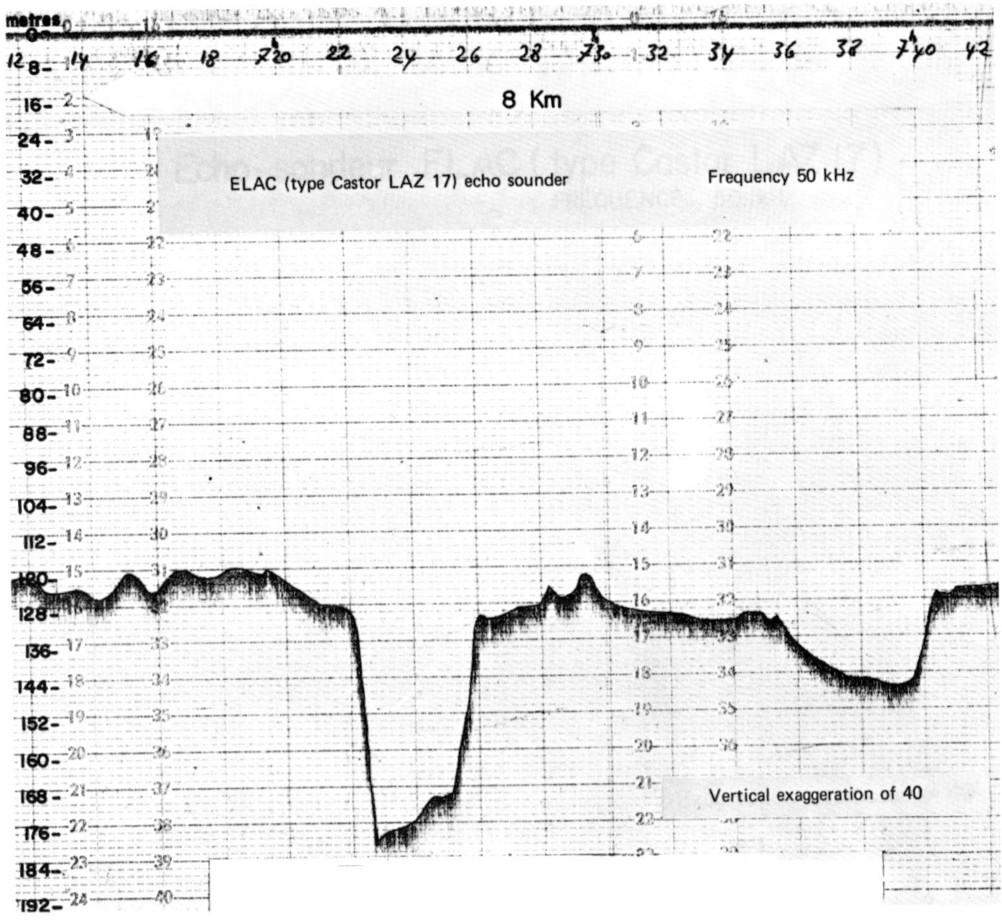

FIG. III.2.1. – Example of echo sounder recording.

III.2.2.2 Corrections to the speed of sound

The speed of sound in sea water (from 1,460 to 1,560 m/s) depends on the temperature (in particular), the salinity and the pressure (depth).

Owing to these highly appreciable vertical and local variations, echo sounders must automatically be calibrated.

In most cases, correction of the speed of sound is allowed for from **calibrations** of the echo sounder frequently performed in situ (preferably twice a day).

This calibration consists in comparing the "depth read" on the echo sounder recording against the "true depth". The "true depth" is materialized by a reflecting metal plate suspended from a carefully graduated steel cable lowered as close as possible to the bottom. Since the measurement is taken with relation to the surface of the water, one must clearly make allowance for the submersion of the transducer.

For each series of measurements, a calibrating curve is then established giving the corrections to be applied for the various depths.

More rarely, correction of the speed of sound is made on the basis of hydrological measurements using bathythermographs (generally) or directly by means of celerimeters and made at the time the bathymetric reading is taken.

For this determination the following are used [2]:

– either nomograms such as that shown in Fig. III.2.2,
– or empirical formulæ, such as :

$$V\,(\text{m/s}) = 1{,}410 + (4.21\,t - 0.037\,t^2) + 1.105\,s + 0.018\,d$$

where:

t = temperature in degrees centigrade,
s = salinity in ppm,
d = depth of water in metres.

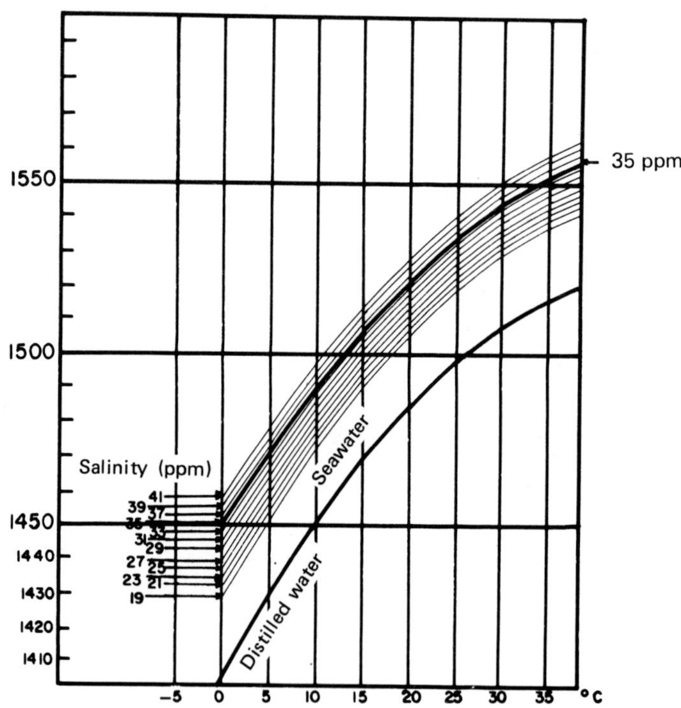

FIG. III.2.2. – Speed of sound in water (on the surface) in terms of temperature and salinity (according to Del Grosso).

III.2.2.3 Correction for the tide

The depths obtained after correction of the speed of sound must be "reduced" by the tide, with in most cases the level of reference being a low water level (lowest water level for the French hydrography datum, *Lowest Astronomical Tide* (*LAT*) for the North Sea, etc.).

The task of obtaining a truly reliable tidal correction sets a delicate problem. In regions where the amplitude of the tide is large, it most certainly represents a major source of inaccuracy in the bathymetry measurement.

The determination of the tide at present in most cases rests on observations ashore and a necessarily imperfect understanding of the system:

- of co-tidal lines (where the tide occurs at the same time),
- and the lines of equal amplitude.

Only use of perfectly reliable tide graphs submerged on the research site provides a satisfactory solution to the problem.

III.2.3 Interpretation of echographs

III.2.3.1 Theoretical interpretation

The exact interpretation of the true relief calls for an understanding of the echo recordings generated by the geometric shapes of the sea bottom [3].

The interpretation is governed by two main factors.

The recorded gradient of the sea bottom is less than the true gradient of the bottom, such that:

$$\sin \theta = \operatorname{tg} \phi$$

where:

$\sin \theta$ is the true slope of the bottom,
$\operatorname{tg} \phi$ is the slope of the echo trace.

Up to about 15°: $\sin \theta \approx \operatorname{tg} \phi$, and the error is therefore practically negligible.

Figures III.2.3 and III.2.4 show the sounding profiles per echo diagrammatically for a sea bottom, for constant and variable gradients respectively.

Any distinct rise of the bottom or any distinct variation of the slope gives a hyperbolic echo trace.

Figures III.2.5 and III.2.6 illustrate examples of echo curves originating restrictively from a rise and a *V* depression.

III.2.3.2 Use of the results: bathymetry

The depths must be subjected in turn to:

- analysis on analog recordings,
- correction and reduction,

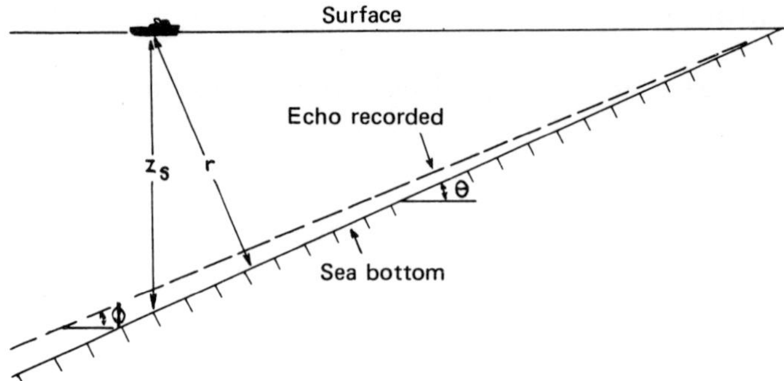

FIG. III.2.3. – Echo sounding profile from the bottom of a constant gradient sea bottom.

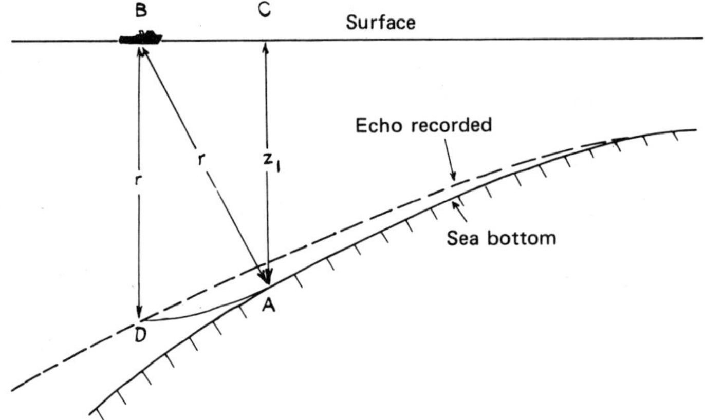

FIG. III.2.4. – Echo sounding profile from a sea bottom with a variable gradient.

– transfer to the planes of position (the position of the echo sounder and not of the radiolocation antenna of the vessel).

The ideal way to choose the depths to be transferred is to note all noteworthy values along the profiles measured at all the gradient change points, so that the gradient can be considered as perfectly regular between two values (**sounding map**).

The following stage consists in plotting contour lines (**bathymetry map**), with equidistance depending:

– first, on the degree of accuracy of the depth readings,
– second, on the density of the profiles made.

This series of operations is carried out:

– either manually,

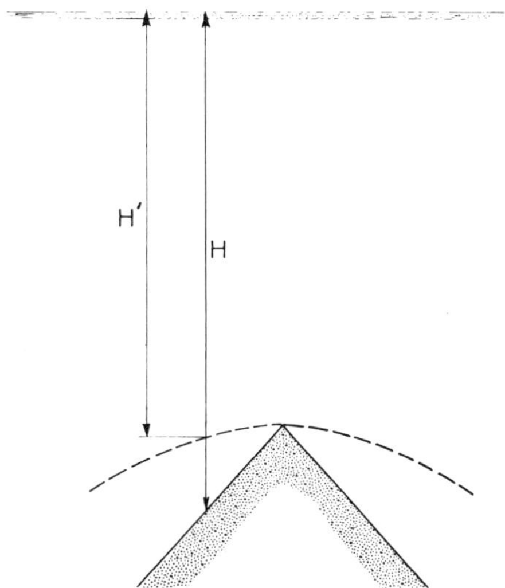

FIG. III.2.5. – Curve of an echo originating from a pointed peak (*H* true depth ; *H'* depth recorded).

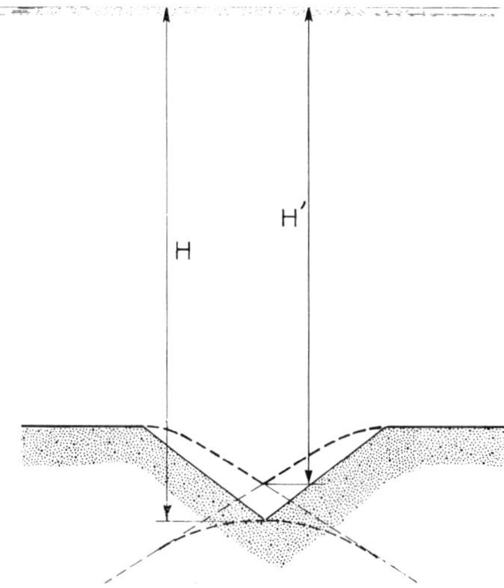

FIG. III.2.6. – Echo curve originating from a "*V*" hollow (*H* true depth ; *H'* depth recorded).

– or automatically (digitization, print-out of numerical elements on punched cards or magnetic tape, filing of these elements with data processing by means of appropriate software, print-out of the final documents in the form of lists and/or tracing board plots).

In the latter case, despite the existence of many algorithms programmed on a computer, the plotting of the contour lines can generally be perfected by manual "smoothing".

III.3 SEA BOTTOM OBSERVATION TECHNIQUES: SIDE-SCAN SONAR AND TELEVISION

Surveying of the topography (morphology and roughness) of sea bottoms is conducted:

– almost invariably by indirect observation by means of a side-scan sonar,

– sometimes, to specify certain details, by direct observation (visual), television and/or photography.

III.3.1 Principle of the side-scan sonar. Formation of the echoes

The side-scan sonar transducer acts both as transmitter and receiver of the ultrasonic signals.

III.3.1.1 Brief description and principle of operation of the side-scan sonar

The system generally consists of (see Fig. III.3.1) [4] [5] [6]:

– a round-nosed cylindrical body towed from the vessel (known as the "fish"), containing one or two (1) transducers (together with the associated electronic circuits),
– a towing cable ensuring the electrical and mechanical links to the towing vessel,
– a one or two rack recorder using either electrosensitive paper or a magnetic tape.

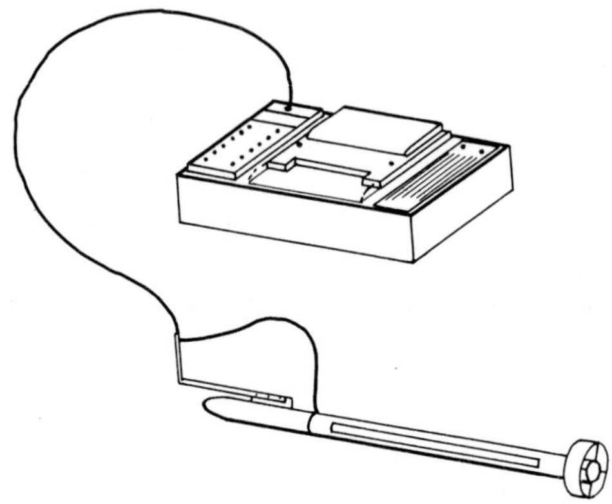

FIG. III.3.1. – Side-scan sonar system.

The side-scan sonar transducer:

– transmits short sound pulses to the water, perpendicular to the direction of travel,
– receives the echoes recorded aboard the vessel (following conversion into electric pulses).

The frequencies used vary from a few tens to about 100 kHz, depending on the particular unit.

III.3.1.2 Formation of the images

The sound pulses transmitted at regular time intervals (the repetition rate essentially depends on the lateral range selected) and the echoes resulting from the irregularities on the sea bottom are recorded as a function of time (two-way trip): clearly, the nearest echoes arrive first, followed by echoes from more distant zones at ever increasing intervals.

(1) The sonar is generally bilateral.

Each group of echoes resulting from a transmission is displayed on the recorder in the form of a trace inscribed cross-wise by the stylus on the recording paper which moves longitudinally (2).

As the vessel advances and the pulses occur one after the other, an image is formed on the recording paper by juxtaposition of the traces (somewhat similar to that obtained on a television screen).

III.3.1.3 Geometry of the ultrasonic beam

The fineness and precision of the recording are a function of the narrowness of the ultrasonic beam, and of the frequency and duration of the pulse transmitted.

The shape of the transducer is selected so as to transmit a fan-shaped beam (see Fig. III.3.2):

– with an angle of a few degrees in the horizontal plane (azimuth),
– with an angle of about 10 to a few tens of degrees in the vertical plane (elevation).

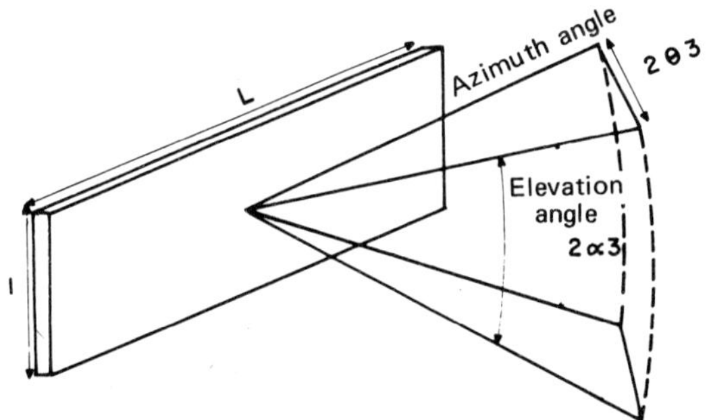

FIG. III.3.2. – Side-scan sonar beam.

The ultrasonic beam can be broken down into the following:

– a primary lobe with an angle defined conventionally as the sector in which the sound intensity is only 3 dB beneath that of the axial (maximum) intensity,
– a number of secondary lobes.

Figure III.3.3 gives a diagrammatic representation of the beam in the horizontal and vertical planes.

The geometry of the primary lobe is defined by the following equations:

$$2\theta_{3\,(\mathrm{radian})} = \frac{\lambda}{L} \qquad 2\alpha_{3\,(\mathrm{radian})} = \frac{\lambda}{l}$$

(2) The recording plot is darker or lighter in proportion to the intensity of the reflected sound signal.

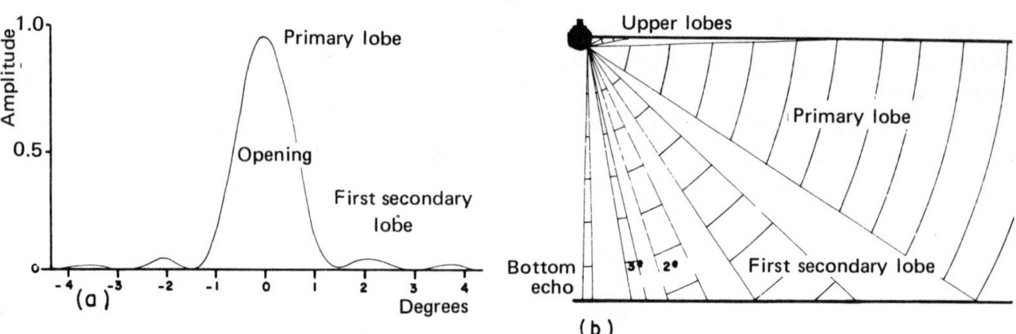

FIG. III.3.3. – Diagrammatic representation of the primary and secondary lobes :
a. In the horizontal plane.
b. In the vertical plane.

where:

L and l = dimensions of transducer,
λ = wavelength.

Even though only the primary lobe is actually used in practice, the secondary lobes present a certain interest. In particular, the sub-vertical lobe:

– gives a section of the bottom of the sea along the path of the vessel,
– enables any echo from an object situated in the water near the vertical of the vessel to be identified (for instance a shoal of fish).

III.3.1.4 Production of the recordings

The regions inscribed in the narrow band $ABCD$ (see Fig. III.3.4) reflect the energy back to the transducer. The initial side echoes recorded will originate from AB and then at increasing lapses of time from EF and lastly CD, these points corresponding to the maximum range of the instrument. By recording the echoes as a function of the travel time, a veritable scan of the terrain is made.

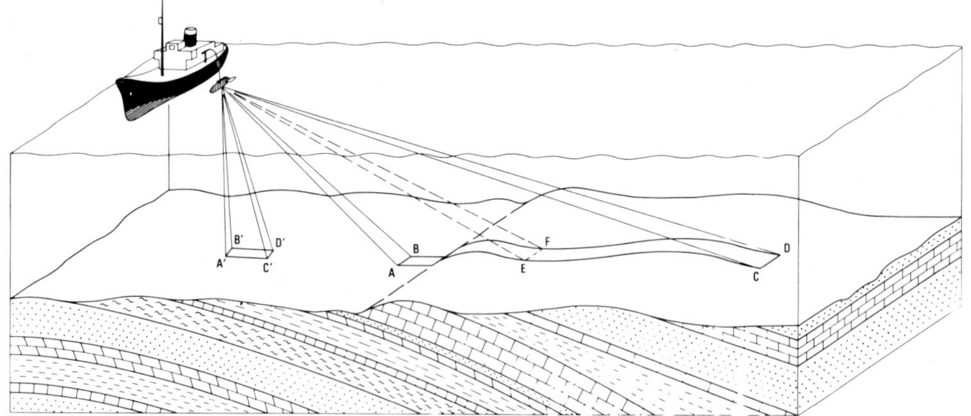

FIG. III.3.4. – Schematic illustration of the geometry of a side-scan sonar beam (only the port beam is shown).

As the vessel advances, the following are obtained:

– first, a continuous bathymetric profile near to the time origin, vertically beneath the transducer,

– second, a relatively continuous image of the bottom as far as the maximum range capability of the instrument.

III.3.1.5 Formation of the echoes. Angle of incidence

The features of the bottom brought to light are:

– either of topographical nature (variation of the angle of incidence),

– or related to the physical characteristics of the soil (variations in the coefficient of reflection or backscattering).

The way in which topographic echoes are formed is shown in Fig. III.3.5. All the folds in the bottom cause the angle of incidence of the acoustic rays to vary and hence also the amount of reflected energy.

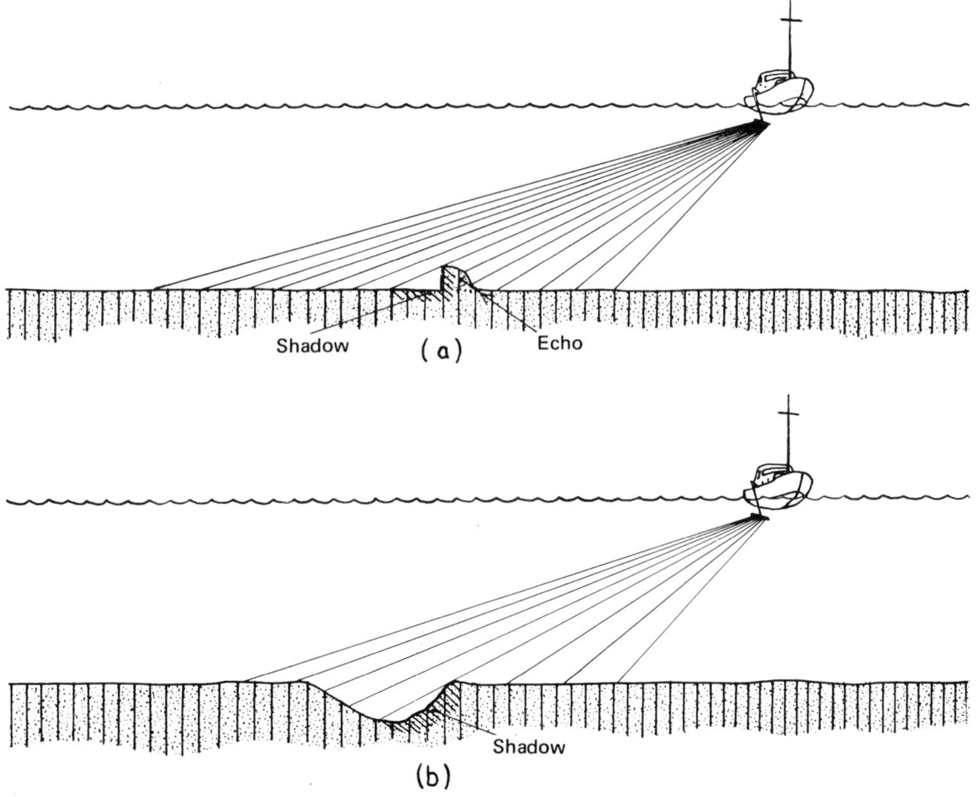

FIG. III.3.5. – Diagrammatic representation of how echoes and shadows are formed:
a. For a relief.
b. For a depression.

The useful part of the recording is that corresponding to angles of incidence of less than 30°, where the coefficient of reflection varies sharply with the angle of incidence. The ideal conditions therefore prevail for detecting variations in the angle of incidence and hence variations in the topography.

A change in the nature of the bottom modifies the intensity of the signal as much or even more than a change in the gradient (especially if the angle of incidence is between 20 and 60°). The reflection coefficient varies considerably when changing from mud to pebbles or rock, while sand lies somewhere in between.

III.3.2 Characteristics of the side-scan sonar

The side-scan sonar is essentially characterized by its range and its longitudinal and transverse resolving powers.

III.3.2.1 Lateral range

The maximum range of a side-scan sonar depends on many factors, the leading ones being:

- the characteristics of the instrument:
  - the pulse duration,
  - the transmission power,
  - the signal/noise ratio,
  - the frequency ($rF^2 = 1{,}300$ is an empirical formula expressing the range in kilometres for an optimum frequency in kilocycles),
- the physico-chemical properties of the medium through which the sound waves are propagated,
- the implementation parameters (Fig. III.3.6):
  - the height of the "fish" above the bottom,
  - the inclination of the axis of the beam from the horizontal.

III.3.2.2 Longitudinal resolution (along the direction of advance of the vessel)

The resolving power is the minimum distance between two points of the surface explored yielding two distinctly separate echoes on the recording. It depends essentially on the pulse duration and the angle of incidence of the sound wave at time *T*.

Owing to the angle of the ultrasonic beam in the horizontal plane (a few degrees), a point object is recorded throughout several transmission cycles (see Fig. III.3.7a).

The longitudinal resolution ΔY depends (Fig. III.3.7b):

- on the angle of the beam θ in the horizontal plane,
- on the recording time *T*, i.e. the distance between the object and the transducer.

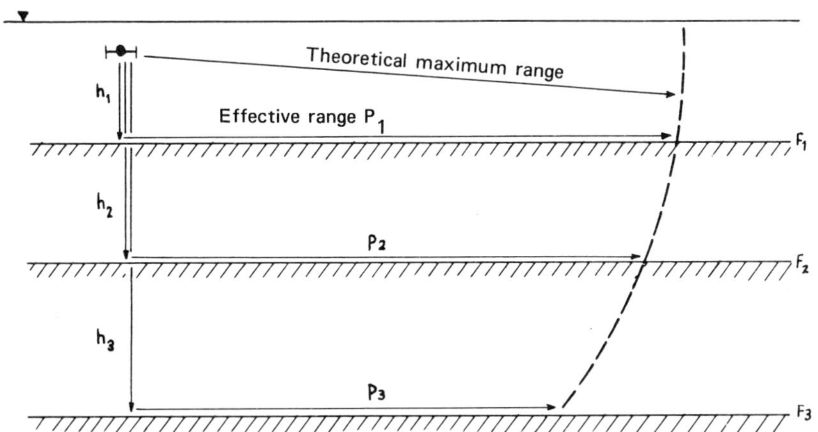

FIG. III.3.6. – Relationship between water depth h and the effective range P.

This is expressed by the relationship:

$$\Delta Y = VT \operatorname{tg} \frac{\theta}{2}$$

where:

V = speed of sound in water.

For example, for values of:

$$T = 0.4 \text{ s (lateral distance 300 m)}$$
$$V = 1{,}500 \text{ m/s}$$
$$\theta = 1.2°$$

the resolution $\Delta Y = 6.28$ m.

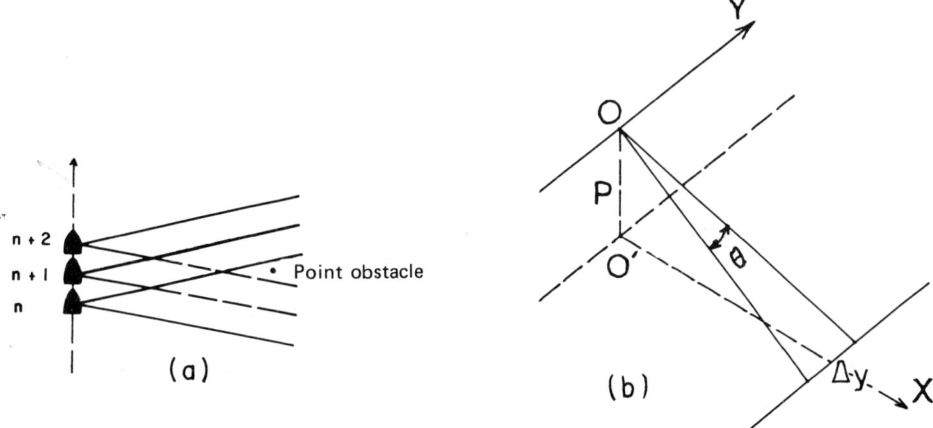

FIG. III.3.7. – Longitudinal resolution.

This longitudinal resolution is only theoretical, since in actual practice the effect of yaw, which is often greater than the angle of the beam in the horizontal plane, is predominant.

III.3.2.3 Transverse resolution (in the direction of scan)

The resolution is generally much better along an axis perpendicular to the direction of motion of the fish. It is all but independent of the depth of the water and depends only on the angle of incidence and the pulse transmitted.

The transverse resolution ΔX can be expressed as:

$$\Delta X = \frac{lVT}{2\sqrt{V^2T^2 - 4h^2}}$$

where:

l = duration of transmitted pulse,
V = speed of sound in water,
T = recording time (expressing the range of the echo),
h = depth of water beneath fish.

For example for values of:

$T = 0.4$ s
$V = 1{,}500$ m/s
$h = 30$ m
$l = 0.1$ ms, or 0.15 m

the resolution $\Delta X = 0.075$ m.

The nomogram given in Fig. III.3.8 shows that the resolution along the direction of scan is all the higher the greater the propagation time (i.e. the closer the angle of incidence is to the horizontal). For a given depth of water beneath the fish, the resolution therefore increases with the distance from the fish.

In practice, many other factors (scatter, refraction, attenuation, etc.,) limit the lateral resolution to 1/1,000th of the maximum range of the instrument.

III.3.3 Distortion of side-scan sonar images

There are various causes for the distortion of side-scan sonar images, including the following:

- the obliqueness of the beams,
- the slope of the bottom,
- the anisotropy of the medium through which the rays propagate,
- the navigating conditions,
- the scales on the recordings.

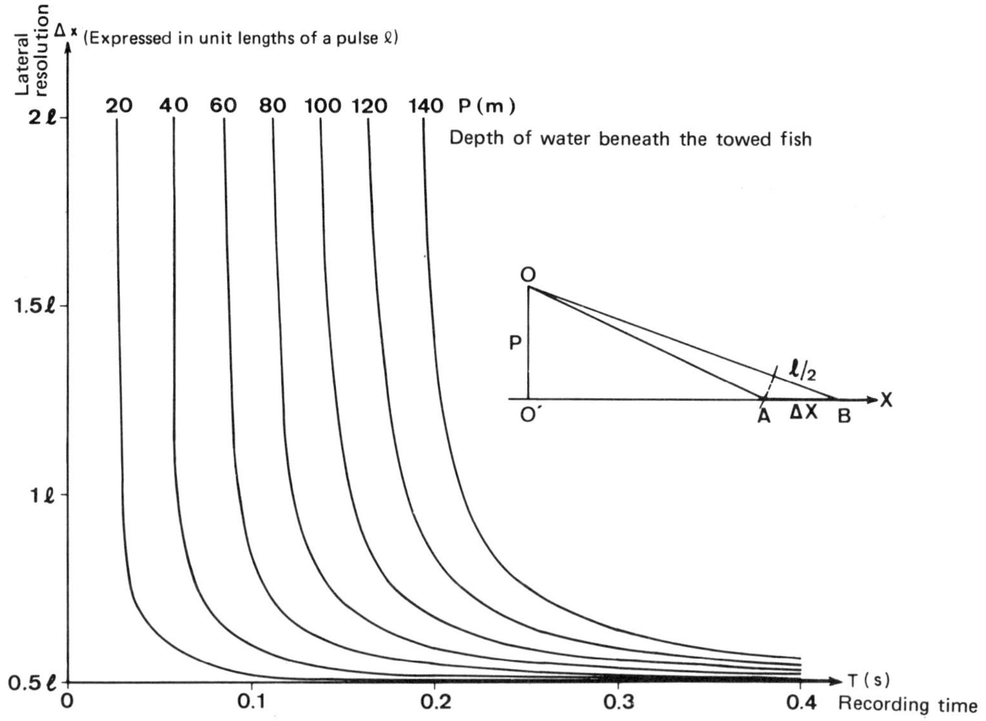

FIG. III.3.8. – Lateral resolution nomogram.

III.3.3.1 Distortion caused by the obliqueness of the beams

The lateral range X scale is not linear with the recording time T, as is shown in Fig. III.3.9.

The image recorded is all the less distorted, the more horizontal the angle of incidence of the beams. Figure III.3.10 shows the $X = f(T)$ curves plotted for various depths of water beneath the fish.

If it is intended to produce an isometric mosaic from the sonar recordings, when putting it together, the initial recording fringe (from $T = 0$) displaying the greatest distortions, must be eliminated.

III.3.3.2 Distortion resulting from the slope of the bottom

It is assumed that the bottom is flat with a slope of α and that the vessel is following a contour line (Fig. III.3.11).

The horizontal distance measured for a given time T from the vertical below the vessel differs depending on whether the beam is directed up the slope or down the slope.

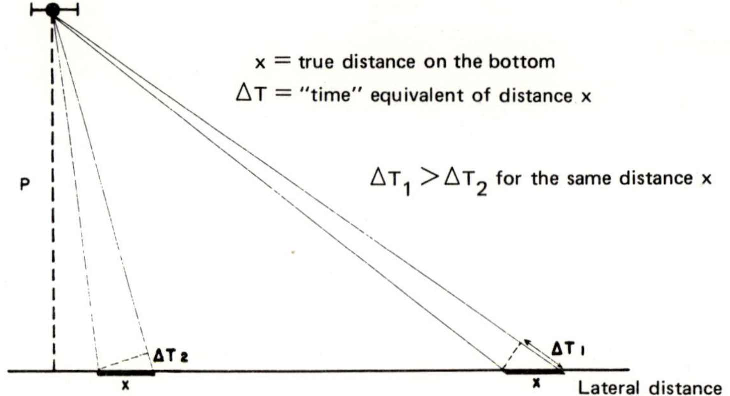

FIG. III.3.9. – Distortion resulting from the obliqueness of the beam.

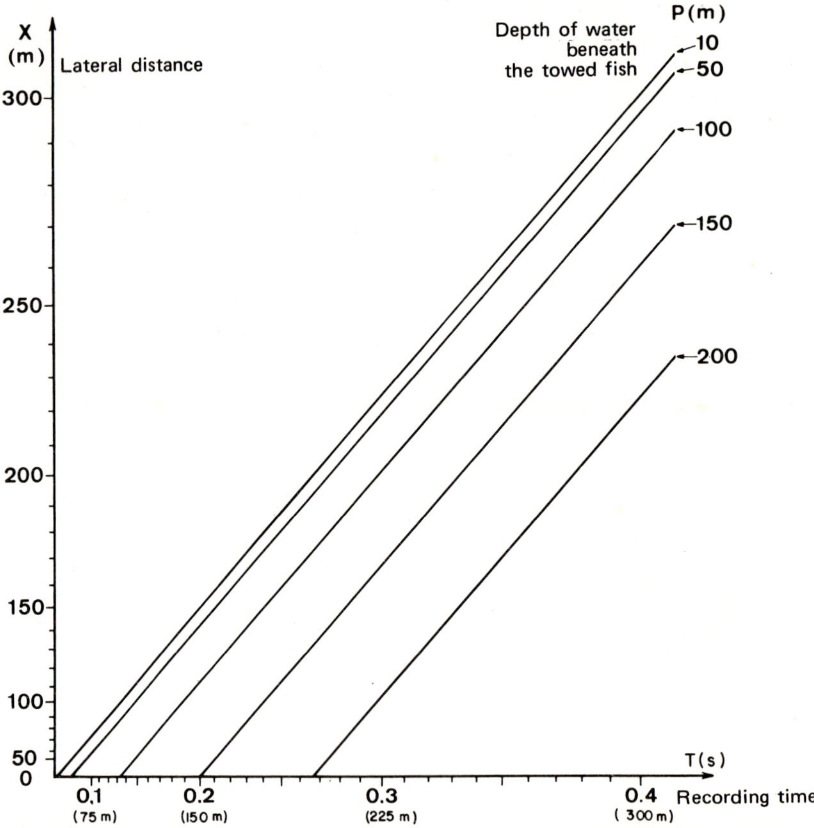

FIG. III.3.10. – Nomogram of distortion caused by the obliqueness of the beam.

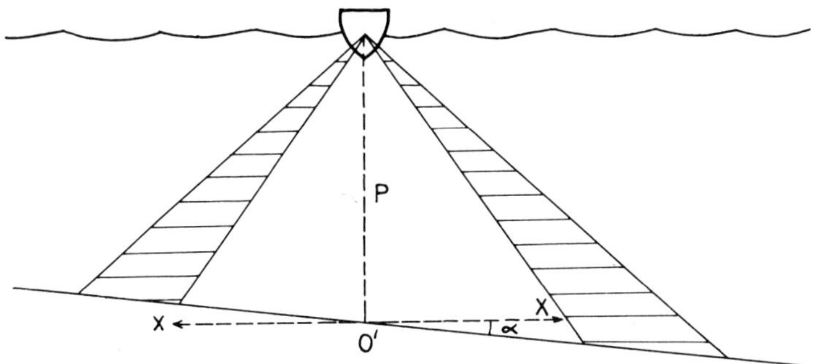

FIG. III.3.11. – Distortion caused by the bottom gradient.

III.3.3.3 Distortion caused by the anisotropy of the medium

A shortening of the maximum range can be brought about by the curve of the sound beams resulting from a velocity gradient in the water crossed:

– in the case of a negative velocity gradient, the distortion phenomenon is shown in Fig. III.3.12,
– in the case of a positive gradient, the sound beams are concave upwards and cannot attain the sea bottom (no bottom echo),
– in the case of a thermocline ([3]), considerable distortions also occur.

All these distortions are difficult to eliminate, since the distribution of the speeds of sound in the water crossed is generally inadequately known.

FIG. III.3.12. – Formation of a shadow zone caused by the curvature of the sound waves (negative velocity gradient).

III.3.3.4 Distortion caused by navigation and the instability of the fish towed

Very careful navigation along the profile governs the quality of the recordings and their quantitative interpretation (*X*, *Y*, *Z* fixing of the echo).

Yaw of the fish can be provoked:

– by the yaw motions of the vessels,
– by slack and tug cycles in the towing, all the more so the shorter the tow line.

(3) Thermocline: a zone of temperature discontinuity in the water.

On the recordings, yaw takes the form of jags that can easily be recognized on structures which are generally approximately rectilinear (rock outcrops, crests of dunes, etc.) (Fig. III.3.13).

The **roll** and **pitch** movements of the fish do not give rise to any appreciable distortions on the recordings.

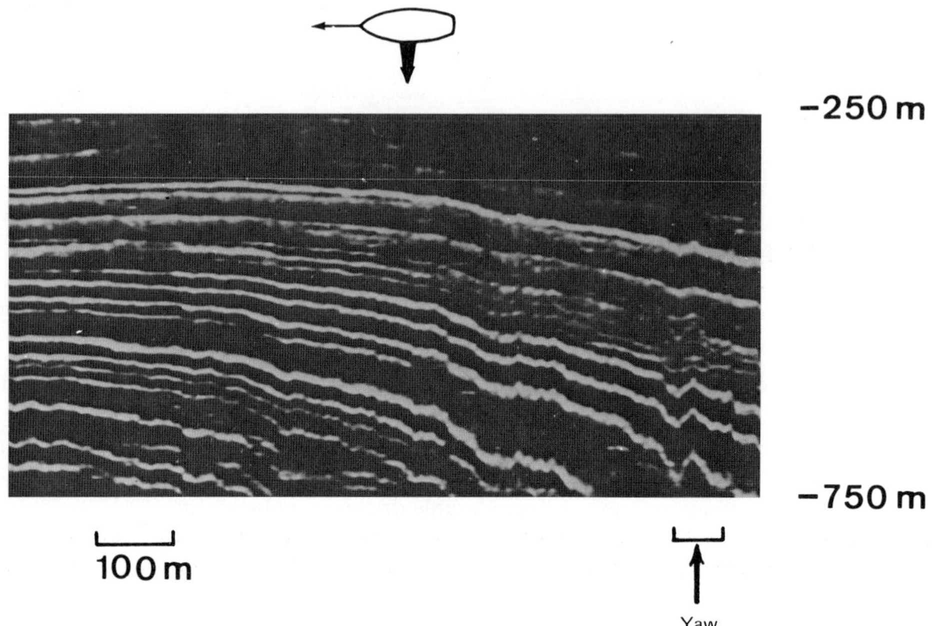

FIG. III.3.13. – Example of distortion resulting from the yaw of the towed fish.

Drift, or the angle between the leading of the vessel (or fish) and the actual path followed, may be caused either by **wind** or **current**:

– in the first case, the drift is not transmitted to the towed fish (Fig. III.3.14a),

– in the second case, the vessel-fish combination will display a drift with relation to the path followed (Fig. III.3.14b), in such a way that the ultrasonic beam is no longer perpendicular to the profile. This results in an error in the fixing of the echo that is all the greater the greater the range. It is therefore advantageous to orient the profiles in such a way as to avoid cross-currents.

III.3.3.5 Scale distortions on the recordings

Owing to the lack of flexibility of most recorders used and the difficulty in keeping the speed of the vessel constant when making the measurements, the range scales in the direction of advance of the vessel and the direction of scan are generally different. The second may be expanded four times or even more, compared to the first.

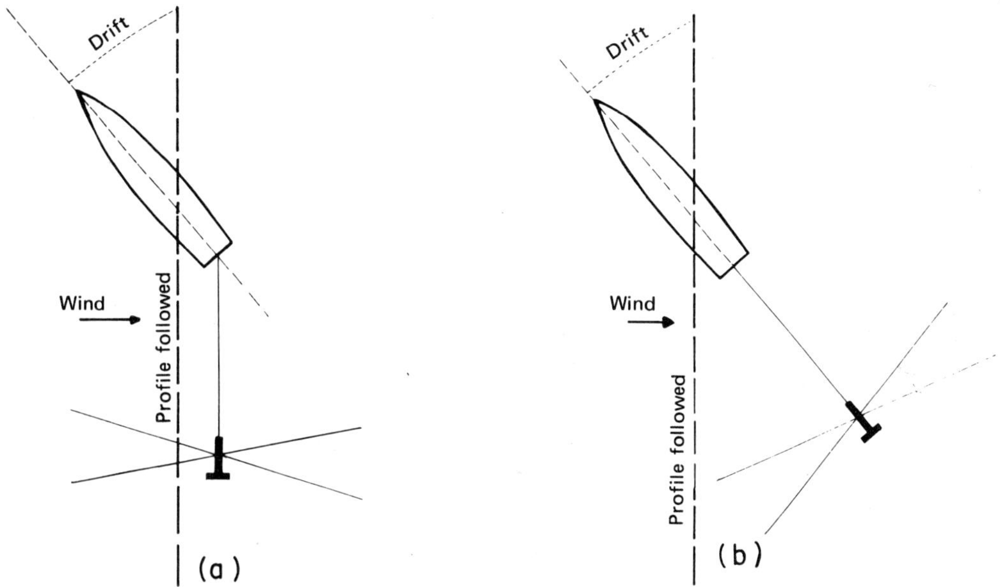

FIG. III.3.14. – Distortion due to drift.

These distortions, which are sometimes quite considerable, must be eliminated if it is intended to assemble together the sonar images in order to form an isometric mosaic. There are several methods of correcting the images, some of which are as yet still in the experimental stage (see Paragraph III.4.2.2).

III.3.4 Implementation of the side-scan sonar

III.3.4.1 Implementation technique

Except for the Transit sonar (*Kelvin Hughes),* secured to the hull of the vessel, side-scan sonars now on the market use a towed fish in which the transducers are housed. The fish is therefore independent of the motions of the vessel.

It is generally easy to use short range sonars owing to the great compactness and lightness of the fish (about 20 kg). It can be launched without a crane (two people are sufficient) and use of a winch is not indispensable. The latter does however provide a certain operational flexibility, particularly in adjusting the depth of submersion and in lifting the fish.

Large range (heavier) sonars require the use of a crane, preferably hydraulic, and a winch. Use of constant tension winch or any other system for absorbing tugs enhances the navigation of the fish.

Stern **towing** would appear the best solution and endows the vessel with greater manoeuvrability (elimination of the risk of entangling the cable in the propeller). The point of attachment of the towing cable must be as stable as possible, and hence must be located as near as possible to the centre of the vessel.

III.3.4.2 Towing speed

The quality of the recording (resolution and distortion) is a function of the speed of the vessel.

The manufacturers' manuals indicate the optimum towing speed in terms of the lateral ranges selected.

Common operating speeds:

– range from 2 to 6 knots for short range instruments,
– and attain 9 knots for large range instruments.

III.3.4.3 Depth of water beneath the fish

The fish must be:

– sufficiently submerged to avoid the reflections which occur on the surface of the water, from the energy transmitted by the upper secondary lobes,
– preferably submerged beneath the thermocline, to avoid troublesome phenomena of distortion.

The fish must therefore lie at an optimum level (depending on the range) above the bottom in order to obtain the most horizontal angle of incidence possible and the best possible resolution.

Depending on whether the bottom is smooth or very broken up, it is accepted that the elevation must be about 10 to 20% of the maximum range. This optimum depth of submersion is obtained:

– either by governing the speed of the vessel, for a given length of the tow line,
– or by adjusting the length of the tow line for a given speed of the vessel.

In the case of a water depth of several hundred metres, the necessary depth of submersion cannot be obtained with lightweight fishes if a reasonable operating speed is to be maintained. This is why most manufacturers have created downward deflectors that can be secured to the cables.

In the case of very great depths, the towing technique is no longer suitable owing to the inaccuracy of location of the fish.

The use of small submersibles provides a remedy to this drawback, but the need to position the submersible by pingers and the cost of the operation unavoidably restrict implementation of this technique.

Remark. It should be pointed out that the best conditions for obtaining topographic echoes are brought about by describing the sonar profiles parallel to contour lines.

III.3.4.4 Advantages of integrated reconnaissance

The results of a side-scan sonar survey are considerably enhanced by using an echo sounder, high resolution seismic reflection and a magnetometer at one and the same time.

Provided a certain number of precautions are taken – in particular the use of separate power supplies – most intruments can be employed simultaneously (however, simultaneous use of a sparker and a magnetometer would appear to be excluded).

III.3.5 Side-scan sonars that can be used

III.3.5.1 Choice of type of instrument

At present, there are two commercially available groups of instruments which can be distinguished by the incompatibility of resolving power together with lateral range.

In order to go into detailed surveying in a narrow zone, a high resolution (about one metre) but low lateral range (maximum 300 m) side-scan sonar will be chosen. In this case, in order to cover a given surface, a greater number of kilometres of profiles will have to be recorded.

In order to cover a large surface, for instance for a preliminary survey, a wide range instrument (500 to 1,000 m) with a lower resolving power (a few metres) will be preferred.

III.3.5.2 Characteristics of various units

Most of the side-scan sonars available fall into the high resolution category (a few decimetres), use high frequencies (about 100 kHz) and have maximum lateral ranges of about 300 to 500 m.

These units meet the requirements of current reconnaissance problems on the continental shelf. Their performances are all very much the same and their lateral ranges cover 50 to 550 m.

The "Geomecanique Sol 110 S" sonar is the only unit capable of reconciling wide lateral range (maximum 1,500 m) with a fair resolution that can be used for large scale or large depth studies.

The characteristics of the main side-scan sonars are summarized in Tables III.3.1 and III.3.2 ([4]). The most commonly used models are the EGG and the Klein.

(4) The "Gloria", a very high range prototype (22 km), is used by the *National Institute of Oceanography*. It operates at a frequency of 6.5 kHz and has a lateral resolution of about 30 m.

TABLE III.3.1

CHARACTERISTICS OF THE MAIN SIDE-SCAN SONARS

Manufacturer and Type	Kelvin Hughes Transit	EGG		Klein	
	MS 43 MS 47	Mk IA	Mk IB	Mk 300	Mk 400
Electroacoustic parameters					
Frequency	48 kHz	105 ± 10 kHz	105 ± 10 kHz	50 kHz	100 kHz
Pulse duration	1 ms	0.1 ms	0.1 ms	0.1 ms	0.1 ms
Beam angle:					
horizontal	1.5°	1°	1.2°	2°	0.75°
vertical	51°	40°	20° or 50°	20°	20°
Beam inclination to horizontal		10°	10° or 20°	10°	10°
Lateral range	275 or 550 m	75, 150 or 300 m	50, 100, 125, 200, 250, 500 m	75, 150, 300 m or 100, 200, 400 m	75, 150, 300 m or 100, 200, 400 m
Dimensions of transducer	112 x 10 x 15 cm				
Fish	Secured to hull		272 Saf-T-Link	300 B	402 A
Dimensions		127 x 10 x 30.5 cm	118 x 11.4 x 30 cm	91.4 x 25.4 x 25.4 cm	122 x 9 x 30.5 cm
Weight (in air)	24 kg + 45 kg (outboard frame)	18 kg	22 kg	41 kg	14 kg
Maximum submersion		360 m (standard) 600 m (spec. version)	600 m (standard)	450 m	450 m
Cable		Reinforced steel	Reinforced steel	Reinforced	Reinforced
Dimensions:					
length		50, 150 or 600 m	50, 150 or 600 m	100 m (standard)	150 m (standard)/500 m
diameter		0.95 cm	0.95 cm	0.64 m	0.95 cm
Weight (in air)		0.35 kg/m	0.35 kg/m		0.52 kg/m
Recorder		259-3	259-3		401
Number of tracks	1	2	2	2	2
Paper:	Wet	Wet	Wet	Wet	Wet
type		Alfax, type A	Alfax, type A		
width	15 or 22.8 cm	28 cm (13 cm/track)	28 cm (13 cm/track)	48 cm (20.3 cm/track)	28 cm (13 cm/track)
Dimensions		28 x 44 x 84 cm	28 x 44 x 84 cm		25.4 x 84.4 x 59.7 cm
Weight		38 kg	38 kg		45 kg
Observations		Towing speed: 2-8 knots	Towing speed: 1-15 knots	Towing speed: 1-8 knots	Towing speed: 0-12 knots

TABLE III.3.2

CHARACTERISTICS OF MAIN SIDE – SCAN SONARS

Manufacturer and Type	ORE		EDO Western 601	Geomecanique-IFP SOL 110 S
	1096	1098		
Electroacoustic parameters				
Frequency	96 kHz	97 kHz	100 kHz	37.5 kHz – 23 kHz
Pulse duration	0.1 ms	0.1 ms	0.1 ms	0.6 ms – 1.2 ms
Beam angle:				
horizontal	2°	2°	1.3°	1.5° – 2.5°
vertical	30°	35°	57°	20° - 25°
Beam inclination to horizontal	10°	15 or 20°	10 or 20°	Adjustable (0 to 90°)
Lateral range	300 m (maximum)	300 m (maximum)	450 m (maximum)	375, 750, 937.5, 1,500 m
Dimensions of transducer	7.6 x 40.7 x 3.8 cm	7.6 x 40.7 x 3.8 cm	45.7 x 6.35 cm (active face)	143.6 x 14 cm (active face)
Fish	196	198	602	
Dimensions		101 x 15.2 x 35.5 cm	165 x 5.7 x 20.5 cm	350 x 50 x 80 cm
Weight (in air)		19 kg	36 kg	450 kg
Maximum immersion	600 m	600 m	600 m	2,000 m
Cable	Reinforced	Nylon (reinforced)	Nylon (reinforced)	Steel (reinforced)
Dimensions:				
length	300 m (standard)	500 m (standard)-300-600 m	75 m (standard)-150 m (standard)	250 m (standard)
diameter			1 cm-1 cm	2.5 cm-3.2 cm
Weight (in air)				2.6 kg/m-3.4 kg/m
Recorder	Unspecified	Unspecified	EDO, types 550 or 552	Mufax (modified IFP)
Number of tracks	2	2	2	2
Paper				wet
type				Muirhead type ROU H
width		28 or 48 cm (20.3 cm/track)		22.8 cm (10 cm/track)
Dimensions				45 x 40 x 50 cm
Weight				30 kg
Observations	Coupled to a penetrator	Towing speed: 0-5 knots	Towing speed: 0-15 knots	

III.3.6 Visual observation of sea floors by television and/or photography

The purpose of visual observation of sea floors is:

– either to calibrate the echoes recorded on the side-scan sonar,
– or to monitor or further specify certain details of the topography or nature of the bottom.

III.3.6.1 Implementation of television

A television camera can be used near the bottom:

– either by means of a tubular frame of the "troïka" type pulled along the bottom at a speed of about 1 knot,
– or by means of a submersible (manned submarine or robot of the "TOM 300" type of the *Comex Seal*) travelling near the bottom.

Observation of the sea bottom by means of a television camera:

– is made either by describing profiles, or by stations,
– calls for a video unit aboard the vessel in order to identify the recordings.

III.3.6.2 Characteristics of the images obtained

The image obtained is 2-3 m wide and has a maximum depth of a few metres and generally makes it possible:

– to obtain details of the topography of the bottom of about 5 to 10 cm resolving power,
– to indicate the true magnitude of the differences in level (up to about 10 cm) which cannot be observed by the side-scan sonar,
– to clearly distinguish the nature of the bottom: sand, mud, rock, etc.

As an example, Fig. III.3.15 shows an image of the sea floor obtained by television from the Télénaute of the *Institut Français du Pétrole* ([5]).

III.4 ANALYSIS AND INTERPRETATION OF SIDE-SCAN SONAR RECORDINGS

The analysis and interpretation of side-scan sonar recordings makes it possible:

– to determine the reliefs or characteristic elements (rock outcrops, sand ridges, etc.) of the bottoms,
– to identify objects: wrecks, wellheads, pipelines, etc., on the bottom.

(5) The "Télénaute techniques" of the *Institut Français du Pétrole* have been developed by *Comex-Seal* and *Thomson-CSF* to build "TOM 300" which is a remote-controlled submersible vehicule for performing tasks of observation, data gathering, sampling and light manipulation.

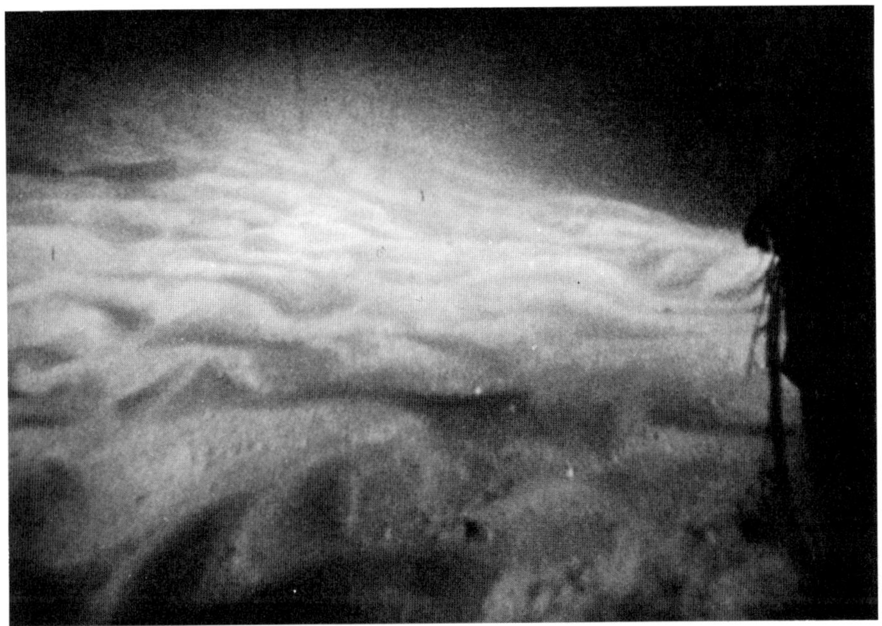

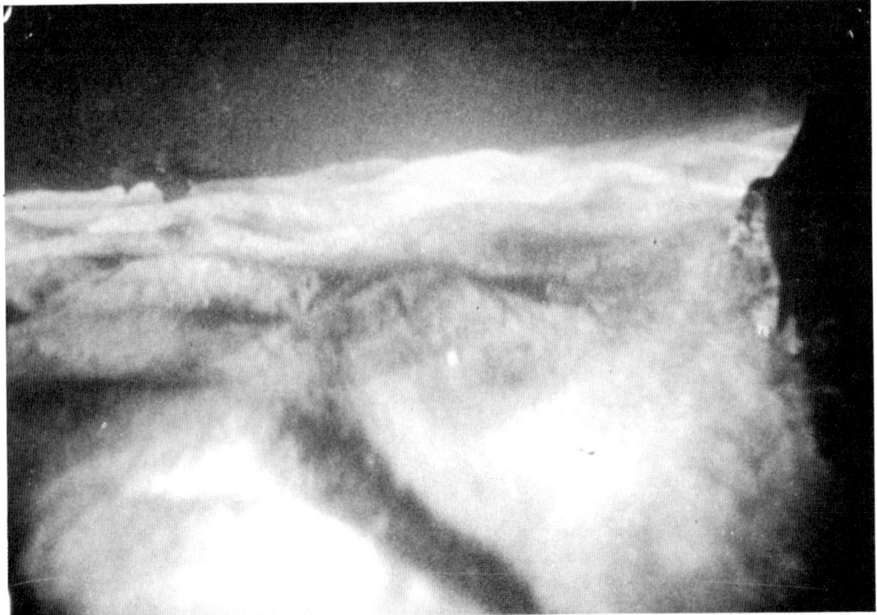

FIG. III.3.15. – Images of the sea bottom obtained by the television camera used on the "Telenaute".

III.4.1 Analysis of the recordings

The formation of echoes and the plotting of recordings have already been described in Paragraph III.3.1.

III.4.1.1 Description of the recordings

Figure III.4.1 shows how acoustic pulses emitted by a bilateral side-scan towed sonar are converted on the analog recording.

For each recording side, starting from the centre, the following are encountered:

– a regular dark line corresponding to the duration of the pulse,

– a light zone corresponding to the passage through the water (H_p = depth of water beneath fish),

– a grey zone corresponding to the response from the bottom scanned by the beam.

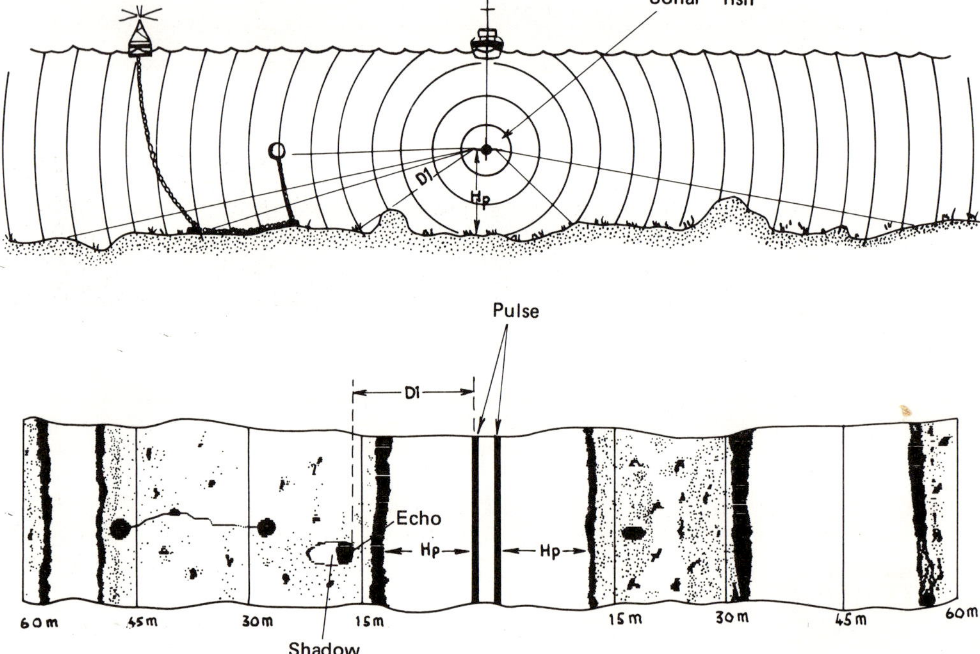

FIG. III.4.1. – Diagram showing a typical recording of side-scan sonar.

The obstacles on the bottom generally take the form of darker zones (acoustic "echoes"). followed, when the obstacle is of any size, by light zones (acoustic "shadow").

Certain spots of different shade on a recording without "shadows" may correspond to different types of soil (and not to reliefs).

On the recordings, the times are automatically converted into ranges, and the speed adopted for the conversion is 1,500 m/s. Where the speed of sound would differ appreciably from this value, the metric scales must be corrected accordingly.

III.4.1.2 Special phenomena

Owing to the geometry of the ultrasonic beam and the requisite conditions for the formation of topographic echoes – as oblique an angle of incidence as possible –, part of certain recordings cannot be validly utilized. The width of this "**blind zone**" varies with the depth of water beneath the fish and the effective angle of the beam transmitted (Fig. III.4.2).

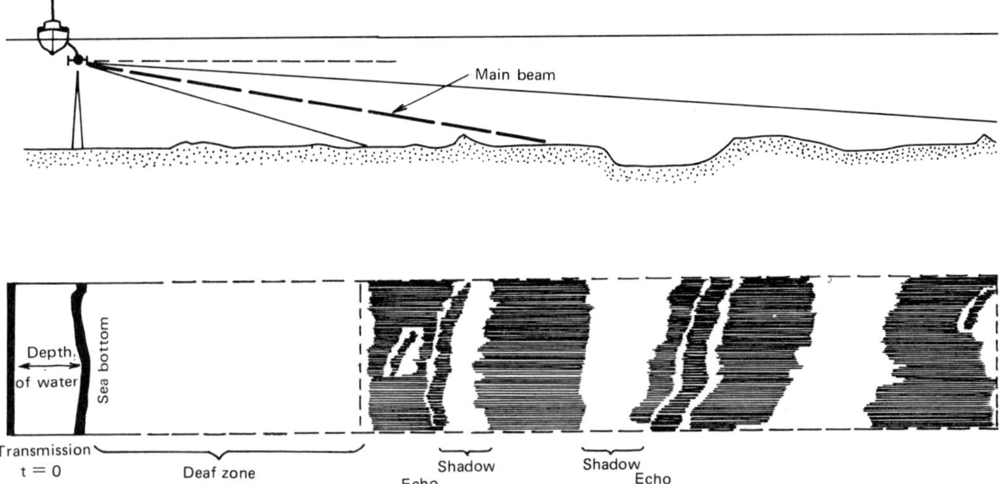

FIG. III.4.2. – Diagram representing the existance of a "blind" zone on certain recordings.

An echo, generally fairly powerful, can be recorded not only by the transducer on the same side as the reflecting obstacle but also in attenuated form by the transducer on the other side of the vessel: this is the "cross talk" phenomenon (real and false echoes) (see Fig. III.4.3).

One therefore constantly has to compare the two channels recorded so that false obstacles caused by such phenomena can be eliminated when interpreting the recordings. Since these echoes are symmetrical around the path followed by the vessel, the more attenuated parasite echo can easily be recognized.

The conditions of implementation – operating speed, navigation of the fish, submersion beneath the thermocline, etc. – determine the reliability and the quality of the recordings.

Depending on the speed of the vessel, the geometric shapes are compressed on the recordings to a varying degree (see Fig. III.4.4).

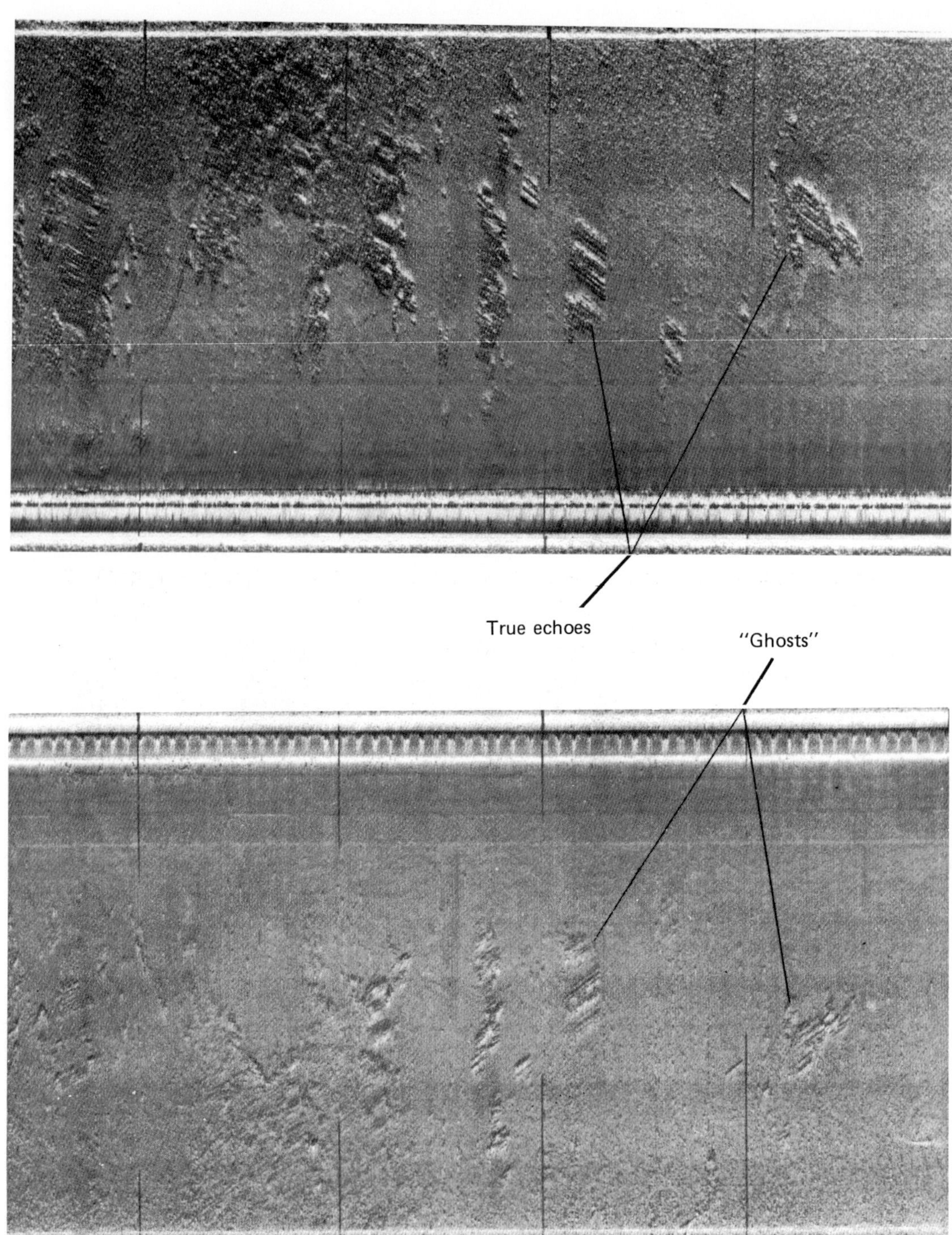

FIG. III.4.3. – Example of "cross-talk".

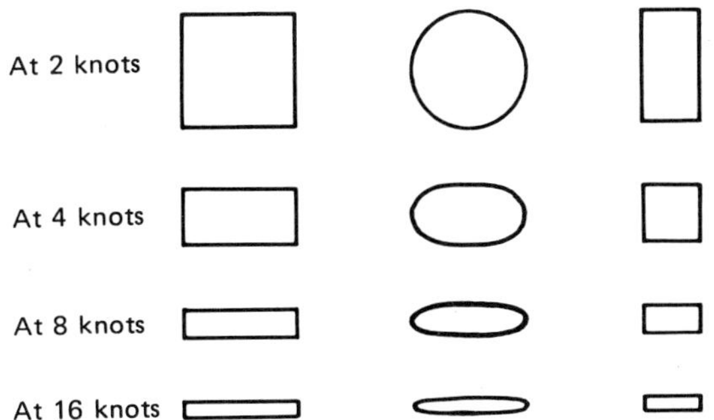

FIG. III.4.4. – Distortions caused by the speed of the ship.

III.4.1.3 Examples of recordings

The definition and resolving power of the various units used which govern the analysis of the recordings are illustrated by the examples given in Fig. III.4.5 to III.4.10.

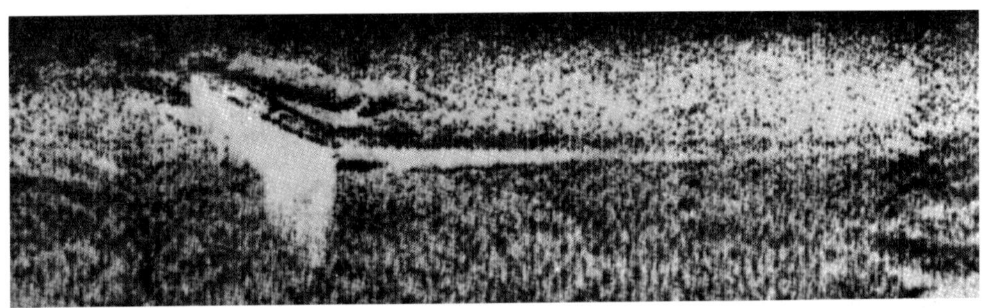

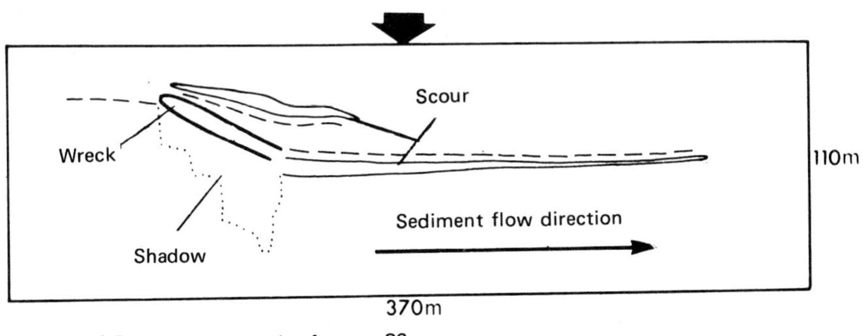

Strait of Dover Depth of water 26 m

Kelvin Hughes sonar transit

FIG. III.4.5. – Side-scan sonar recording showing a wreck.

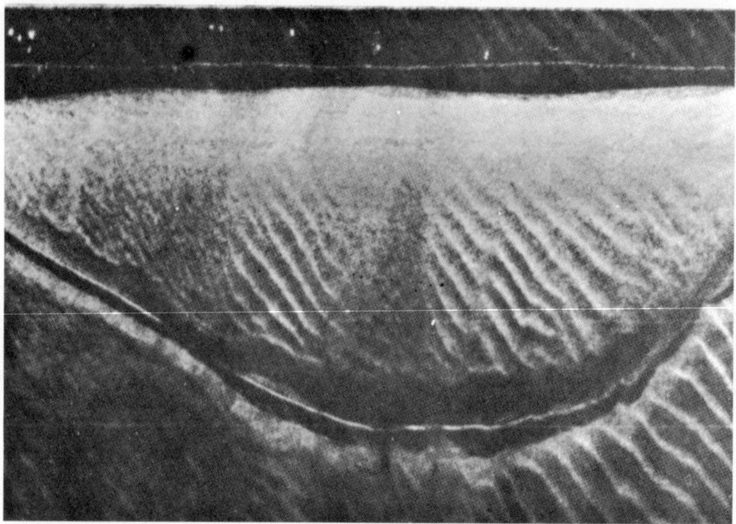

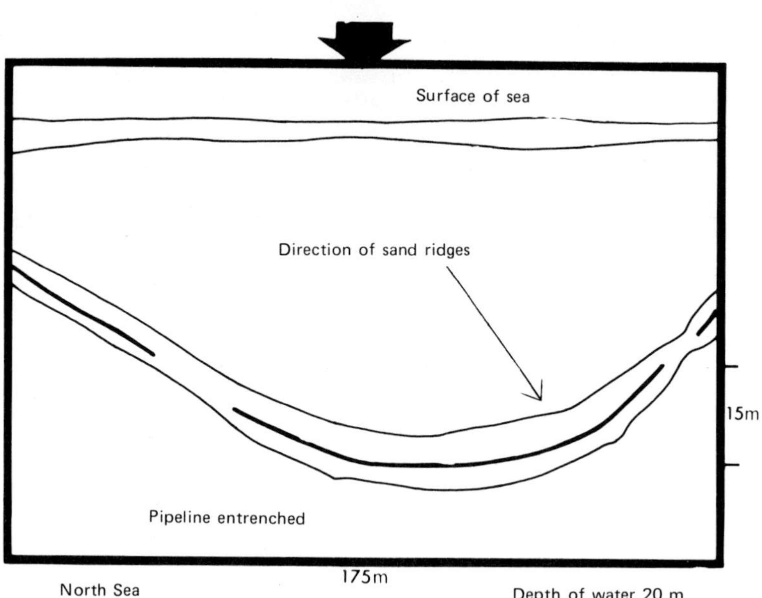

FIG. III.4.6. – Side-scan sonar recording showing a pipeline buried in a trench.

FIG. III.4.7. – Side-scan sonar recording in a granite outcrop zone.

The reader will doubtless note that the analysis of the recordings makes it possible, in addition to "visualizing" objects, outcrops, ridges, etc., to deduce also the direction of advance of the sediments (see Fig. III.4.5, III.4.9, and III.4.10).

III.4.2 Interpretation of the recordings

Interpretation of side-scan sonar recordings has two aims:

– first, to delineate the terrains of various kinds forming the sea bottom,

– second, to pinpoint the obstacles revealed by the echoes and (as regards the main ones) to calculate their height.

III.4.2.1 Determination of the position and height of an obstacle

Figure III.4.11 shows the geometry of the system formed by the submersed side-scan sonar and the sea bottom.

The various basic terms used are defined as follows:

– depth of water (P_e) = vertical distance between the surface and the bottom,

– depth of submersion (P_p) = vertical distance between the surface and the sonar fish,

– depth of water beneath fish (H_p) = vertical distance between the sonar fish and the bottom (PP'),

– horizontal distance (D_h) = horizontal distance between the point on the bottom directly beneath the fish and an object on the bottom,

– lateral range (D_l = oblique distance between the fish and an object on the bottom (measured on the recording),

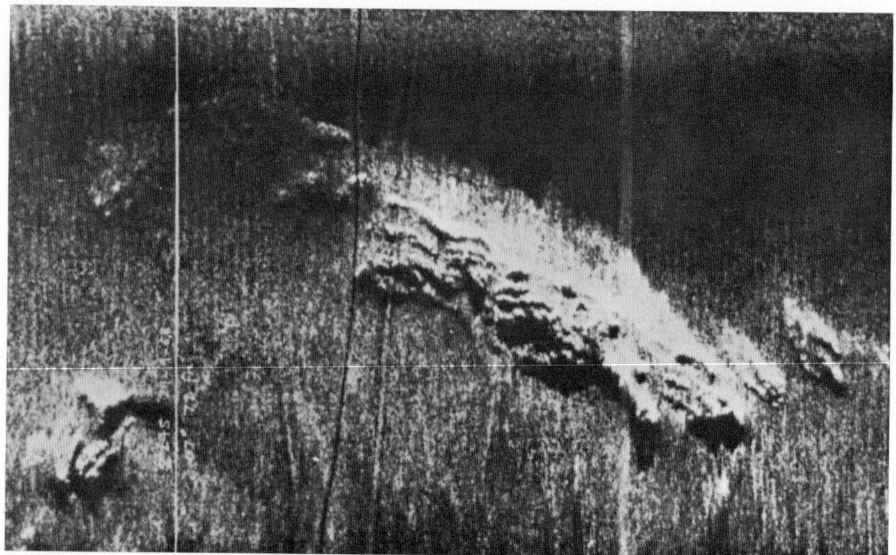

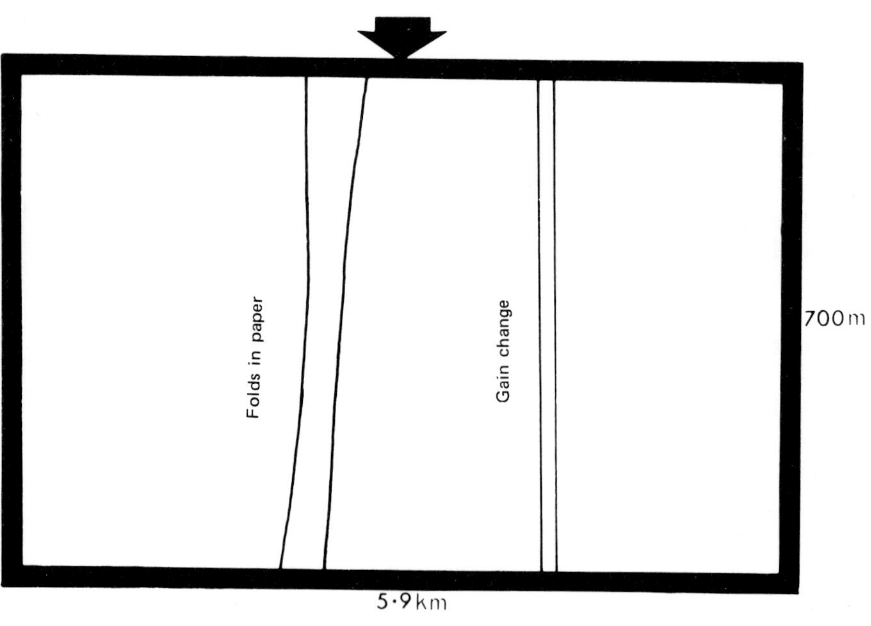

FIG. III.4.8. – Side-scan sonar recording showing an isolated outcrop of rock.

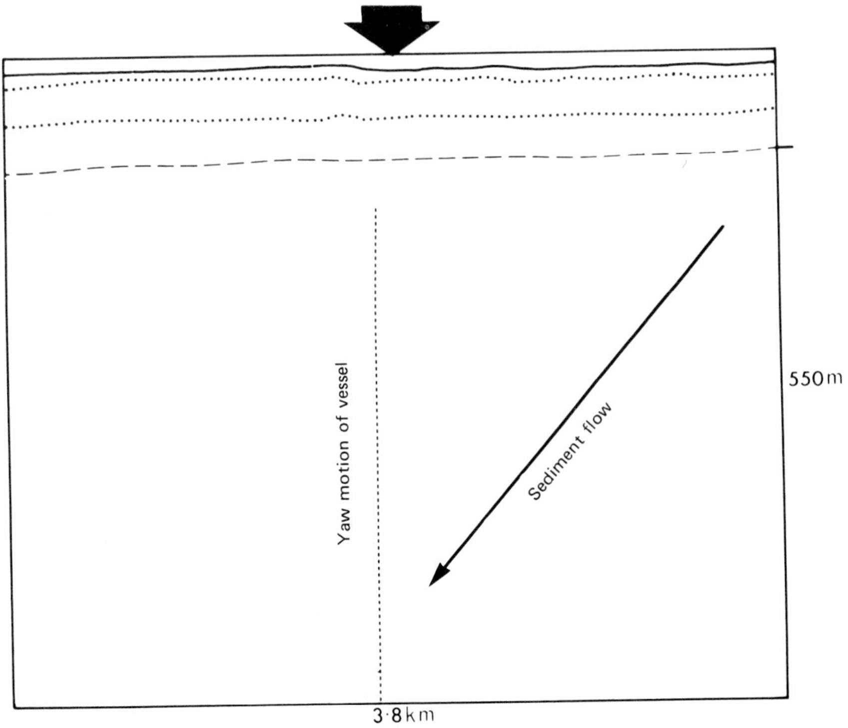

English Channel Depth of water 30 m

FIG. III.4.9. – Side-scan sonar recording showing a system of furrows.

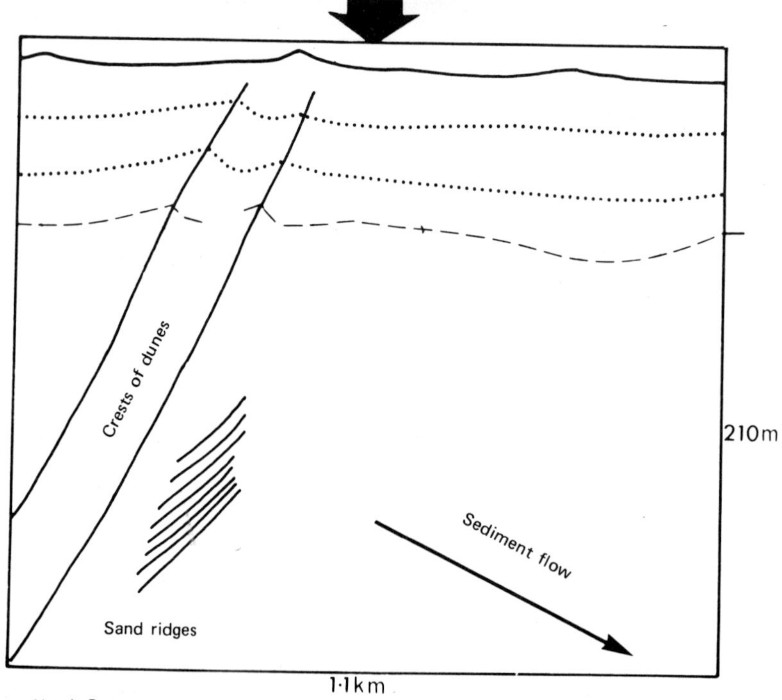

FIG. III.4.10. – Side-scan sonar recording in a zone of sand ridges.

– height of obstacle (H_0) = distance between the top of the obstacle and the bottom at the horizontal plane directly beneath the fish (AA') and the farthest point of the acoustic shadow (O) measured on the recording.

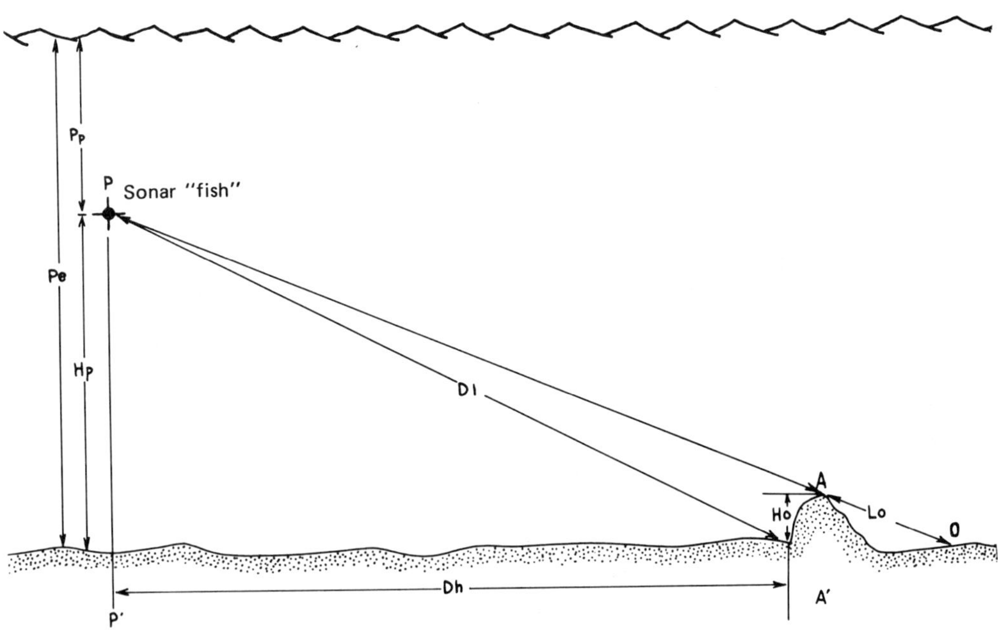

FIG. III.4.11. – Geometry of "side-scan sonar-sea bottom" system.

The **longitudinal position** (along the path followed by the vessel) of an obstacle is determined with relation to the pulses corresponding to topographically recorded points.

On the recording (see Fig. III.4.12):

– n and $n + 1$ are two successive pulses,
– a is an obstacle between these two pulses.

On the bottom:

– N and $N + 1$ are the two topographically recorded points corresponding to n and $n + 1$,
– A' is the orthogonal projection of obstacle A on path $N - N + 1$.

The longitudinal range NA' is given by the equation:

$$NA' = na \frac{N - N + 1}{n - n + 1}$$

Naturally, one has to allow for the "antenna correction", i.e. the horizontal distance between the radiolocation antenna and the fish.

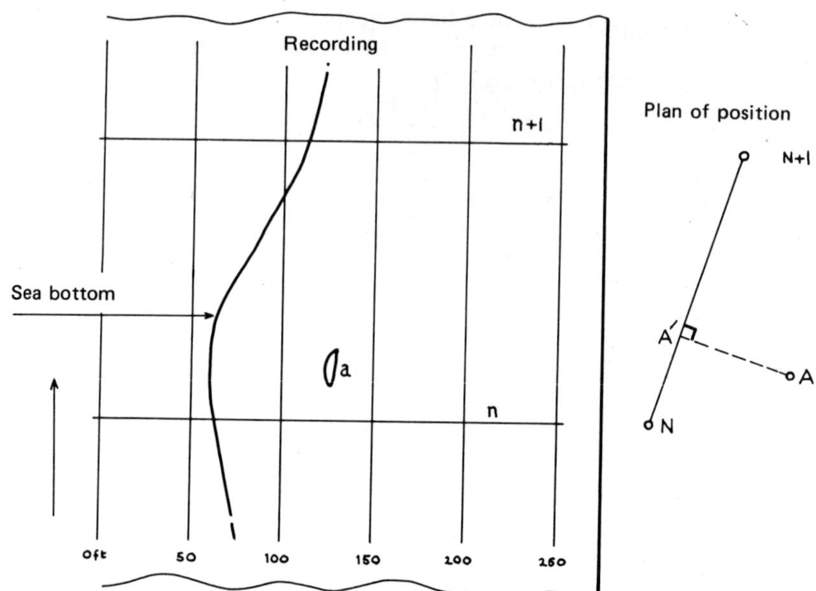

FIG. III.4.12. – Determination of longitudinal position of obstacles.

The **transverse position** is obtained by converting the oblique ranges measured on the recording into horizontal ranges.

On the assumption as an initial approximation that the bottom can be assimilated to a horizontal plane, then (see Fig. III.4.11):

$$D_h = \sqrt{D_l^2 - H_p^2}$$

Knowledge of NA' and D_h locates obstacle A.

The height H_0 of the obstacle is obtained by resolving two triangles POP' and AOA' (see Fig. III.4.11):

$$H_0 = \frac{L_0\, H_p}{D_l + L_0}$$

III.4.2.2 The mosaic

Each time the totality of elements recorded enables fixes to be made, this is done manually, takes time and is liable to errors, so it is of advantage to produce sonar mosaics which in pratice form charts to scale.

This operation breaks down into two stages.

Correction to scale (in blocks generally of 1 km):

– of the longitudinal scale, depending on the speed of the vessel and the payoff speed of the recording paper,

(a) Exaggeration of vertical scale: 7

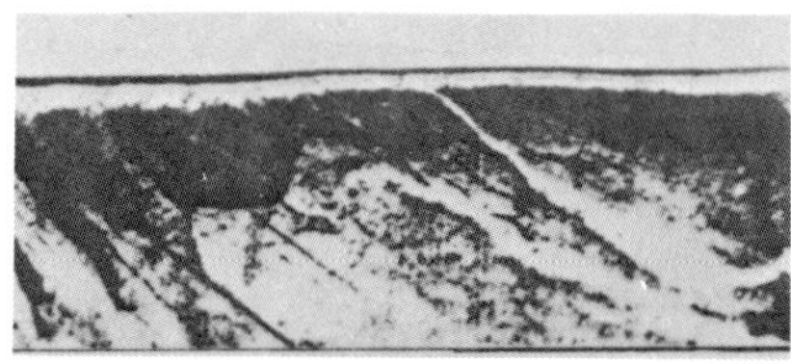

(b) Scale 1/1

FIG. III.4.13. – Example of original (a) and anamorphosed (b) recording.

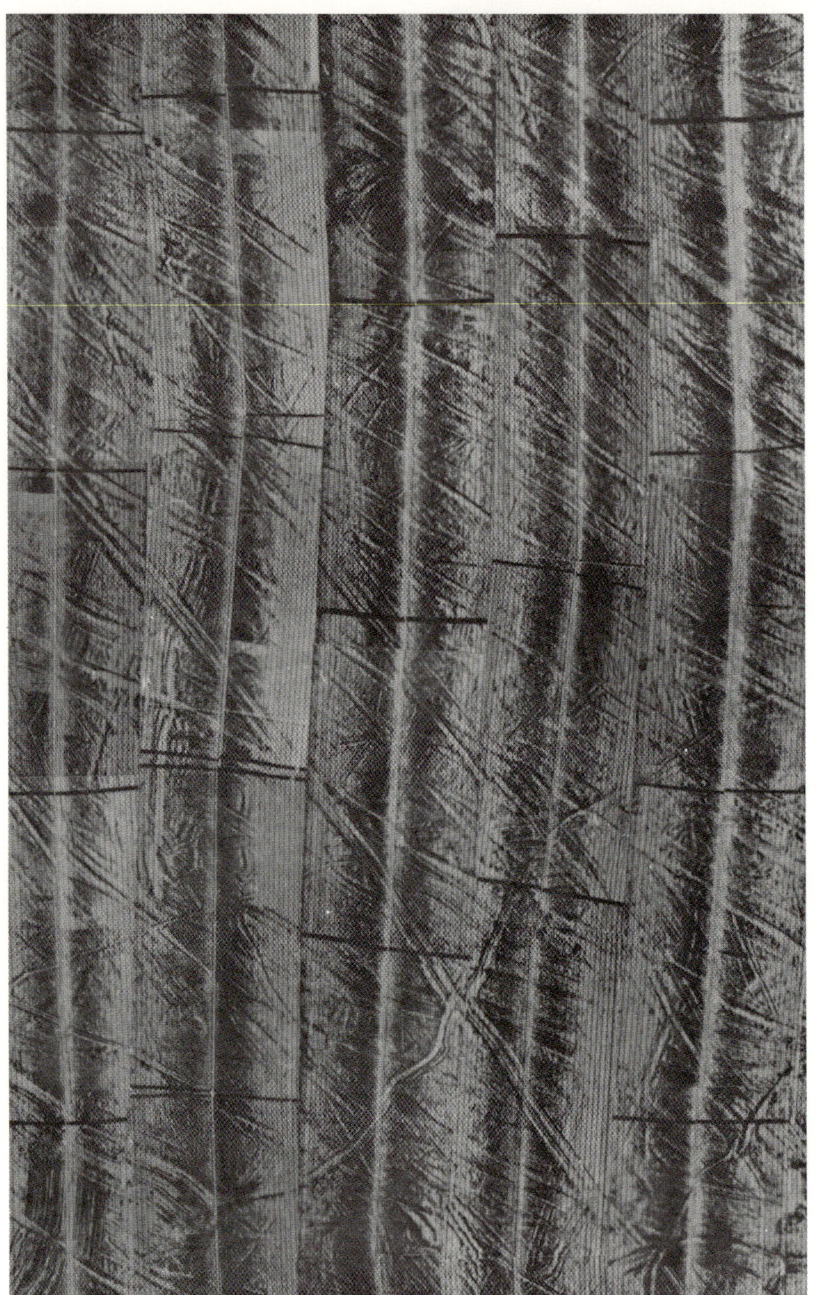

FIG. III.4.14. – Example of mosaic (scale 1 : 10,000) showing a "network of scours" cut into a soft bottom by icebergs.

– of the transverse scale, depending on the width of the paper and the scale adopted for the recording,

is performed by an optical method known as "anamorphosis" (used by the *Bureau d'Etudes Industrielles et de Coopération de l'Institut Français du Pétrole (BEICIP))* (6).

Figure III.4.13 shows the results of an optical anamorphosis whereby equality of scale has been ensured by compressing the scale in the direction of illumination (transverse) seven times with relation to the scale in the direction of advance of the vessel (longitudinal).

The photographic reproduction of the documents obtained is performed at a given scale (commonly 1: 10,000). By assembling these documents together, the mosaic is formed.

Such isometric mosaics reveal the true shape and orientation. On broken up bottoms, they alone enable all the topographic and lithological features to be fixed in both X and Y coordinates and delineated exactly. Figure III.4.14 gives an example of a mosaic.

BIBLIOGRAPHY

[1] HAYES (J.W.) – "Echo-Sounding and Seismic Profiling." *Offshore Services,* April 1977, p. 55, 57, 58, 60, 62, 63.

[2] URICK (R.J.) – *Principles of Underwater Sound for Engineers.* Mac Graw-Hill, 1967.

[3] KRAUSE (D.C.) – "Interpretation of Echo-Sounding Profiles". *Rev. Hydrogr. Intern.,* Monaco, Vol. 39, n° 1, p. 65-123.

[4] CHOLET (J.), FONTANEL (A.) et GRAU (G.) – "Etude du fond de la mer à l'aide d'un sonar latéral". *XIe International Photogrammetry Congress,* Lausanne, 8-20 July 1964.

[5] FLEMING (B.W.) – "Side-Scan Sonar : a Practical Guide", *The International Hydrographic Review,* Vol. LIII, No. 1, January 1976.

[6] KLEIN (M.) – "Side-Scan Sonar", *Offshore Services,* April 1977, p. 67, 68, 71, 72, 75.

(6) There are other scale correction methods :
– photographic (*National Institute of Oceanography*),
– cathodic (*Bath University of Technology*),
– from magnetic tapes.

CHAPTER IV

Seabed Exploration by high Resolution Seismic Prospecting

CONTENTS

INTRODUCTION

1. "High resolution continuous seismic reflection" (or continuous seismic sounding) is the widest-used and most economical method for studying the first hundred metres of soil beneath the sea floor.

The method enables the geometry, structure and configuration of the geological strata to be determined. However, in the prevailing state of techniques, seismics **alone** does not make it possible to make any affirmation:

- as to the nature of the soils,
- and yet less, as to their physical and mechanical properties.

While certain interpretations sometimes justify a presumption as to the state of consolidation of the soils (owing to the degree of penetration, for instance of signals with a given frequency and energy), these assumptions must necessarily be verified by core samples or in situ geotechnical measurements.

2. Preliminary recording of seismic profiles on a marine site makes it possible:

- to fix the locations of the geological and geotechnical soundings (drillings/core drillings and in situ measurements) as a function of the variations in the configuration of the subsoil,
- to reduce the number of these soundings,
- to extrapolate where necessary the results of core drillings and in situ measurements.

3. All seismic techniques currently applied for the reconnaissance of marine soils use the continuous **reflection** method. The refraction method is applied only when seismic reflection proves to be inoperative or the results obtained do not yield the expected accuracy.

4. Several types of devices are used in "high resolution seismics" [1] [2] [3] [4] [5] [6]. The main of them are:

- sediment sounders (or penetrators),
- boomers,
- sparkers.

These devices are characterized by their transmission frequency and consequently the penetration of the signal and its resolving power (or definition):

- the penetration is inversely proportional to the transmission frequency,
- the resolving power (and reflective quality) decreases with the penetration and increases with frequency.

IV.1 SEDIMENT SOUNDERS

Sediment sounders, also known as *pinger probes, mud penetrators or bottom sounders*, are devices derived from the echo sounder and used for the surveying of surface strata (a few tens of metres deep) of unconsolidated marine floors, in order to determine their sedimentary structure and possibly their nature [7].

IV.1.1 Principle and characteristics of sediment sounders

The sediment sounder is a low frequency echo sounder. It can be implemented in different ways.

IV.1.1.1 Principle of sediment sounders

A transducer type transmitter with a frequency of about 5 kHz sends a pulse to the bottom.

The pulse, after reflection from the bottom and the initial reflecting surfaces beneath the bottom (to a penetration of a few metres to about 30 to 40 m, depending on the degree of consolidation of the sediments), is recorded:

- either by the transmitter acting as receiver (the most frequent case),
- or by a separate receiver (streamer) operating as a receiver.

The speed of propagation through the sediments in the bottom being unknown, the recording is displayed in terms of the two-way time.

IV.1.1.2 Characteristics of sediment sounders

The parameters defining a sediment sounder are identical to those given for echo sounders.

The nominal frequency, which is lower than in echo sounding, generally lies between 1 and 7 kHz.

The directivity is an increasing function of frequency: for instance, the directivity of a transducer used in sediment sounding is less than that of an echo sounder.
The usual sediment sounder has a beam angle of 20 to 50° (see Fig. IV.1.1).

The resolving power generally decreases as the energy provided increases.
Owing to the poorer directivity in sediment sounding, one has to increase the energy provided in order to obtain a useful energy comparable to that of echo sounding (covering the same area of the sea floor).

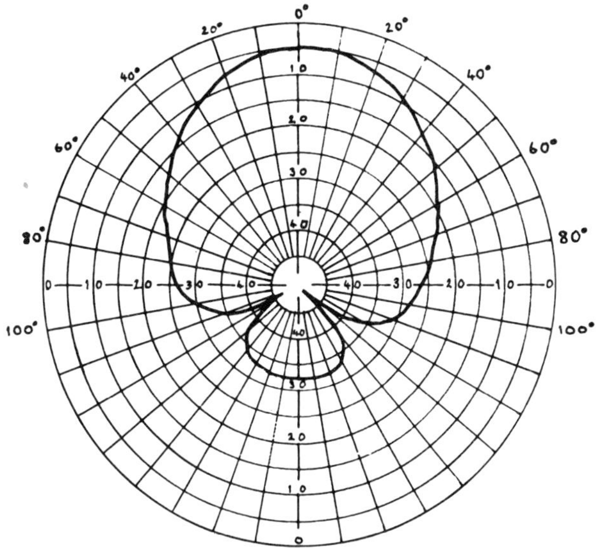

FIG. IV.1.1. – Directional diagram of a sediment sounder (EDO Western 515, 3.5 kHz).

This increase in energy generally results in an elongation of the signal and hence a lower resolving power, which reaches no more than 30 cm under the most favourable conditions. It should be noted that the increase in energy is quickly limited by the cavitation phenomenon.

IV.1.2 Implementation of sediment sounders and units used

IV.1.2.1 Implementation of sediment sounders

A sediment sounder can be used:

– either by attaching the transducer to the hull of the vessel, as for echo sounders,
– or by mounting the transducer in a fish towed behind the vessel at a shallow depth in order to gain independence from the movements of the vessel.

The efficiency of sediment sounders decreases as the depth of water increases. For great depths, it may therefore be of advantage to work nearer to the bottom in order to reduce energy losses.

The limiting conditions of implementation given by the manufacturers of sounders are:

– speed of vessel: 6 knots,
– state of sea: 4 to 5.

In reality, it is often necessary to operate at a lower speed.

TABLEAU IV.1.1

CHARACTERISTICS OF SEDIMENT SOUNDERS

Manufacturer	EGG	ORE	EDO Western	Raytheon	CESCO	Thomson CSF
Type	Pinger probe 229	Sub-bottom profiling system : model 1,036	Model 515	RTT Transducer TC-7 1,000 Transceiver PTR 106 A	SONIA Mark I or II	TEF
Frequency	5 kHz	3.5-7 kHz	3.5-7 kHz	7 kHz	3 kHz	3.5-6 kHz
Power		10 kW	2 kW/10 kW	2 kW	10 kW	1 kW
Pulse duration	0.4 ms Energy = 0.4 J Firing rate = 20 pulses/s		0.5 ms	0.1-1 ms	0.4 ms	Adjustable from 0.1 ms
Transmission level (ref.: 0.1 Pa at 1 m)	98 dB	114 dB	112/116 dB			107 dB
Directivity (at – 3 dB)		55° for 3.5 kHz 40° for 5 kHz 30° for 7 kHz	20/40°	36°	23°	
Dimensions	(L) 213 cm x (w) 30 cm x (h) 30 cm	Transducer = 152 x 46 x 36 cm Transceiver = 43 x 43 x 18 cm	Transducer = 120 x 120 x 44 cm Transceiver = 48 x 35.5 x 36 cm	Transducer = ϕ 43 cm x (h) 20 cm Transceiver = 48 x 43 x 13 cm	$\phi = 1$ m	$\phi = 21$ cm, $w = 26$ cm
Weight	55 kg	Transducer = 122 kg Transceiver = 23 kg	Transducer = 75 kg	Transducer = 16 kg Transceiver = 25 kg	1,300 kg	Transducer = 10 kg
Recorder: Paper Dimensions Weight	EGG 254 Wet paper 28 cm 100 x 60 x 20 cm 95 kg	Gifft 4000 Wet paper 48 cm 70 x 40 x 20 cm 30 kg	PESR Dry paper 48 cm (w) 79 cm x (h) 9.4 cm x (depth) 51 cm 43 kg	DE 719 RTT Paper 18 cm 46 x 39 x 23 cm 21 kg		EGG 254
Implementation	Towed fish (15 m cable)	Towed fish	Towed fish (variant adaptable to hull)	Adaptable to hull	Secured to hull	Towed fish
Observations	Assembly of 4 transducers of 5 kHz	Other models: – 1.032 (3.5-7 kHz) 4 transducers mounted on swivel frame – 1.038 (1.4 kHz)	Other models: 1.5-3.5-7 kHz 2 kW 15/30°		Auxiliary frequencies = 1.5-6 kHz	

IV.1.2.2 Units used

Table IV.1.1 gives the characteristics of sediment sounders currently in use.

These devices make it possible under optimum conditions to attain:

– a penetration of about 30-50 m in loose sediments,
– a resolution of about 30 cm.

In consolidated sediments (for instance dense sands), penetration is no more than a few meters.

The ORE, EDO, and Raytheon models yield the most satisfactory results.

The SONIA model also gives good results, but its complex implementation resulting from its dimensions limits its use.

Sediment sounders are used:

– either for detail studies,
– or in conjunction with sparker or boomer type devices to obtain a good penetration along with a satisfactory resolution.

IV.2 BOOMERS (AND THE UNIBOOM)

The boomer or thumper is an electromechanical source invented by *EGG.*

IV.2.1 Principle and characteristics of the boomer

IV.2.1.1 Principle of the boomer

The boomer consists of:

– an induction coil against which an aluminium plate is applied by a system of springs,
– a bank of capacitors (connected to a sparking circuit) producing electrical discharges through the coil at regular intervals.

With each discharge, the eddy currents induced in the conductive plate cause it to move violently away from the coil. The initial movement of the plate triggers the acoustic pulse.

IV.2.1.2 Characteristics of the boomer and Uniboom

The signature of a 1,000 J boomer is given in Fig. IV.2.1. The duration of the signal is about 5 ms.

The spectrum for this boomer ranges from 200 to 2000 Hz.

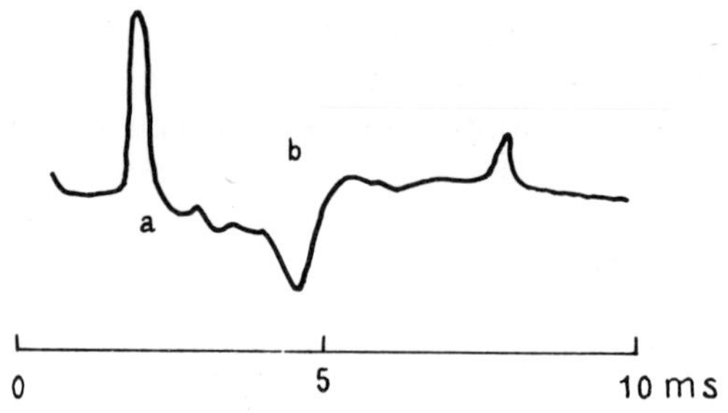

FIG. IV.2.1. – Signature of a 1,000 J boomer.

From the standpoint of energy distribution, the figure reveals:

– a very high amplitude of the initial pulse peak (*a*),
– a peak of negative amplitude (*b*) extending the signal.

This secondary peak is caused by the cavitation which arises behind the plate in the depressurized zone.

In the **Uniboom** system, the secondary pulse is eliminated by providing an elastic diaphragm on the inner face of the plate from the depressurized side. This diaphragm then absorbs part of the energy and thus limits the cavitation.

The **duration of the Uniboom signal** is limited to about 0.2 ms (Fig. IV.2.2).

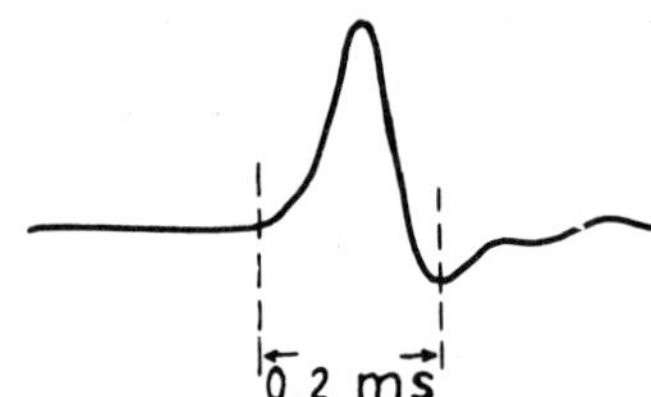

FIG. IV.2.2. – Signature of a Uniboom.

The **frequency spectrum** ranges from 500 to 10,000 Hz on the average (the frequency decreases slightly as the energy output increases).

The resolving power:

– of the **boomer** proper is not less than 2 m, owing to the considerable length of the signal,
– with the **Uniboom**, it can theoretically get down to 30-40 cm (comparable to the best sediment sounders).

IV.2.2 Application of boomers. Equipment

IV.2.2.1 Application of boomers

The boomer is attached to a fish towed by the vessel.

The limitation in the energy of boomers (5,000 J per coil) is caused by the cavitation effect resulting in wear of the plate. The cavitation is reduced by submerging the boomer to a greater depth (the higher hydrostatic pressure raises the cavitation threshold).

The Uniboom, which is not affected by cavitation, is used from a small catamaran towed on the surface by the vessel.

However, owing to the fact that the unit is not submerged, it can only be used in calm seas (maximum force 3 to 4).

IV.2.2.2 Equipment: boomers and Uniboom

Table IV.2.1 summarizes the technical characteristics:

- of 500 and 1,000 J boomers,
- of the Uniboom (100, 200 and 300 J at 3,500 V).

TABLE IV.2.1

TECHNICAL CHARACTERISTICS OF BOOMERS
(Constructor *EGG*)

	Boomer		Uniboom		
Energy (J)	500	1,000	100	200	300
Duration of signal (ms)	0.4	0.8	0.2	0.2	0.2
Transmission level ref: 0,1 Pa at 1 m (dB)	110	114	95	104	107
Spectrum (Hz)	300-3,000	200-2,000	700-14,000	500-10,000	400-8,000
Firing rate (pulses/s)	2.5	1.2	6	4	2
Resolution (m)	2		0.3		
Dimensions (m)	2.64 x 0.66 x 0.30		1.58 x 0.84 x 0.59		
Weight (kg)	74		90		
Implementation	Code		Attached to a catamaran, towed on the surface		

In practice, the Uniboom (commonly known as the boomer) is at present the only model used.

For an energy of 100 J, the transmission rate can reach 6 pulses/s.

The characteristics of use are approximately the following:

- penetration = about 75 m,
- resolving power = 0.40 m.

IV.3 SPARKERS

In sparkers, the seismic shock signal is emitted by the spark generated by the discharge from a battery of capacitors.

IV.3.1 Principle and construction of sparkers

IV.3.1.1 Principle of sparkers

After having first been charged up to a voltage of several thousand volts, a battery of capacitors coupled to a sparker circuit is suddenly discharged at regular intervals between two or several immersed electrodes.

The spark emitted evaporates the surrounding water, creating an energy-charged bubble of gas, the expansion of which releases a seismic shock signal.

IV.3.1.2 Construction of sparkers

The simplest emitter consists of two or three electrodes laid out linearly inside a metal frame acting as the ground. The disadvantage of this device is that it generates bubbles of such size that their resonance frequency interferes with the useful spectrum.

This defect can be remedied by building multi-electrode sparkers (with up to $n = 900$ electrodes). When the energy is divided up among n electrodes, smaller bubbles are formed and have a higher frequency that falls outside of the useful spectrum [8].

IV.3.2 Characteristics of sparkers

IV.3.2.1 The "signature" of the sparker

The "signature" (i.e. characteristic pattern) of the sparker (Fig. IV.3.1) shows:

- a sharp initial pulse followed by,
- secondary pulses resulting from the bubble effect.

The duration of the bubble effect increases with the energy of the emission source.

The initial period of the bubble effect obeys a law proportional to $W^{\frac{1}{3}}$, where W represents the electrical energy applied.

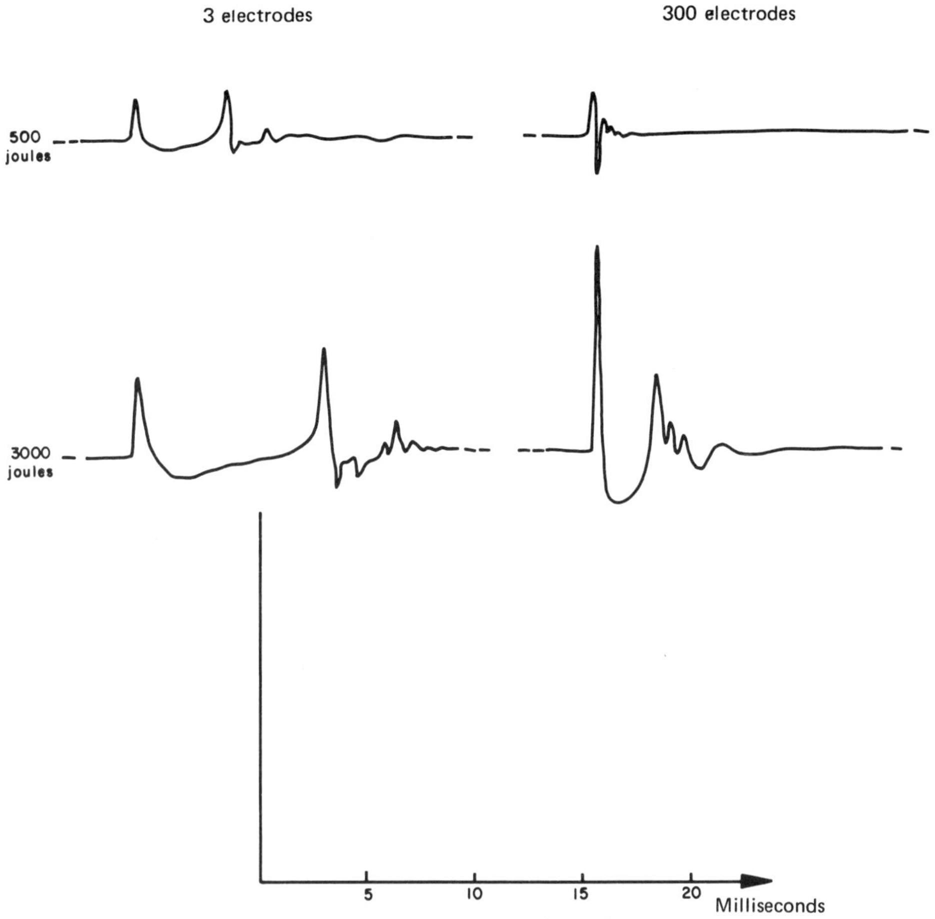

FIG. IV.3.1. – Signature of a sparker.

In the case of a **single-electrode sparker**, the initial period is:

$$T_1 = kW^{\frac{1}{3}}$$

where:

W = electrical energy applied,

K = a coefficient which is a function of the depth, electrical charging voltage, diameter of the electrode and contact surface area with the water.

In the case of **a sparker with n electrodes**, the energy applied W is distributed among the electrodes. The initial period of each bubble then becomes:

$$T_n = K\left(\frac{W}{n}\right)^{\frac{1}{3}} \qquad \text{i.e} \qquad T_n = T_1 n^{-\frac{1}{3}}$$

The period therefore decreases as the number of electrodes increases.

The pressure (i.e. the **amplitude** of the signal) corresponding to the initial pulse peak is also theoretically proportional to $W^{\frac{1}{3}}$

– in the case of **a single electrode**, the pressure is given by the equation:

$$P_1 = kW^{\frac{1}{3}}$$

– in the case of *n* **electrodes**, this pressure should become:

$$P_n = nk\left(\frac{W}{n}\right)^{\frac{1}{3}} = n^{\frac{2}{3}} P_1$$

In practice, this formula has been corroborated for a small number of electrodes, but the pressure quickly tends towards a limit and then falls off for a large number of electrodes ($n > 100$).

A compromise has to be found between the duration and the amplitude of the signal. With a power supply of 4,000 V, this compromise lies between 2 and 10 J per electrode:

– below 2 J, the spark may not occur,
– above 10 J, the power of resolution decreases.

IV.3.2.2 Penetration of the signal and power of resolution

The *EGG* Sparkarray (3 or 9 electrodes) has an emission power varying from 500 to 24,000 J. For a power of 1,000 J, the frequency spectrum lies between 100 and 1,000 Hz. At water depths of 150 to 200 m, its performances are:

– penetration = 150 m,
– resolution = about 6 m.

The *IFP/Geomecanique* sparker, equipped with 200 electrodes and developing an energy of 1,000 J (4,000 V power supply), has the following performance:

– penetration = 150 m,
– resolution = 2 to 3 m.

Actually, relatively low energies are used in most cases (between 50 and 500 J), yielding:

– penetrations of about 50-100 m,
– a resolution of about 1.5 to 2 m.

IV.3.3 Implementation of sparkers. Equipment used

IV.3.3.1 Implementation of sparkers

Sparkers are relatively easy to use.

The transmitter and receiver (streamer) are submerged to a shallow depth (about 0.50 m) and towed either alongside or astern of the vessel.

The optimum operating speed is 4 to 5 knots with a maximum state of sea of force 5.

IV.3.3.2 Equipment used

Today, a wide variety of sparkers is available: they are characterized by their energy expressed in joules (or watts per second in American literature):

– the energy applied varies from a few tens to several thousands of joules. In high resolution seismics, it is always necessary to operate at an energy less than 1,000 J,

– voltages vary from 4,000 to 20,000 V.

Table IV.3.1 summarizes the leading characteristics of commonly used systems.

TABLE IV.3.1

	EGG Sparkarray (Modular system)	IFP-Geomecanique 2 m long	Common arrays
Voltage (V)	4,000	4,000	
Energy (J)	between 500 and 24,000	1,000	between 50 and 500
Number of electrodes	3 or 9	200	
Frequency (Hz)	between 100 and 1,000	between 100 and 2,000	
Penetration (m)	150	150	50-100
Resolution (m)	≈ 6	≈ 2-3	≈ 1.5-2

Remark. During a survey in 1976, in morainic soils, pingers, air-gun, miniflexichoc, surface and deep towed boomers and sparkers where tested with analogic and mono or multichannel digital recorders. These tests led to a comprehensive comparative study of the performances (penetration and definition) of most of off-the-shelf high resolution equipments [6].

IV.4 ANALYSIS AND INTERPRETATION OF HIGH RESOLUTION SEISMIC RECORDINGS

The result of continuous seismic recording is an unprocessed data document suggestive of a geological section.

Analysis of this document to detect the various specific phenomena liable to occur in reflection seismics is needed before any attempt can be made to interprete it.

IV.4.1 Description of recordings

Figure IV.4.1 shows a typical continuous seismic sounding recording with definition of the various lines.

IV.4.1.1 Lines recorded

On the recording (Fig. IV.4.1), the following various lines can be seen from top to bottom:

– a rectilinear black line representing the transmission pulse,

– a series of parallel lines corresponding to the direct path of the sound from the transmitter to the receiver, the irregularity of these lines results from the variations in distance between the transmitter and receiver, particularly during turns, this direct arrival of the signal at the receiver does not exist when the transmitter and the receiver are the same (sediment sounders),

– a series of traces representing the sea bottom and the various underlying reflective surfaces.

IV.4.1.2 Horizontal and vertical scales

The recording also shows:

– a series of vertical lines corresponding to the pulses, i.e. the various points surveyed topographically,

– a series of horizontal lines, marking the time scale.

The horizontal scale does not bear a ratio of 1 to the vertical scale.

The horizontal scale is about 1: 5,000 for typical recording conditions, namely:

– speed of vessel = 4 to 5 knots,

– paper advance speed = 2 to 3 cm/min.

The vertical scale is graduated in two-way milliseconds (TWT): this is generally from 100 to 300 ms for a paper width of 11 in (28 cm), thus varying from 1: 250 to 1 : 1,000.

IV.4.1.3 Examples of recordings

The penetration and resolution of the different units used, which govern the analysis and interpretation of the recordings, are shown by the various examples shown in Fig. IV.4.2 to IV.4.7.

The examples are given in decreasing order of power of resolution and, correspondingly, increasing order of power of penetration (sediment sounder, Uniboom, 500 J and lastly 1,000 J sparker).

IV.4.2 Analysis of specific phenomena

The principle of specific phenomena on continuous seismic sounding recordings has been described earlier under "Elements of Seismic Reflection".

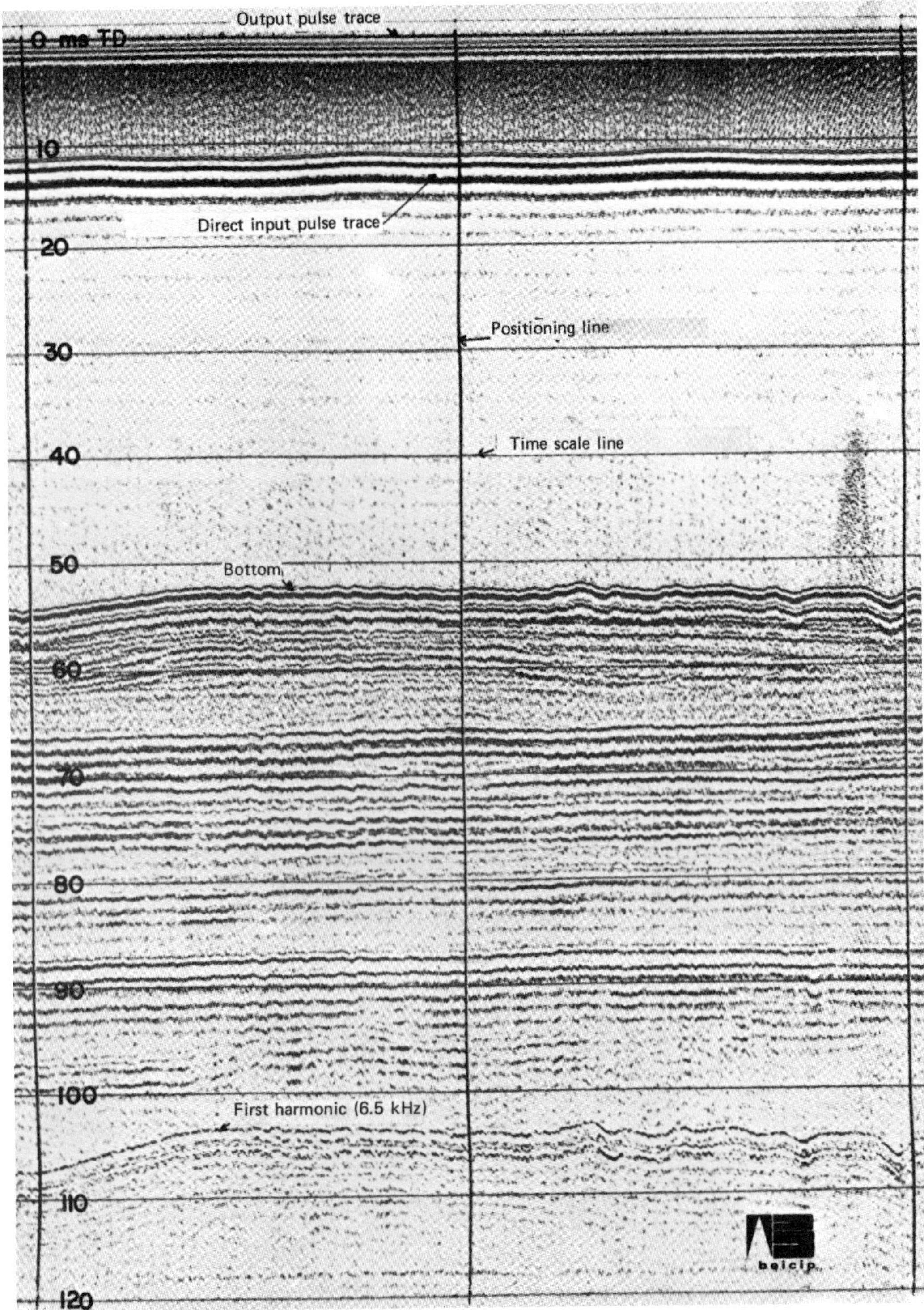

FIG. IV.4.1. – Description of a seismic sounding recording (200 J boomer).

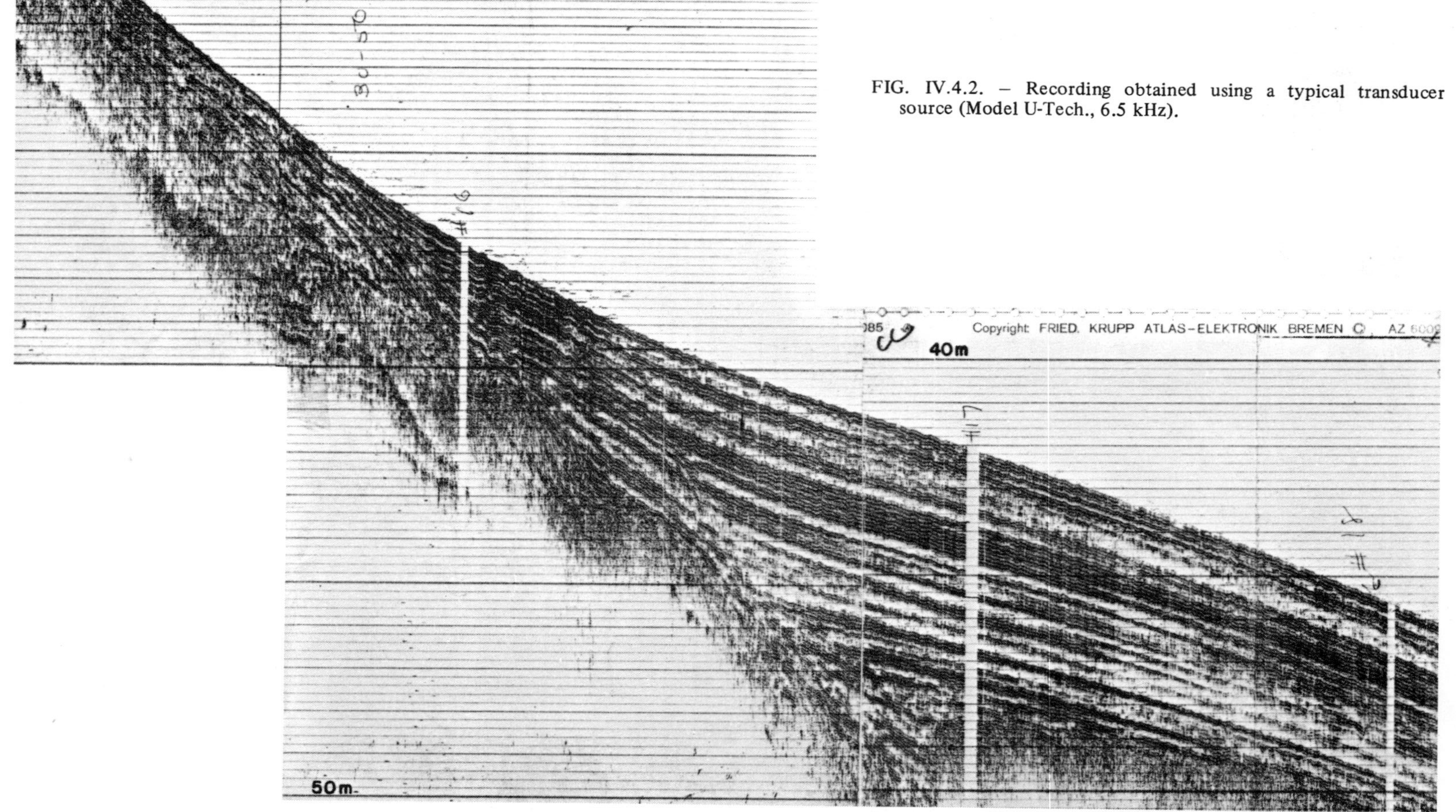

FIG. IV.4.2. – Recording obtained using a typical transducer source (Model U-Tech., 6.5 kHz).

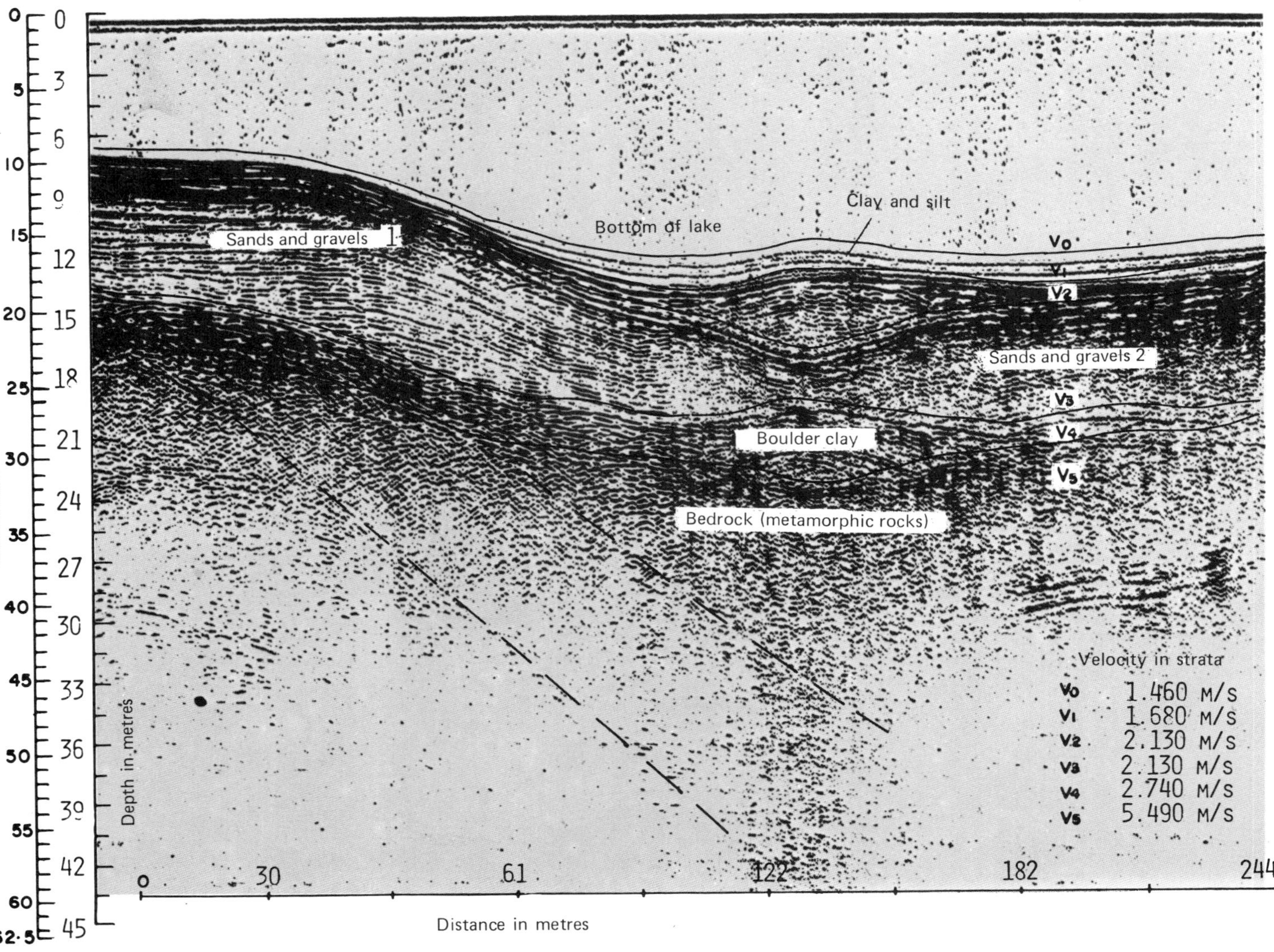

FIG. IV.4.3. – Recording obtained using a Hydrosonde ED 10 seismic source.

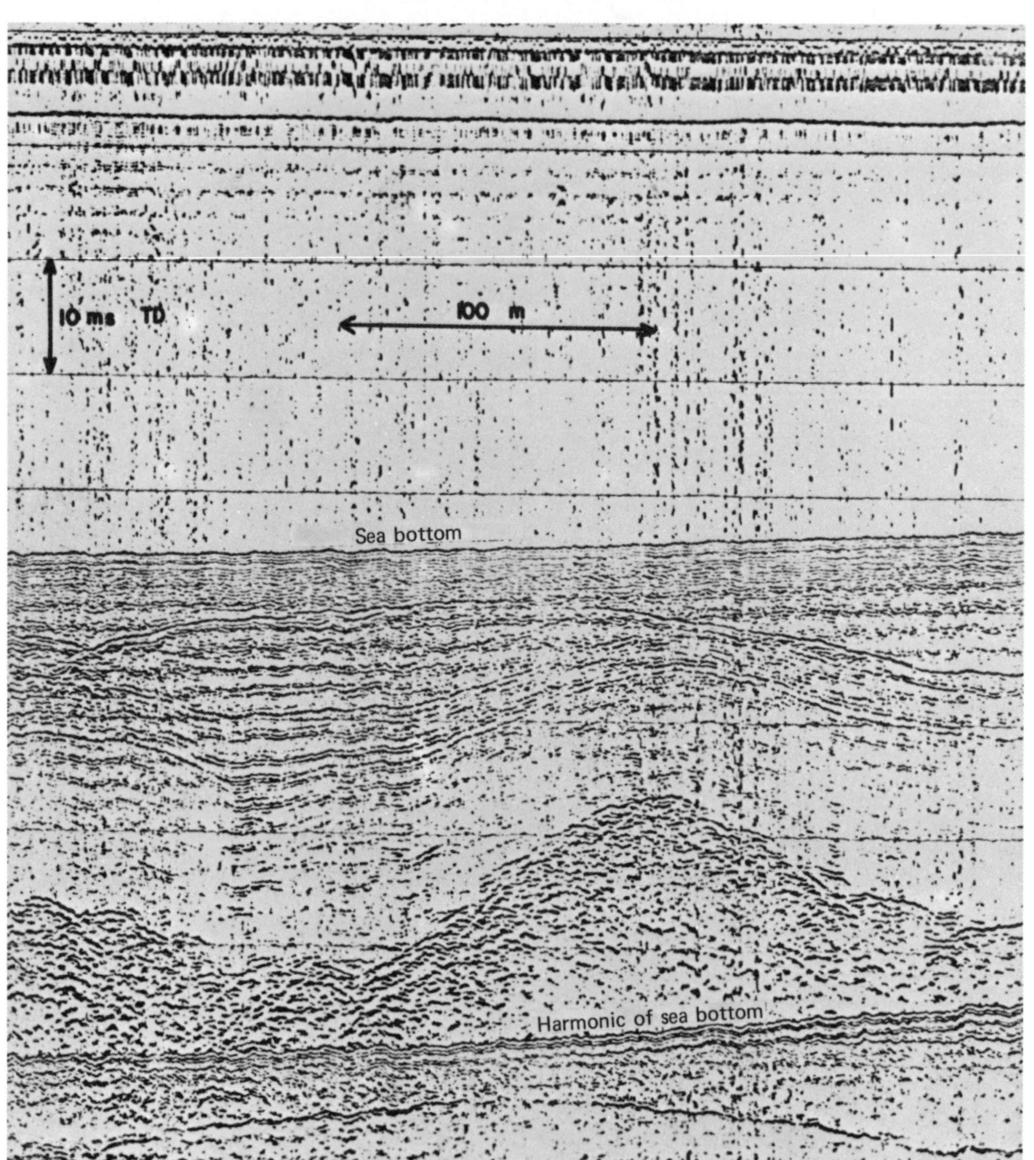

FIG IV.4.4. – Recording obtained using a boomer (Uniboom).

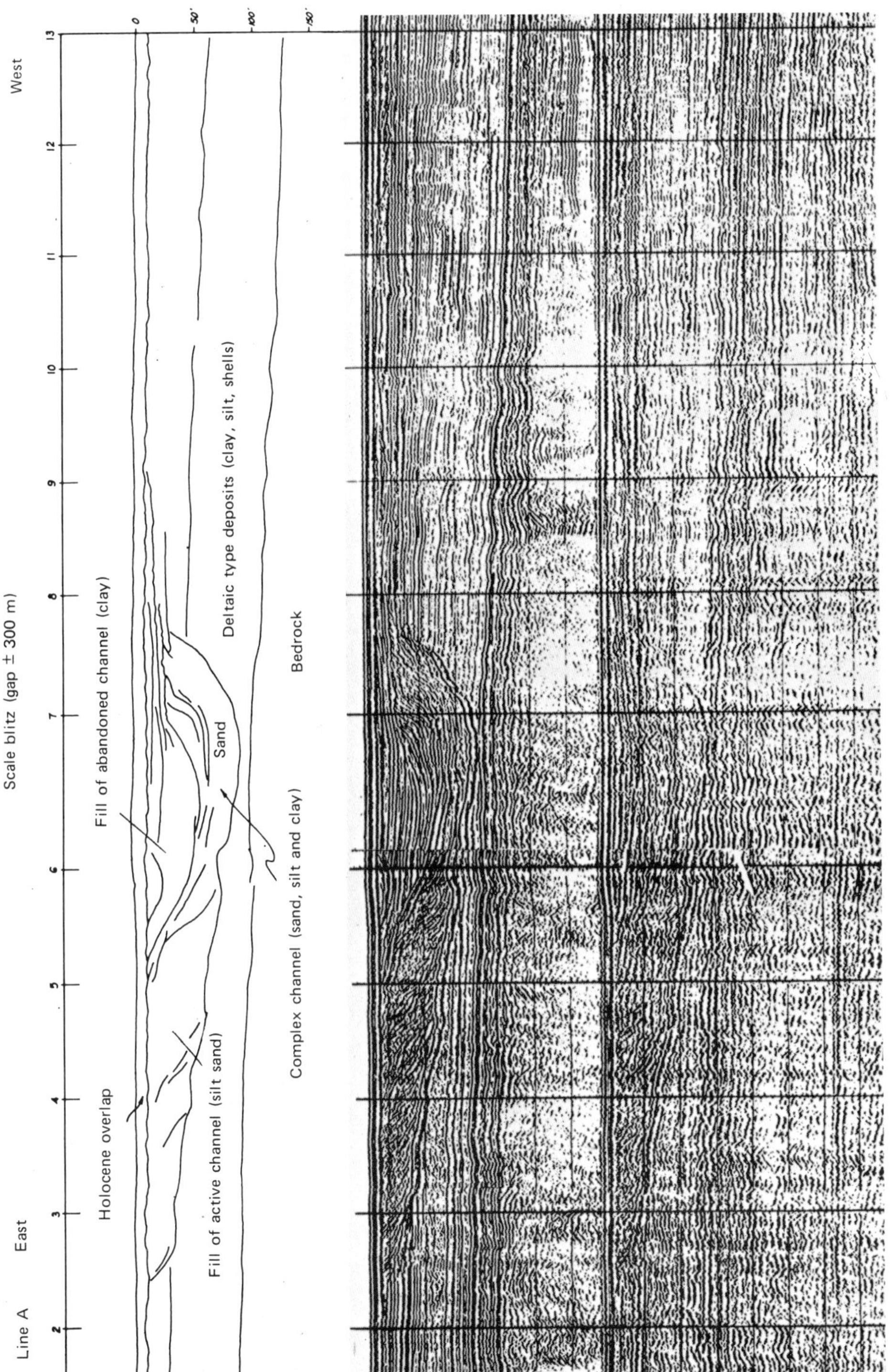

FIG. IV.4.5. – Recording obtained using a sparker (Gulf of Mexico).

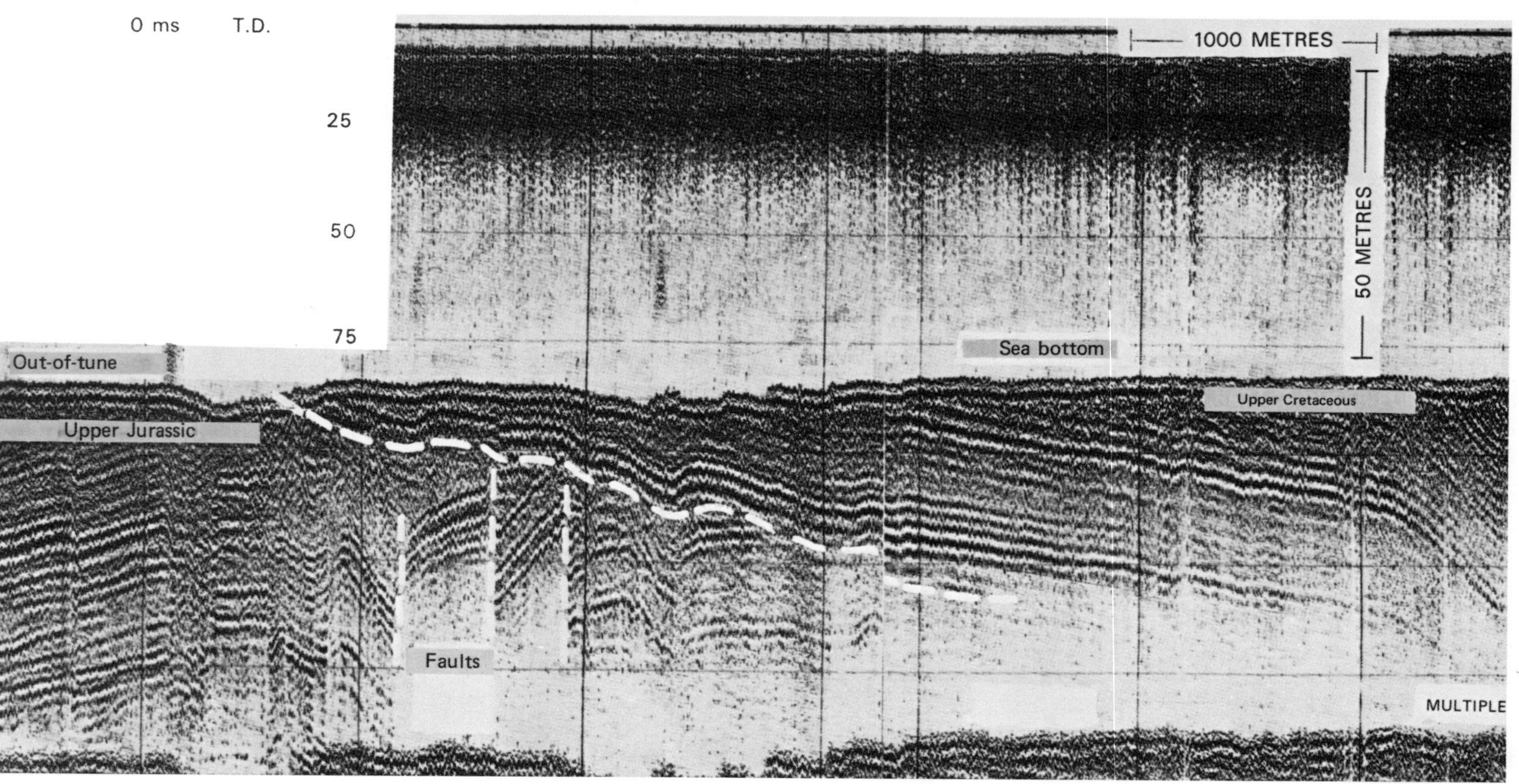

FIG. IV.4.6. – Recording obtained using a sparker (English Channel).

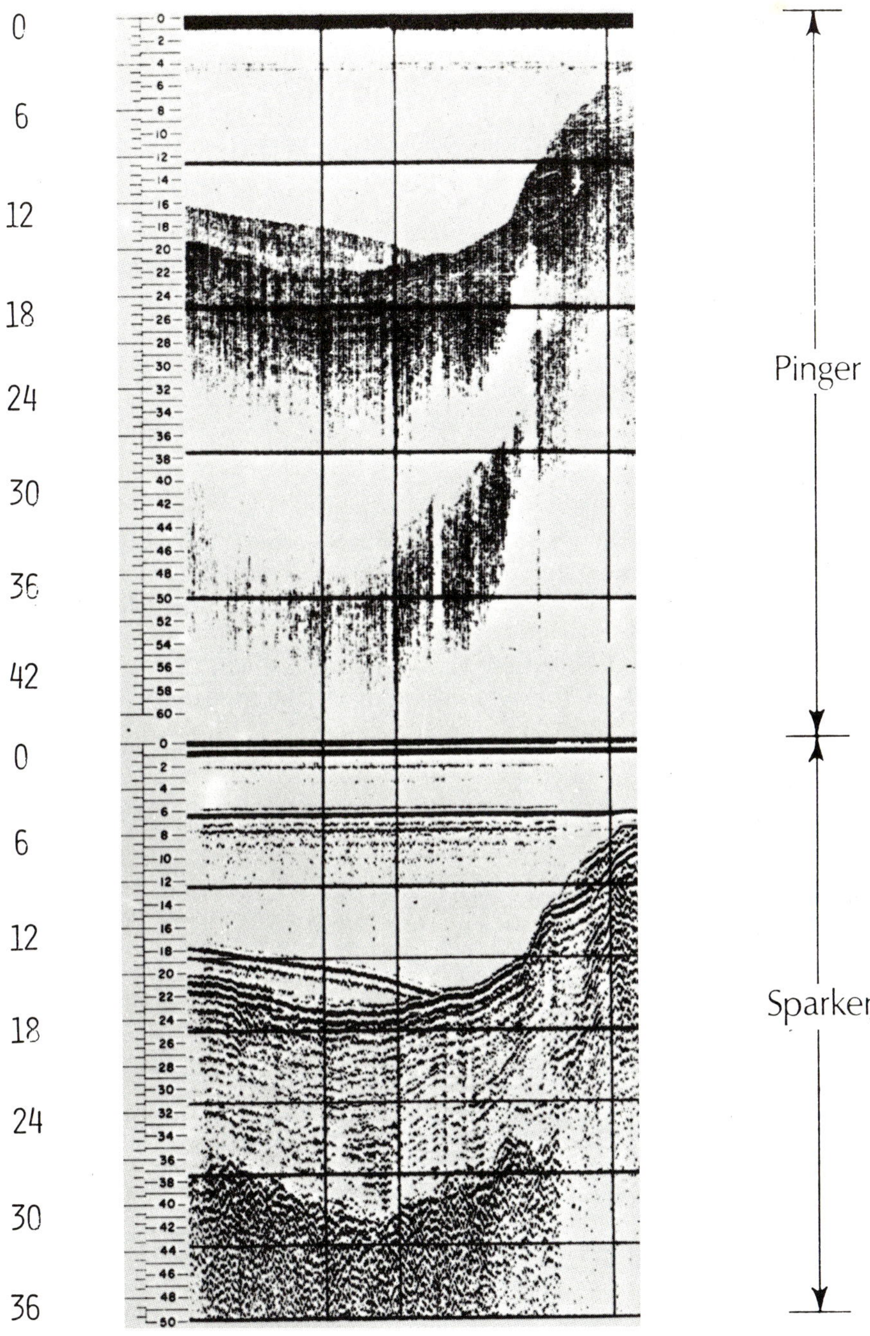

FIG. IV.4.7. – Example of "two-track" recording.
Top : transducer type source (pinger).
Bottom : seismic source (sparker).

IV.4.2.1 Reverberation and "pedalling"

Reverberations or "ghost" **reflections** are the result of reflection of the seismic wave from the air-water interface.

The waves propagate in water with low attenuation, the bottom and surface representing two highly reflective horizons. These physical conditions give rise to the "pedalling" phenomenon.

In most cases, the sources generate **residual oscillations** of the initial pulse.

The upshot of this combination of phenomena is an elongation of the signal transmitted, which, when recorded, results in a thickening of the reflectors and in particular of the energy reflector formed by the sea bottom.

IV.4.2.2 Multiple reflections

Seismic waves reflected by the interfaces of the various media encountered, again propagate to these interfaces after reaching the surface of the sea (which itself is a good reflector). The result is **multiple reflections**, commonly known as "**multiples**".

Since the multiple reflections have crossed twice (or *n* times) the two-way path, they appear on the recordings twice (or *n* times) as low as the primary or single reflections (see Fig. IV.4.8).

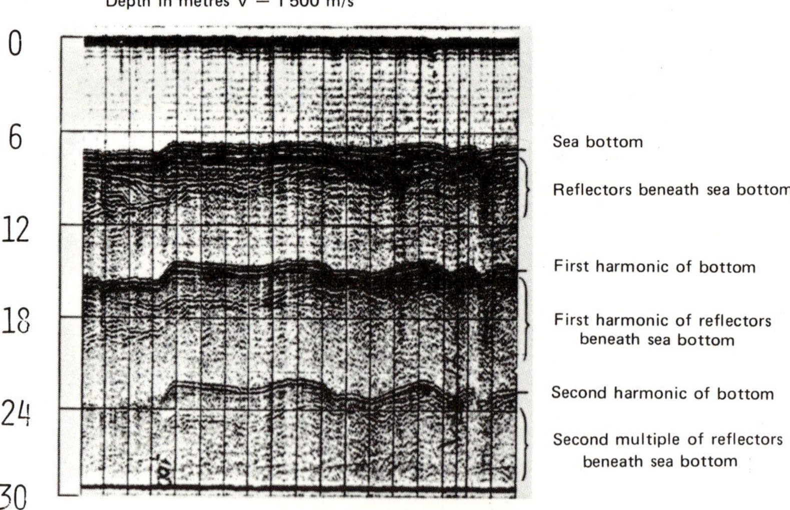

FIG. IV.4.8. – Example of multiple reflections.

The reflections with the highest energy, in particular the sea bottom, result in the most clearly distinct "multiples".

This phenomenon is particularly obvious and bothersome where the depth of water is shallow, since the "multiples" can then mask the primary reflections.

IV.4.2.3 Diffraction

The phenomenon of the diffraction results on a recording in hyperbolic echoes, only one branch of which is sometimes distinguishable.

Diffraction is brought about by specific points with the property of reflecting the incident rise in all directions. Such for instance is the case of isolated erratic blocks, sharp edges (reefs, etc.), eroded surfaces, faults, and so on.

IV.4.3 Interpretation of recordings

There are several stages to interpreting seismic recordings:

– elimination of the parasite acoustic shapes ("multiples", diffraction or scatter hyperbolas),
– identification of the reflectors,
– reconstruction of the maps,
– geological interpretation.

IV.4.3.1 Elimination of parasite figures

The detection of the "multiples" is the first operation to be performed. Depending on the particular case:

– detection may be immediate (clearly evident),
– detection calls for point-by-point comparison of the "possible multiple to-primary reflection" distance and the "primary reflection-to-surface" distance.

In certain cases, it may be very difficult to follow a "multiple".

Beneath the initial bottom "multiple" and hence beyond the equivalent of the depth of water, the plot is very faint and is only possible for high energy or highly sloped horizons.

The problem set by diffraction hyperboles can be eliminated only by digital recording.

In actual practice, diffraction phenomena are rarely a hindrance. Sometimes they may even enable geological accidents to be detected:

– an almost vertical alignment of the diffraction figures probably indicates a fault,
– an almost horizontal alignment probably indicates an erosion surface.

IV.4.3.2 Identification of the reflectors

Study of the **shape of the reflectors** represents the direct analysis method. It consist in processing the recording like a geological profile and drawing immediate conclusions therefrom. For instance:

– rectilinear (or planar) reflectors generally represent loose, non-tectonized uniform sediments (often of the Plio-Quaternary),

– curved reflectors can be explained by the sedimentation itself, but more often represent distortions of tectonic origin,

– discontinuous or complex reflectors indicate not only tectonic distortions, but also phenomena such as erosion, back-filling, discordance, etc.

Study of the shape of reflectors thus provides an approach to the tectonic development of the series considered.

The nature of the reflectors is indicated by the appearance of the reflection (a function of amplitude, frequency, signature, etc.):

– the essential reflectors are recognised by their remarkable appearance and lateral persistance,

– variations in nature invariably correspond to variations in the lithological and sedimentological properties of the reflectors.

The plot of the reflector is extended throughout the length of the profile (correlation by continuity). The same procedure is applied to the other profiles, starting from a well identified intersection with the first profile.

For the plot of other reflectors for the study as a whole, assistance is obviously drawn from the possible conformity between reflectors.

The plot of the reflectors enables pinchouts, gaps, faults, etc., to be detected.

IV.4.3.3 Reconstitution of the maps

For each of the reflectors recorded, the following can be accomplished at each point:

– the two-way path time can be measured,

– the **isochronous maps** (curves of equal trajectory time for a horizon) can be plotted.

The problem of transformation of the isochronous maps:

– into **isobar maps**, namely curves of equal depth of a horizon,

– and **isopach maps**, namely curves of equal thickness of a formation (this transformation is known as plotting),

presupposes knowledge of the speed of the sound wave through the strata crossed.

The speeds in the various strata are:

– either measured in calibration drillings,

– or estimated from data concerning the formations encountered (speeds in loose formations –clays, silts, sands, etc.– vary from 1,600 to 2,000 m/s).

The depth of the reflectors are then calculated, knowing the measured or estimated speed V, in accordance with the relationship:

$$h = \frac{V \cdot t}{2}$$

The "sea bottom" reflector can also be calculated and restored (generally with $V = 1{,}500$ m/s) in order to obtain the bathymetry.

The time t will be corrected to allow:

– for the "transmitter-receiver" range (correction for obliqueness),

– for the submersion of the transmitter and the receiver.

IV.4.3.4 Geological interpretation

The interpretation method consists of:

– defining the fields between two identified reflectors which display homogeneous seismic characteristics,

– analysing these characteristics in sedimentological terms,

– establishing the dimensional characteristics of the elements thus defined (lateral extension, erosions, discordances, etc.),

– consolidating all the data thus obtained into stratigraphic terms, allowing for the geological history of the surrounding regions.

A **highly exhaustive** analysis in continuous seismic sounding implies:

– the simultaneous use on the same profile of several units of distinct characteristics, so as to derive the best resolution/penetration ratio and hence the optimum litho-seismic image,

– the generation of a network of profiles adapted to the geological problem to be solved, i.e. with meshes less than the wavelength of the accidents sought.

The search for optimum exploitation of the results must consolidate all the available data:

– information deduced from the bathymetry and/or other geophysical methods,

– consideration of the geological continuity, etc.

However, uncertainty still generally remains owing to:

– either the existence of stratigraphic and tectonic accidents,

– or the absence of clear correlationships with the geology of the surrounding regions (terrestrial sites or already known marine sites).

One should recall that seismic reflection permits neither determination of the propagating speeds through the terrains crossed nor, a fortiori, their identification.

It is therefore invariably appropriate to cross-relate the interpretation of seismic reflection to a certain number of samples at various depths (surface core drilling and depth core drilling). The preliminary interpretation of the recordings will furthermore make it possible to choose the best positions for the core samples intended.

BIBLIOGRAPHY

[1] MARKÉ (P.A.). – "The Development and Use of Acoustic Energy Sources for Marine Seismic Profiling", *IERE Conf. on Electronic in Oceanography,* Southampton 1966.

[2] INGHAM (A.E.). – *Sea Surveying* (Vol. I and II). John Wiley and Sons, London, New York, 1975.

[3] AMAR (R.). – Recent French Sea-bed Reconnaissance Methods and Techniques", *Ind. Petrol.. Europe Gaz-Chim.,* Vol. 41, No. 4, pp. 51, 53-56, April 1973.

[4] HAYES (J.W.). – "Echo-Sounding and Seismic Profiling", *Offshore Services,* April 1977, p. 55, 57, 58, 60, 62, 63.

[5] BURNS (F.M.). – "Deep-Tow and Digital Techniques". *Offshore Services,* May 1977, p. 47-48.

[6] DES VALLIERES (T.), KUHN (H.), LE MOAL (R.) and DUVAL (J.). – "Test of Various High Resolution Seismic Devices in Hard Bottom Areas". *OTC*, Houston, May 1978, Paper OTC No 3221.

[7] LOWELL (F.C. Jr.) and DALTON (W.L.). – "Development and Test of a State-of-the-Art Sub-Bottom Profiler for Offshore Use", *OTC,* Houston, 1971, Paper OTC No. 1340.

[8] CASSAND (J.) and LAVERGNE (M.). – "Etinceleurs multi-électrodes à haute résolution", *Geophys. Prospecting,* Vol. 18, No. 3, p. 380-388, September 1970.

CHAPTER V

Seabed Exploration Coring Devices and Techniques

CONTENTS

INTRODUCTION

1. A considerable number of drilling and coring units and techniques have been developed for several years now for the reconnaissance of marine soils. The depth of investigation reached varies from a few metres to over 100 m depending on the technique used. The quality of the cores recovered depends on the method implemented and the nature of the soils.

While not detracting from the many geological advantages of knowledge of the soils, the present chapter dwells more specifically on the geotechnical aspect which governs the type and importance of the structural foundations to be set up on the site.

2. The various coring units and techniques used on the continental shelf fall into six categories, depending on the method of implementation, operation, maximum penetration reached, etc.:

– drilling and wireline coring from a surface support (ship or possibly drilling platform); the method uses various coring variants: most often by percussion, fairly rarely by rotation and for a short time now by push into certain consolidated soils,

– drilling and wireline coring using submerged sounding units set on the bottom and operated by divers; this method is limited by the depth of water to which divers can economically be used, namely about 40 m,

– vibrocoring achieved by means of hydraulic or electrical vibro-drivers; the penetration depth can vary from a few metres to 10-15 m,

– "flexocoring", derived from the "flexodrilling" method implemented either from a frame on the bottom, or a specially equipped vessel,

– coring by means of sounding units laid on the bottom and remote-controlled, the development of these units is still encountering major technological difficulties, and they are still of inadequate reliability and excessive cost,

– surface coring by means of a gravity corer or stationary piston corer (of the Kullenberg type); the penetration reached varies from a few metres to about 10 or more metres in soft sediments.

3. The main laboratory tests on core samples for:

– identification of the soils,
– measurement of the shear strength,
– estimation of the settlement,

are briefly summarized in Chapter I "Concepts of soil mechanics".

V.1 DRILLING AND WIRELINE CORING FROM A SURFACE SUPPORT

The technique of drilling and wireline coring is at present the most widely used method for the reconnaissance of soils for geotechnical purposes. Several contractors use the method, including *McClelland Engineering Inc., Fugro-Cesco, National Soils Services Inc., Dames and Moore,* etc.

The core sampling barrel is made to penetrate into the soil by percussion or push. The rotary wireline coring method, little used from a surface support (see Paragraph V.1.7), is applied particularly by means of the submerged sounding units described in Paragraph V.2.

V.1.1 "Percussion" or "push" wireline coring method

The wireline coring method is applied [1] [2] [3] [4] [5] [6]:

– either mostly from supply vessels specifically equipped for this purpose (Fig. V.1.1),
– or from small drilling vessels 75-80 m long in the North Sea particularly (Figs. V.1.2a and V.1.2b, see Section VI.1.3),
– or sometimes from drilling platforms (jack-up or semi-submersible types).

FIG. V.1.1. – Supply vessel equipped with WABCO 1,500 (Failing 1,500) drill.

(a)

(b)

FIG. V.1.2. – Soil reconnaissance vessels.
(a) "Surveyor" (*Heerema*)
(b) "Mariner" (*Heerema*)

V.1.1.1 Principle of method

Reconnaissance of the soils is achieved by the following in turn (Fig. V.1.3):

– by drilling down to the level selected for taking the sample,
– by wireline coring through the guide tube formed by the drilling string.

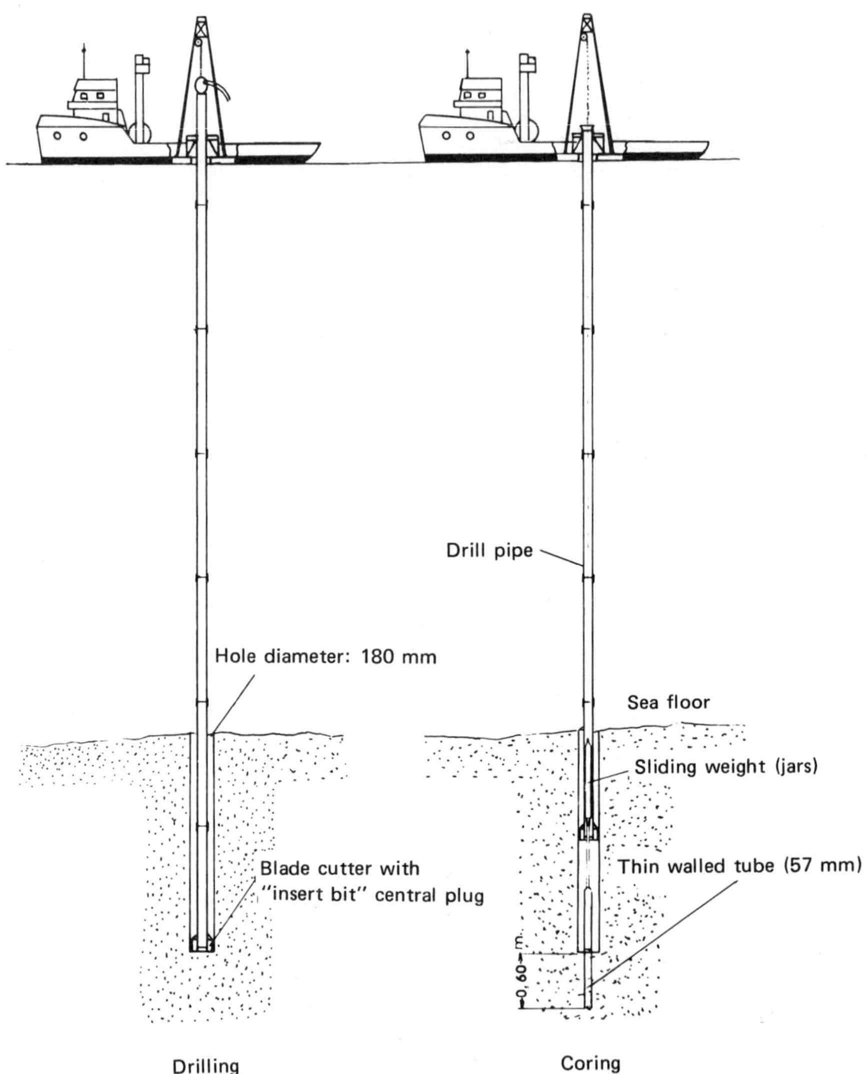

FIG. V.1.3. – Details of the wireline soil sampling procedure

In the case of **percussion** coring (Fig. V.1.4), the sequence of operations is as follows:

- the drilling string and bit are raised by about 2 m,
- the "insert-bit" forming the central plug of the drilling bit is raised by wireline,
- the wireline of the core barrel suspended from a sliding weight is lowered,
- the core barrel is driven into the soil by dropping the wireline actuated sliding weight,
- the sliding weight and the core barrel are raised by wireline.

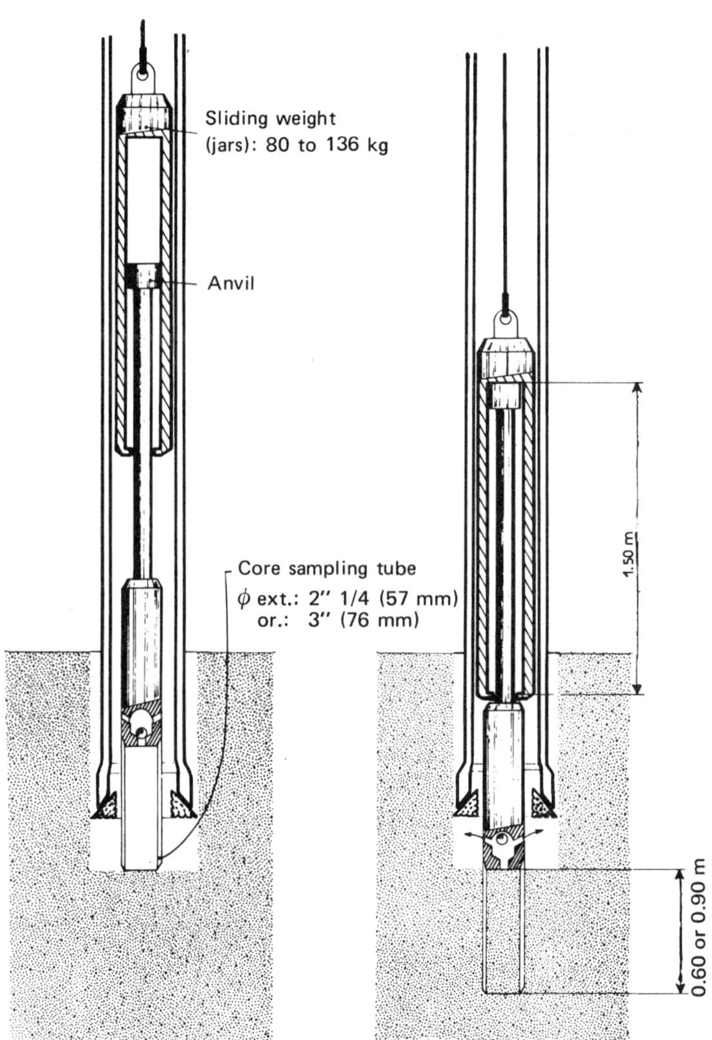

FIG. V.1.4. – Driving device by blows of core sampling tube.

In the case of **push** sampling (Fig. V.1.5), which is still in its early stages of development, the sequence of operations will probably be as follows:

– the drilling string and bit are raised by 1 to 2 m,
– the "insert-bit" is raised by wireline,
– the core barrel is lowered by the wireline and then clamped below the drilling bit,
– the core barrel is driven into the ground by the weight of the drilling string,
– the core barrel is raised by wireline.

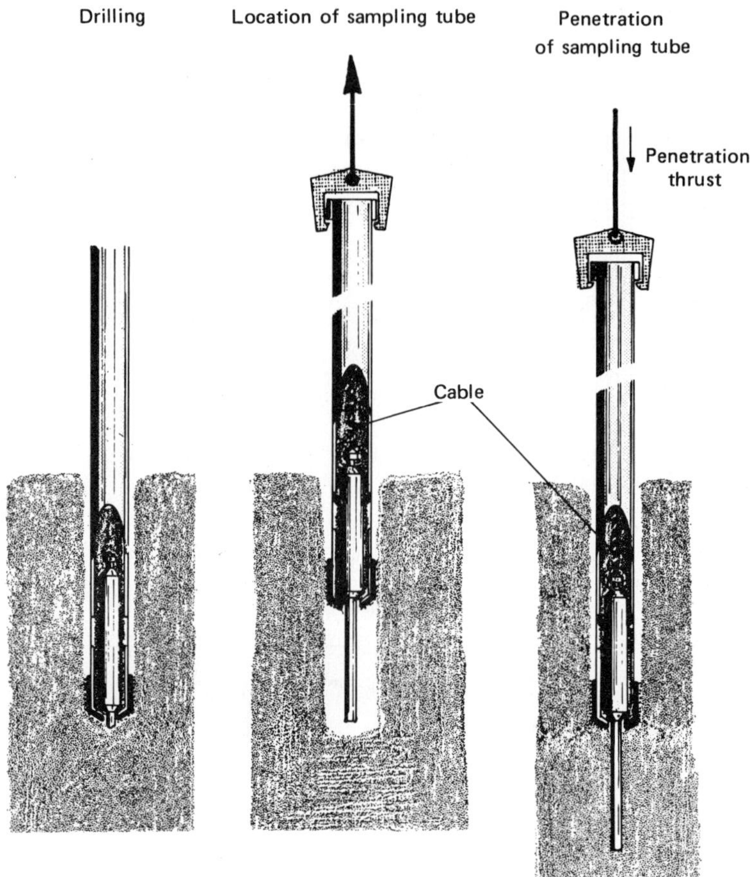

FIG. V.1.5. – Driving device by push of core sampling tube (push sampler)

V.1.1.2 Equipment used

The equipment used for wireline core sampling essentially comprises:

– a Failing 1,500 type drilling rig (modified for operation at sea) or simple power tongs (BJ, Bowen, etc.) (Fig. V.1.6) for turning the drilling string,

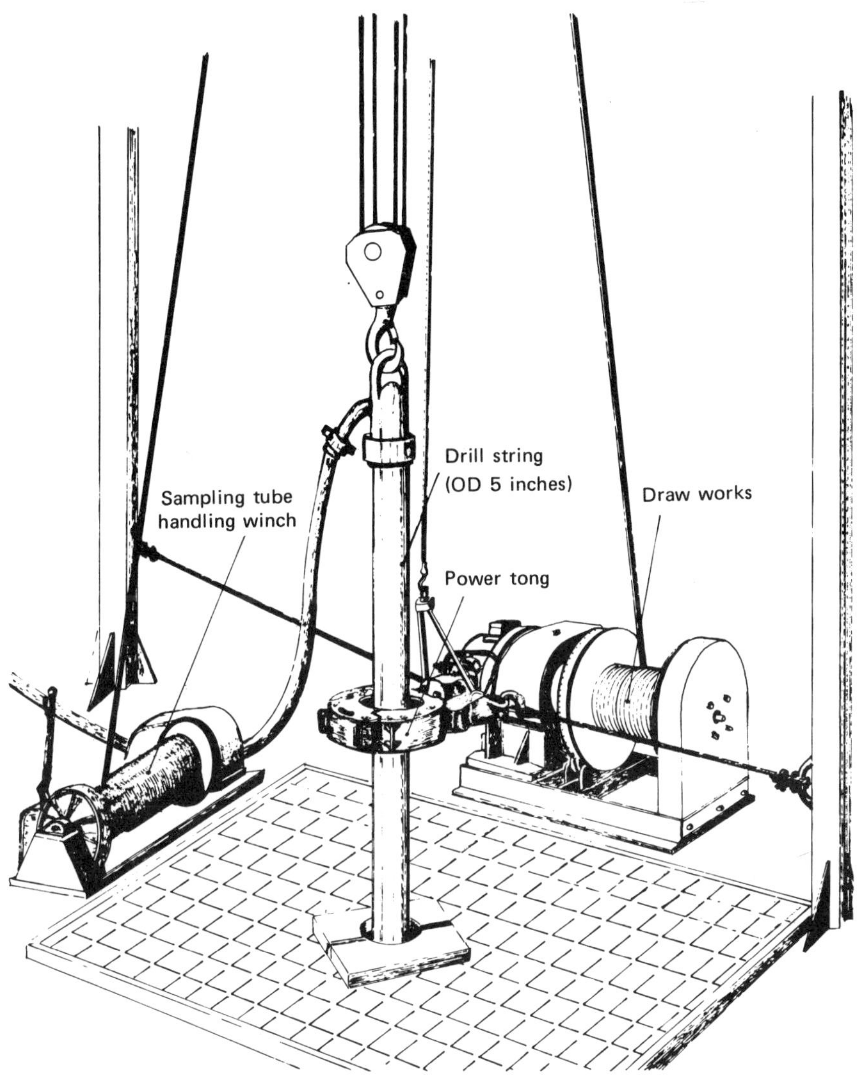

FIG. V.1.6. – Diagram of drilling equipment with power-tongs.

– a 4″ or 5″ drilling string consisting of elements of different lenghts (0.90, 1.50, 3, 4, 5, 6 and 9 m) running through the central well of the ship; the drilling string is equipped with a 7″, 8 1/2″ or 9″ drill bit with tungsten carbide cutters,

– a drilling fluid circulation pump,

– a drawworks,

– a winch actuating the sampling tube.

The Failing 1,500 drilling rig is generally skid mounted. Its characteristics are:

– total weight = about 8 t,
– height of mast = 8.20 m (or 14.80 m),
– mast lifting force = 9 t,
– drive motor = Diesel,
– maximum length of drilling string = 450 m (1,500 ft).

The power tongs offer the advantage on the drilling rig of reducing operating times (by eliminating the need for the kelly).

V.1.2 Characteristics of drilling and wireline core sampling method

V.1.2.1 Drilling characteristics

The load on the bit during drilling is the result of:

– the weight of the drilling string,
– the weight of the 6″ diameter drill-collars about 10 m long (or the hydraulic pressure in the case of a Failing 1,500 type drill).

The rotating speed differs with the process used:

– speed of the turntable of a Failing 1,500 unit = 73, 127 or 220 rpm,
– driving speeds when using power tongs = 25 to 100 rpm (average 60 to 80 rpm).

The torque developed at 60 rpm is about 900-1,000 m × kg.

The flow-rate commonly used is about 500 l/min.
An open fluid circuit is used, discharging onto the sea bottom.

The drilling fluid, which invariably consists of seawater with an addition of natural clay, bentonite, etc., has a density:

– of about 1.20 to 1.30 in soft soils or soils only slightly consolidated,
– from 1.05 to 1.10 in the hard soils of the North Sea.

V.1.2.2 Core sampling characteristics

The core sampler consisting of a thin wall barrel is equipped at the top end with a special device for securing the wireline clamping system. The penetration is obtained:

– either by dropping a sliding weight raised by wireline (percussion),
– or by the weight of the drilling string (thrust) (see Section V.1.1.1).

The following are **the dimensions of the sampling tubes**:

– outside diameter:
  – commonly 2 1/4" (57 mm),
  – more rarely 3" (76 mm),

– length:

– commonly 0.60 m,
– more rarely 0.90 m.

The dimensions of the samples taken are:

– diameters = 54 or 72 mm,
– maximum lengths = 0.60 or 0.90 m,
– actual lengths:
– 20 to 30 cm in hard soils for the "percussion" method,
– approximately 60 cm in the case of the "thrust" method.

Percussion driving of the sampler into the soil is brought about by means of a sliding weight:

– weighting 80 kg (2 1/4" tube) or 136 kg (3" tube),

– falling from a height which can be adjusted between 1.50 and 3.00 m. In actual fact, allowing for the vertical motion of the vessel, the elasticity of the cable and the considerable vibrations, the dropping height varies and is apparently less than the theoretical height set.

The number of blows of the sliding weight needed to drive the sampler through a given distance provides a qualitative estimate:

– of the cohesion of the soils (clayey formations),
– or its relative density (sandy formations).

In highly consolidated formations, the number of blows is limited to 30 for a penetration of about 60 cm, to avoid the possibility of breaking the core barrel wireline during extraction.

Driving of the sampler into the soil by push is achieved under the effect of the weight of the drilling string.

This method has several advantages over the percussion method, namely:

– the greater length of the samples taken,
– minimum disturbance.

However, in very hard formations buckling of the core barrel can prevent it from passing through the drill bit and string, thus necessitating raising of the whole drilling string.

The sampling sequence can vary with the specifications required, since the interval between two core samplings varies with the depth, nature and heterogeneity of the soil. In practice, a sample is taken:

– almost continuously through the first 10 m,

– then at ever increasing intervals: from 1.50 to 5 m below a depth of about 40-50 m (provided the soil is homogeneous).

The penetration into the formation varies according to the purpose sought:

– less than 50 m to verify the homogeneity of surface strata,
– commonly 50 to 100 m in hard or very hard soils,

– often 100 to 150 m in certain soils of low consolidation or heterogeneity.

The maximum penetration may be 150 to 200 m.

V.1.3 Offshore support and implementation

Satisfactory implementation of the method is governed by the choice of a stable offshore support.

V.1.3.1 Naval support

The naval support used to implement the drilling and wireline sampling method essentially depends on the geographic region concerned, i.e. the oceanographic conditions.

In average sea conditions for water depths of less than 200 m, supply ships are frequently used:

– 40 to 55 m long,
– equipped for drilling with a central well (often 0.5 to 0.6 m in diameter),
– comprising a 4 point anchoring system (1,400 to 2,300 kg anchors).

In conditions of difficult seas, implementation of the method calls for the use of stable supports adapted to the site to be reconnoitred:

– vessels at least 75 to 80 m in length such as the "Surveyor", "Mariner", "Ferder" used in the North Sea, Mediterranean, off the African coast, etc. (see Section VI.1.3.1),
– jack-up and in particular semi-submersible platforms, which are being used more and more frequently as supports in the North Sea in particular, for exploitation drilling,
– catamarans about 40 m long, such as the "Duplus",
– vessels specifically fitted out for drilling and core sampling in the Gulf of Alaska.

In deep seas over 200 m in depth, a dynamic-positioning vessel must be used. Several vessels of this type are at present in service: the "Sealab" (about 100 m), "Arctic Surveyor" (80 m), etc.

V.1.3.2 Link between bottom and surface and positioning tolerances

The rigid connection between bottom and surface consists of the 4 or 5″ drill string. During coring, the corer is not secured to the drill string. The flexible wireline connection does not completely isolate the core sampler from the vertical motion of the vessel.

The rigid bottom-to-surface link cannot allow for any movement in the vessel of more than 3 to 5 m without risk of twisting the drill string and subsequently preventing any use of the sampling tube.

If the efficiency of heave compensator is inadequate, the motion of the vessel caused by waves leads to fatigue in the drill string which in turn may result in breaking, which generally occurs near the bottom.

V.1.3.3 Equipment and space needed on the vessel

The drilling and coring equipment needed on the vessel essentially comprises:

– the drilling equipment or rotation winch,
– the drilling and wireline coring drawworks,
– the drilling fluid circulation pump and tanks.

The storage comprises:

– the drilling rods,
– the bags of drilling mud products.

The ship almost always has a laboratory for analyzing the cores immediately (water content, shear strength, etc.).

The layout for all this equipment and these products requires a clear area of about 200 to 300 m^2.

V.1.3.4 Manpower requirements

Continuous operation calls for two teams of technicians for drilling and laboratory measurements.

Each team will consist of:

– two technicians for the drilling operations,
– two technicians for the wireline operations: operating control, assembling and dis-assembling of the core barrel,
– two operators for preparing the drilling mud,
– one technician for the laboratory measurements,

i.e. a total of 7 per shift.

V.1.4 Oceanographic conditions

V.1.4.1 Water depth

The maximum water depth compatible with application of the method depends on the support and anchoring system used.

In the case of funicular anchor system vessels, the following are commonly used:

– supply-boats for depths of 50-100 m, and specialized vessels (75-80 m long) for depths of 100-150 m,
– the maximum depths would be about 150 and 180-200 m respectively.

In the case of a drilling platform the maximum depths are those at which such supports can be used, namely:

– a maximum of 100 m for jack-up platforms,
– 100 to 150 m and more for semisubmersible platforms.

In a case of a vessel equipped with a dynamic positioning system, the depth of water is limited only for operational reasons (length of drilling string). The "Sealab" and "Arctic Surveyor", for instance, are designed for depths of 800 to 1,000 m.

V.1.4.2 Waves, wind and current

The maximum acceptable wave amplitude for drilling and wireline coring operations clearly depends on the support used:

– in the case of supply boats, drilling cannot be carried out in wave amplitudes of more than 1.5 m,
– in the case of vessels such as the "Surveyor", "Mariner", etc., the wave amplitudes must not exceed 2.0 to 2.5 m and the winds must be less than 6 on the Beaufort scale,
– in the case of drilling platforms, the wave amplitudes can reach 5 to 6 m without hindering operations.

The speed of the current often governs the anchoring possibilities:

– anchoring of supply boat implies a current of less than 2 to 3 knots,
– anchoring of vessels 75 to 80 m long (such as the "Mariner" or "Surveyor") would appear possible with currents up to 3 or 4 knots; these vessels are capable of laying their own anchors, and anchoring time varies from 3 to 5 h provided the current is less than about 3 knots.

V.1.5 Performance of the drilling and wireline percussion coring method

After the vessel (or platform) has been anchored, the operation consists in:

– insulating the string of drilling rods,
– drilling a 7", 8 1/2", etc., diameter hole,
– sequential sampling of cores at set intervals.

V.1.5.1 Drilling and wireline coring rates

Among the characteristics of the methods, contractors sometimes indicate that the reconnaissance will not require more than one day per borehole.

Under ideal conditions, in a water depth of not more than 100 m, it is indeed possible to drill a 50 or 100 m borehole with 25 or 50 cores in one or two days, i.e. an average of one core sample per hour.

In reality, the time taken for drilling and wireline coring varies considerably, depending on:

– the support used,
– the weather and oceanographic conditions; in the North Sea, a supply-boat would be usable only very little,
– the nature of the formations.

Whenever the weather and oceanography conditions permit (wave amplitude less than 2 m), the drilling and wireline coring speed in the compact sands and consolidated clays of the North Sea can attain 60 m/d on two shifts. In reality, the average rate is often no more than 25 to 30 m/d in highly consolidated clays.

The operating rate from a semisubmersible platform with wave amplitude of 5 to 6 m has been known to reach 60 to 70 m in a 36 h period, i.e. an average speed (including coring) of about 2 m/h.

V.1.5.2 Time lost through weather conditions

In the particular case of the North Sea, reconnaissance of soils from a vessel 75 to 80 m long (of the "Mariner" type) is:

– practically confined to the period from March to October, with a working time of no more than 30 to 40%,
– rarely possible during the period from November to February; the working time is about 10% maximum and is practically nil in certain winters.

The actual performance of the reconnaissance method must therefore be calculated allowing for probable lost time throughout the campaign as a whole.

V.1.6 Geotechnical characteristics of the cores sampled

The geotechnical quality of the cores sampled using wireline methods varies widely with the nature of the soils and the sampling conditions.

V.1.6.1 Types of soils and recovery

Wireline coring with small diameters (54 or 72 mm) allows the sampling:

– in soft muds or clays, of highly disturbed cores which are nonetheless used to determine the shear strength,
– in consolidated or highly consolidated clays, of moderately disturbed cores, though generally fairly representative of the mechanical characteristics of the soils,
– in loose sands, very highly disturbed samples, used only for lithological identification.

The length of the cores, which is often a maximum when the "thrust" method is used, varies very considerably when the "percussion" method is used:

– in soft soils ($c_u < 5$ t/m²) ($c_u < 50$ kPa), the length of the core is practically that of the core barrel,

– in cores of average consolidation ($c_u \approx 10$ to 20 t/m²) ($c_u \approx 100$ to 200 kPa), the wireline coring method generally proves satifactory, the percentage of recovery is high and disturbance relatively limited,

– in highly consolidated clays, of the overconsolidated type in the north of the North Sea (c_u up to 50 t/m² (500 kPa) and more), the length of the core sampled by percussion rarely exceeds 20 to 30 cm and disturbance may be considerable owing to the high coefficient of friction against the walls of the core barrels,

– in loose sands, the sample may be taken by drilling with a drilling mud of fairly high density, so the specific gravity measured is generally not representative of the specific gravity in place.

V.1.6.2 Disturbance and geotechnical qualities of the cores

The disturbance of cores sampled by wireline coring is explained at one and the same time by the following:

– the method of insertion of the core barrel (in particular by percussion),

– the small diameter of the cores (often 54 mm) (see Fig. V.1.7),

– the relaxation of the stresses subsequent to sampling (inherent in any coring method), causing the samples to expand,

– the disturbance to the top section of the core caused by the action of the drilling bit and circulation of the drilling mud,

– the development of chemical and biological reactions in the soil caused by the preservation of the cores at temperatures above the temperature at the sea bottom (5° C in the North Sea, for instance),

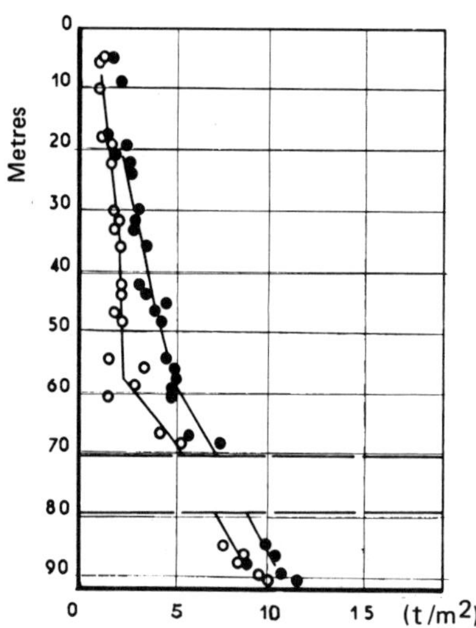

Sample taken by tube on 2 1/4 inch (57 mm) wire-line

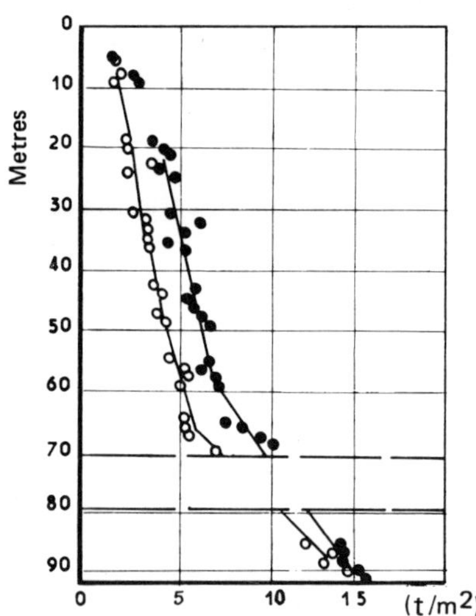

Sample taken by 3 inch (76 mm) tube, with piston

o Unconfined compression test

• Laboratory scissometer

FIG. V.1.7. – Influence of the core sampling tube on the shear strength of a clay.

– transport and handling in the laboratory.

In the specific case of the overconsolidated clays of the North Sea, disturbing is also explained by:

– the high level of the frictional forces against the walls of the core barrel on penetration and extraction of the core,
– frequent frothing which occurs at the lower end of the coring tube.

The lower and upper ends of each core must therefore not be used to determine the mechanical properties.

Disturbance can be highlighted immediately with the first measurements of the undrained shear strength s_u made aboard the vessel by:

– unconfined compression,
– "fall cone" penetration,
– pocket penetrometer,
– vane or "Torvane".

For instance, after penetration of the "fall cone", it is almost invariably observed that:

$$s_{u\,(\text{at centre})} > s_{u\,(\text{on outer surfaces})}$$

V.1.6.3 Drilling and coring logs

The load on the bit varies during drilling and the torque is not measured.

The rate of progress therefore constitutes the only drilling log. In the case of the highly consolidated formations on the North Sea, the following average speeds are observed:

– 5 to 6 m/h in dense sands,
– 1 to 2 m/h in very highly compact clays ($c_u > 50$ t/m^2) ($c_u > 500$ kPa).

The dynamic penetration log of the corer ("percussion" method) would therefore appear difficult to use, especially in the case of highly consolidated clays:

– the number of blows of the sliding weight applied to the core barrel is determined empirically by the operator in the light of the rate of drilling penetration, at each level considered,
– the energy applied by impact is not constant,
– the maximum number of blows is generally no more than 35, by convention.

V.1.7 The "rotary" wireline coring method from a surface support

V.1.7.1 Principle and description of the method

The "rotary" wireline coring method from a surface support is used only infrequently for reconnaissance of soils at sea, though it is nonetheless applied *Geocon* (Canada), *Preussag* (RFA), *Ingemer* (France), etc.

The Longyear type corer, widely used ashore, is set up on the vessel.

Drilling is done continuously by means of a "double coring device" of the Longyear type (see Section V.2).
The top of the inner tube is automatically capped by an "overshot" system which is pulled up by wire line inside the drilling string.

V.1.7.2 Possible improvements to the method

As currently applied, rotary wireline coring involves considerable disturbance of the cores sampled in loose soils owing to circulation of the fluid and the small diameter of the coring tube.

The method could certainly be improved considerably by adapting core barrels equipped with a retractable cutting tool (of the Mazier type, see Section V.2) enabling cores at least 100 mm in diameter to be taken.

V.2 CORING USING SUBMERGED ROTARY DRILLS OPERATED BY DIVERS

Application of the wireline coring method by means of submerged rotary drilling machines set up on the bottom and operated by divers, is limited in practice to depths of about 40 m. However, in addition to harbour tasks, this method is still very widely used for studying the siting of petroleum structures, particularly in the Persian Gulf.

V.2.1 Principle and description of rotary drilling machines

The rotary drill is used for core drilling, generally of the Wirth type. The various models which exist differ particularly in their dimensions and capacities (see Figs. V.2.1 and V.2.2) [7].

V.2.1.1 Principle of Wirth type drilling machines

The Wirth type drilling machine (see Fig. V.2.1) is:

– set on the sea bottom,
– driven by two hydraulic motors supplied by a central hydraulic power plant installed aboard the vessel,
– operated by divers in depths of generally less than 40 m,
– used for wireline coring or making in situ measurements.

This type of drilling machines reduces shore drilling problems to those ashore.

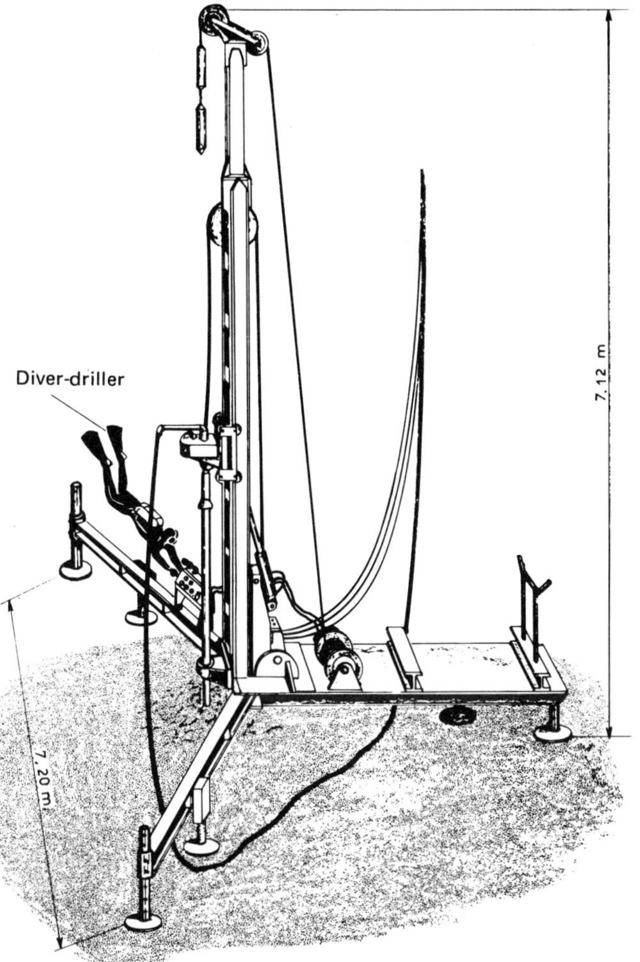

FIG. V.2.1. – Wirth type B1 A drill.

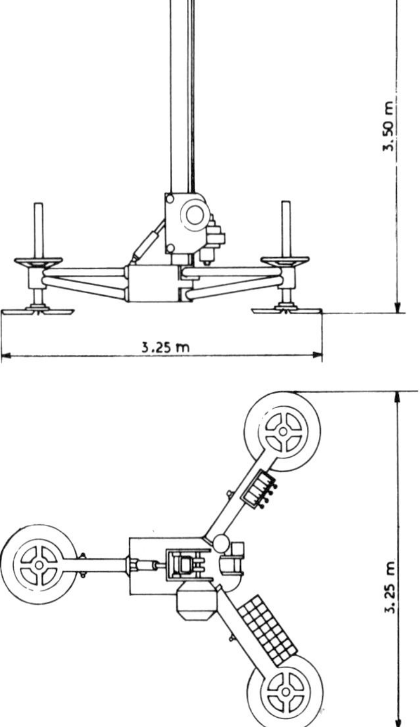

FIG. V.2.2. – SM 150 drill.

The wireline double corer method is applied by means of various types of equipment (see Fig. V.2.3):

– Wirth, Longyear, Craelius, Diamant Boart, for coring in hard or consolidated formations in particular,
– Mazier, mounted on Longyear, for coring in sandy terrains in particular.

All these units comprise:

– a reinforced outer barrel fitted at the bottom with a rotating core bit; this barrel penetrates into the soil under the combined effects of rotation and vertical thrust,
– a core barrel secured inside the rotary string; this core barrel is mounted on ball bearings to prevent it from rotating and to limit disturbance of the cores taken.

In the case of loose sediments, where conventional rotary, coring cannot be used, the Hageman corer, with its inside barrel vibrating by means of a mechanical pawl system, can be used.

After the core barrel has been filled, the diver lowers into the drilling string on "overshot" system on a wireline, consisting of a type of grapple which automatically fits onto the head of the tube.

The core barrel is lifted up through the drilling string to the surface for analysis and storage.

V.2.1.2 Description of equipment used

The models BO 1 or B1 A Wirth drilling machines set on the bottom essentially consist of:

– a frame resting on the sea bottom on a tripod and three adjustable jacks,
– a rotary head with two independent hydraulic motors (model BO 1) or coupled motors (model B1 A),
– an inclinable mast hinged at the bottom with an effective height of 3.35 m (BO 1) or 4.80 m (B1 A),
– a translation mechanism operated by means of a jack and hydraulic locking system,
– a hydraulic capstan winch with a tractive effort of 1 to 2 t (depending on the type), an adjustable speed (0 to 2.5 m/s) and 100 m of wireline for operating the core barrels,
– drilling strings 1.50 m long,
– a pressure and flow control and monitoring console.

The effective mast height enables two 1.50 and 3 m rods coupled together to be raised.

The surface equipment on the vessel consists of:

– the central hydraulic unit powered by an air-cooled Diesel engine developing about 50 hp; this unit has a throughput of about 50 l/min at a maximum pressure of 180 to 200 kg/cm^2 (18 to 20 MPa),
– the mud pump.

The umbilical hose between the bottom and the surface consists of flexible hydraulic lines, tested to 3 times the service pressure, i.e. at least 600 kg/cm^2 (60 MPa).

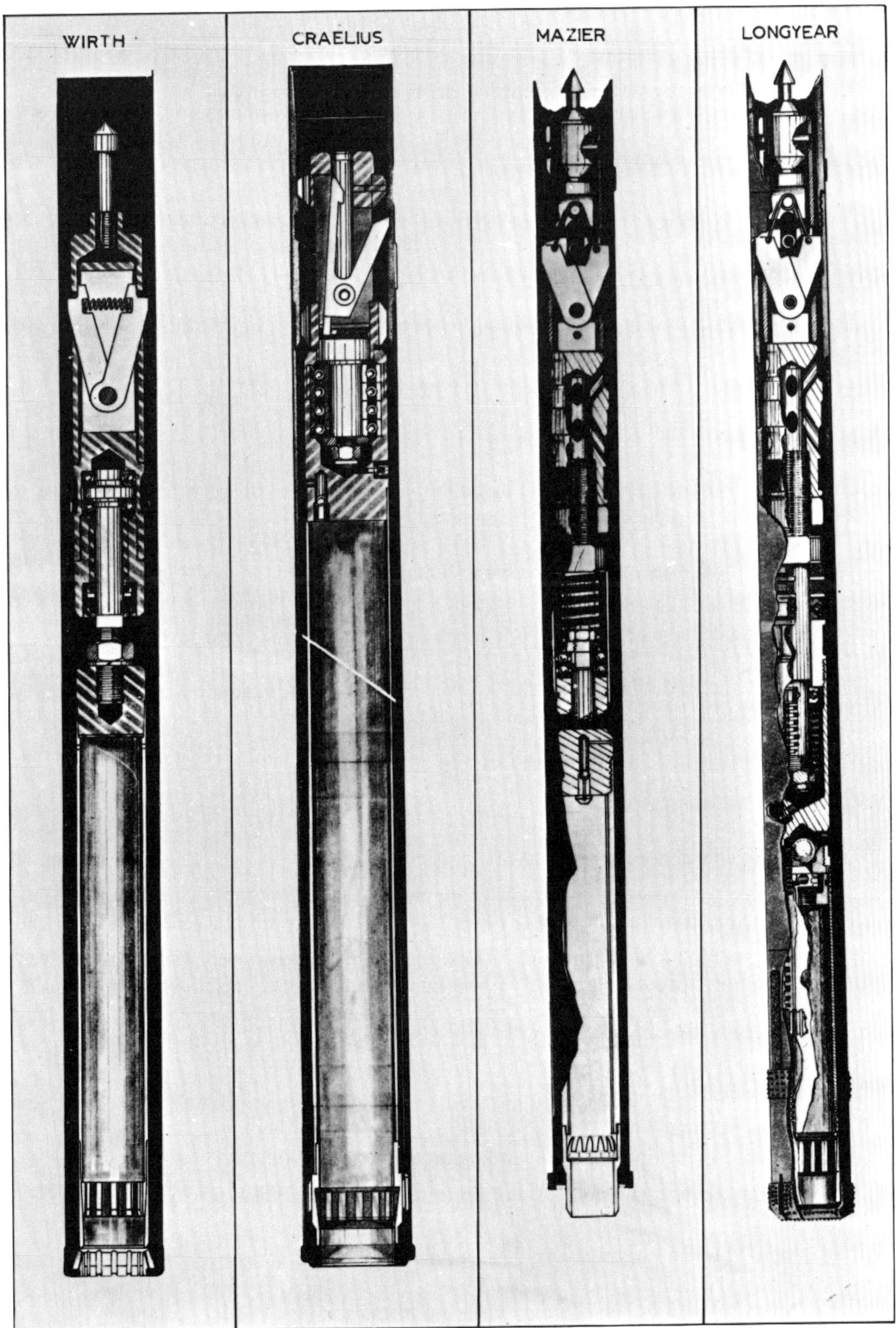

FIG. V.2.3. – Cable-extractable rotary samplers.

V.2.2 Characteristics of use of rotary drills operated by divers

V.2.2.1 Drilling characteristics

The load on the bit is applied hydraulically by means of jacks. Maximum thrust is about 2 to 4 t, depending on the model (2.2 t for the BO 1 sounder and 3.8 t for the B1 A sounder).

There is a continuously variable **rotation speed** (driven by the hydraulic motor) from 0 to about 400 rpm.

The torque developed by this type of sounder is about 200 to 400 m x kg at 50 rpm, depending on the model.

The flow rate of the drilling fluid is deliberately restricted to about 80-120 l/min in order to reduce disturbance of the cores sampled.

The drilling fluid, depending on the formations, consists of:

- seawater in consolidated soils,
- mud with a density of about 1.10 in loose sediments.

The pullout force of the drilling string varies from 3 to 6 t, depending on the particular model on the drill.

V.2.2.2 Coring characteristics

The "double corer" consists mainly of:

- a rotating outer barrel equipped with a diamond core bit,
- an inner core barrel secured to the inside of the rotating string (see Section V.2.1.1).

The Mazier corer mounted on the Longyear system and equipped with a cutting tool protruding below the core bit mounted on its end enables the terrain to be "punched" and protected form the circulating fluid. The tool retracts when it encounters a hard or rocky terrain. This technique improves the rate of recovery in sandy formations.

The dimensions of the corers are:

- typical barrel length = 1.50 m,
- barrel diameters which differ depending on the type of corer.

Table V.2.1 gives the diameters of the barrels and the cores sampled (see Fig. V.2.3).

The sample is taken continuously in successive downward steps of 1.50 m of the core barrels.

The recovery rate varies considerably with the formation (see V.2.5).

TABLE V.2.1

Diameters of corer and core (mm)	Wirth corer HQ	Craelius corer	Longyear corer		Retractable Mazier corer mounted on Longyear
			PQ	PQ3	
Core bit.	122.6	150	122.6	122.6	122.6
Outer barrel:					
OD	119	139.7	117.5	117.5	117.5
ID.	113	127	103.2	103.2	103.2
Inner barrel:					
OD	72	111	95.3	92.2	92.2
ID.	68	107	88.3	84.9	88.9
Core	68	105	85	83	82

In principle, **penetration** attains:

- 30 to 50 m with the PQ (85 mm),
- 50 to 100 m with the HQ (68 mm).

In actual practice, penetration does not commonly exceed:

- 20 to 30 m with PQ,
- 50 to 60 m with HQ.

V.2.3 Offshore support and implementation

No special vessel is needed to implement Wirth type sounders.

V.2.3.1 Type of vessel used

On a sheltered site, a pontoon with minimum dimensions of 20 × 8 m can be used, equipped with 4 winches and a minimum of 100 m of anchor line, accompanied by a tug.

In the open sea, in order to reduce down-time, the following would be appropriate:

- either a supply-boat,
- or a trawler:
  - with a minimum length of 30 m,
  - with a free deck area of 8 × 5 m,
  - equipped with 4 winches (if possible at constant tension) and a minimum of 150 m of anchor line.

V.2.3.2 Link between bottom and surface and positioning tolerances

The **semi-flexible** link between the bottom and the surface consists of (see Fig. V.2.4):

- the cable carrying the drilling machine,

– the flexible supply lines to the hydraulic motors,
– the core barrel hoist wireline.

The semi-flexible connection and its implementation by divers requires the vessel to be located within a radius of 10-20 m at the most above the drilling point.

V.2.3.3 Equipment, handling facilities and space necessary

The equipment needed for drilling and core sampling has already been described (see Section V.2.1.2).

The weight and dimensions (with the frame folded and the most raised) of type BO 1 and B1 A drilling machines are given in Table V.2.2 below:

TABLE V.2.2

	Wirth corer BO 1	Wirth corer B1 A
Frame folded and mast raised:		
– height (m)	5.80	7.12
– span (m)	6.20	7.20
Effective height of mast (m)	3.35	4.80
Weight (t)	1.8	3.5

Implementation of Wirth drills calls for the following hoisting facilities:

– either a loading boom,
– or a gantry or crane (in the case of a supply-boat).

This handling equipment should have the following characteristics:

– hoisting capacity = 3 t (BO 1) or 4 t (B1 A),
– clearance beneath hook = about 5 m,
– outboard overhang = 4 to 5 m.

The minimum space needed on the vessel is about 40-50 m^2, for:

– conveying the drill from one coring point to another,
– the central hydraulic power plant,
– the mud tank and storage of the mud products,
– handling of the core barrels,
– the diving cylinder recharging compressor,
– the decompression chamber.

V.2.3.4 Required manpower

Implementation of an submerged drilling machine requires the simultaneous use of the following:

– one diver-driller on the bottom,
– two technicians on the surface.

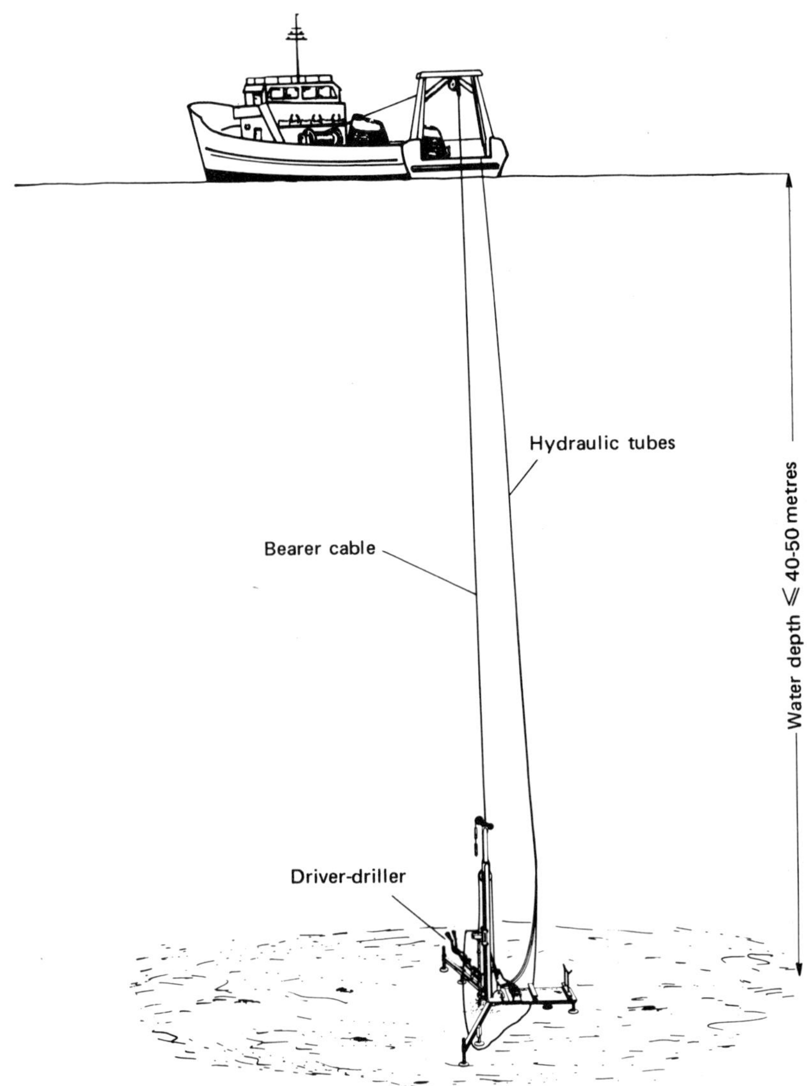

FIG. V.2.4. – Implementation of the Wirth drill.

Extended operation of the drill calls for the successive intervention of numerous diver-drillers, the number of which varies, depending on the depth of water, namely:

- 20 to 30 m = five to six divers,
- 30 to 40 m = at least eight divers.

As an example, in a depth of 30 to 35 m:

- the time on the bottom can attain 1 h 30 min (with only one descent every 24 h),
- decompression takes 1 h.

With a one-person decompression chamber, and with eight divers, a cycle of 1 h 30 min drilling plus a pause of 30 min can be repeated 8 times every 24 h.

V.2.4 Oceanographic conditions

V.2.4.1 Water depth

The maximum acceptable water depth is about 35 to 40 m; the time on the bottom for one diver is then less than or equal to 30 min.

For depths of over 40 to 45 m, the procedure is no longer economically applicable, in view of the need for too many diver-drillers.

V.2.4.2 Waves and current

The device can be used from a supply-boat only in waves of less than 1.5 to 2 m.

The supply-boat cannot be anchored in currents of over 2 to 3 knots.
Furthermore, the work of the divers and the stability of the drill machine on the bottom also imply very weak currents.

V.2.5 Performance of submerged drilling machines implemented by divers

V.2.5.1 Core extraction rates

The core extraction rates vary with the depth of water, nature of the formations and depth of penetration of the core drilling.

Under good weather conditions, a 30 m borehole in a water depth of 30 m can be expected to take about 30 h.

In general, core extraction rates are in the region of:

- 6 to 7 m/d under a water depth of 20 m,
- up to 15 m/d in soft formations.

V.2.5.2 Recovery rate

Just as for drilling on land, the recovery rate is linked to the type of corer and the nature of the formation:

- in rock and highly consolidated soils, the rate of recovery would attain 80 to 90%,
- in sands and soft clays, the rate of recovery is generally no more than 30 to 50%.

The use of non-return valves and drilling muds adapted to the resistance of the formation would also theoretically enable recovery rates of 80% to be reached in loose sediments.

V.2.6 Geotechnical characteristics of the cores sampled

Just as for any coring procedure, the geotechnical quality of the cores depends first and foremost on the nature of the formations encountered.

V.2.6.1 Types of soil

Wirth, Longyear and Craelius rotary wireline corers are well suited:

– for the geological reconnaissance of all types of formations,
– for the geotechnical reconnaissance of rock (limestone coral rock in the Persian Gulf, for instance) or consolidated clays.

The Mazier rotary corer (Soletanche type) with a retractable cutting tool is particularly well suited for taking cores in sandy formations.

The vibrating corer (Hageman) is best for coring loose sediments.

V.2.6.2 Limits and possibilities of use of the rotary wireline corer

In the case of loose sediments, the rotary corer is poorly adapted to technical reconnaissance owing to:

– the high speed of the corer; it is generally agreed that the disturbance of cores from soils of low consolidation (cohesion < 5 to $10\ t/m^2$) (< 50 to 100 kPa), is invariably considerable if the rotary speed is more than 15 to 20 rpm,
– the disaggregation of the cores as a result of circulation of the drilling fluid, particularly with sands; the use of drilling muds in unconsolidated sediments improves the recovery rate, but cannot prevent disturbance.

In a case of rock or consolidated clays, the rotary wireline coring procedure enables cores to be extracted which are acceptable for geotechnical analyses.

V.2.6.3 Advantages of special "cutting tool" and "vibrating" corers

The retractable cutting tool corer can be particularly recommended for sandy soils with very low cohesion:

– the cutting tool protruding from the core bit protects the soil from the drilling fluids,
– the use of a suitably adapted drilling mud improves the recovery rate even further.

The vibrating corer is applicable in particular to clayey soils with low consolidation:

– penetration takes place without difficulty,
– disturbance is quite limited.

Remark. Wirth type sounders are used:

– either for core extraction,
– or for the use of in situ measuring instruments: static or dynamic penetrometer, standard penetration test, vane, pressuremeter, Lugeon permeameter, etc.

V.2.7 Vriens Diving Company device with diving bell

The *Vriens Diving Company* (Netherlands) has built a device for reconnoitering soils from a diving bell [8].

V.2.7.1 Description of device

The device essentially consists:

– of a ballasted cylindrical base 6 m in diameter,
– of a diving bell with inside diameter of 2 m.

The device is lowered by cables from a catamaran and weighs:

– 35 t in air,
– 30 t in water (this weight can be increased by ballasting to 50 t).

The unit is designed for reconnaissance down to a maximum depth of 200 m.

V.2.7.2 Functions envisaged

This type of device enables commonly used soil reconnaissance techniques to be applied. The main techniques envisaged are:

– coring by means of a double corer ("sleeve" sampler) with a diameter of 100 mm, developed by *Soil Mechanics Laboratory* of Delft; the penetration into the ground under thrust (down to a maximum of 20 m) and the method of extraction of the core lead to minimum disturbance of the soil,
– conventional in situ measurements using the static penetrometer and vane,
– electrical logs for determining the in situ porosity (and specific gravity),
– measurements of the permeability.

This device, which a priori offers many advantages, nonetheless appears delicate to use owing to its weight and dimensions.

V.3 CORING BY MEANS OF VIBROCORERS

There are a considerable number of hydraulic, pneumatic or electric vibrocorers. The powers developed, the dimensions of the corebarrel, the depths of penetration reached, etc., vary considerably from one unit to the next. It is not possible to give a detailed description of each appliance, so information is to be confined to:

– description of the operating principle and leading characteristics of hydraulic vibrocorers such as the Frabeltra vibropercussion corer, developed by *Institut Français du Pétrole* and built by *Geomecanique* [9] (see Figs. V.3.1 and V.3.2) or the TLM annular vibro-corer [10] (see Fig. V.3.3),

– tabular listing of the various hydraulic, pneumatic or electrical units at present in existence, together with their characteristics of use.

V.3.1 Principle, description and operating conditions of hydraulic vibrocorers

Vibrocorers are submerged devices driven by an electric motor and operate by vibration or by vibropercussion (generally double effect) and are adapted for:

– coring in loose formations,
– driving of piles or conductor pipes.

V.3.1.1 Brief description

Vibrodrivers operating by vibration and percussion consist of three main parts:

– a vibration generator consisting of 2 (or 4) out-of-balance weights rotating in opposite direction and driven by a hydraulic motor,
– a device for converting the vibrations into percussions (hammer and anvil system) by means of cylinders and springs,
– a system for clampling onto drilling strings by means of hydraulic jacks.

The hydraulic motor is supplied from a central hydraulic power plant (on the vessel) driven by a Diesel engine. The variable flow-rate pump (up to 300 to 400 l/min) operates at a maximum pressure of about 250 kg/cm^2 (25 MPa).

The bottom-to-surface link consists of a set of flexible tubes which are used at pressures of from 200 to 250 kg/cm^2 (20 to 25 MPa) and comprise:

– the supply lines,
– the control lines.

FIG. V.3.1. – Frabeltra vibrodriver.
(Institut Français du Pétrole process)

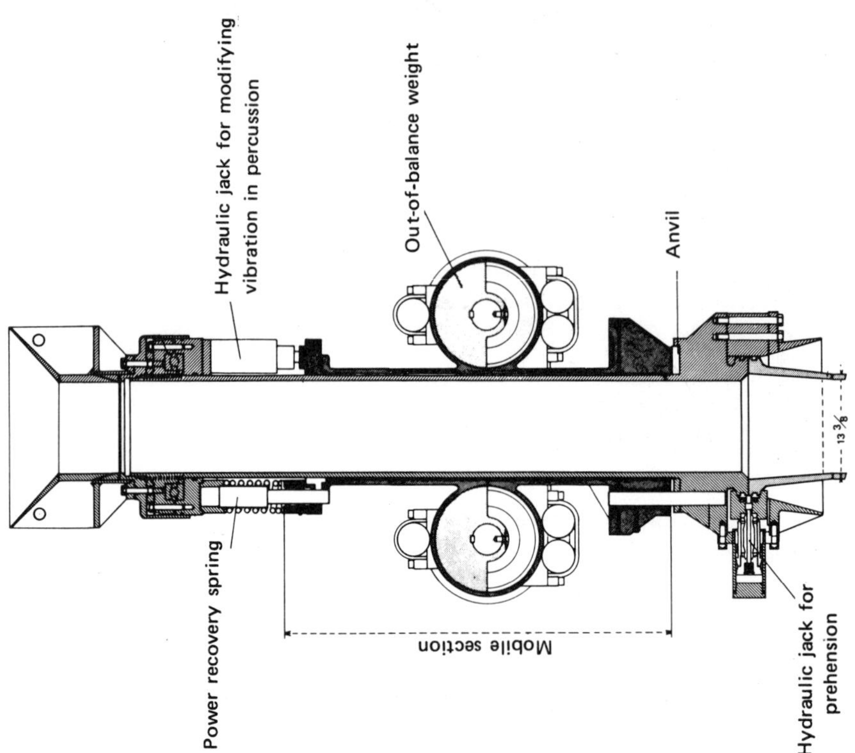

FIG. V.3.2. – Diagram of operating principle of the Frabeltra vibrodriver.
(Institut Français du Pétrole process)

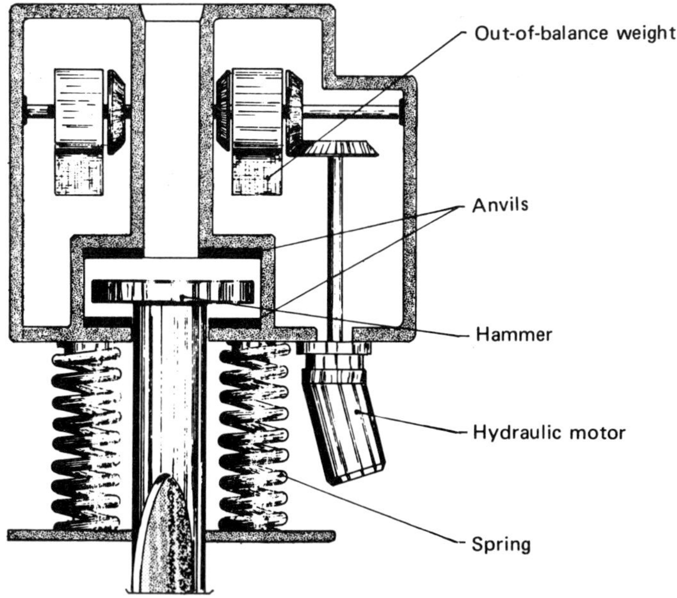

FIG. V.3.3. – Diagram of operating principle of the TLM annular vibrodriver.

V.3.1.2 Operating conditions

Changeover from "vibration" to "percussion" operating mode is controlled hydraulically from the vessel:

– by applying pressure to the jacks, the "vibration generator to tube clamping system" link is rendered rigid; the tube is driven by "vibration",
– by releasing the pressure of the jacks, the "vibration generator to clamping system" link is made flexible; with each cycle, the moving section of the machine strikes the anvil of the tube clamping system; the tube is then driven by "percussion".

Vibrodrivers are generally double-acting and are therefore capable of operating equally well:

– in the normal driving direction,
– in the "breakout" direction.

V.3.2 Characteristics of use of hydraulic vibrodrivers

The characteristics of hydraulic vibrodrivers differ considerably with the particular model considered. As an example, the operating characteristics of two vibrodrivers are given below.

V.3.2.1 Operating characteristics

Steady-state operation of a vibrodriver is typified:

– for vibration, by a rotary speed V_c and consequently a centrifugal force F_c,
– for vibropercussion, by a rotary speed $V_p < V_c$ and a driving (or breakout) force F_b much greater than F_c.

Table V.3.1 summarizes the steady-state operating characteristics of the Frabeltra vibropercussion driver [1] and the TLM annular vibrodriver.

TABLE V.3.1

		Frabeltra vibropercussion driver	**TLM annular vibrodriver**
Weight of unit (t)		4	1
Operation on "vibration"	Rotating speed (rpm)	Maximum = 1,240	600 to 1,500
	Centrifugal force (t)	Maximum = 30	1.5 to 9
Operation on "vibropercussion"	Rotating speed (rpm)	Maximum = 1,020	Steady-state = 750 to 1,100
	Driving force (t)	Maximum = 80 to 100	Maximum = 15 to 18

V.3.2.2 Coring characteristics

The coring operation takes place **continuously** in a single run:

– either to refusal,
– or until total penetration of the tube,

depending on the nature of the formations encountered.

The dimensions of the core barrels are:

– length of barrel elements = 3 to 9 m,
– diameter of tubes = from about 4 1/2" (TLM) up to 13 3/8" (Frabeltra).

The barrels are equipped with non-return valves for loose sediments.

The diameter of the cores varies from about 100 to 300 mm, depending on the unit (TLM or Frabeltra) and the clamping device.

The core barrel is broken out of the ground:

– if possible by pulling on the operating wireline in order to limit disturbance of the core,

---

(1) Of the type developed by *Institut Français du Pétrole* and built by *Geomecanique (Technip Geoproduction)*.

– if necessary by counter-driving on the vibropercussion driver; this certainly increases the amount of disturbance of the core and can also considerably reduce the recovery rate.

Penetration depends on the type of unit, the method of implementation and the nature of the soil:

– with the Frabeltra vibropercussion driver (diameter 13 3/8″), penetration in sands is often about 10 m,
– with the TLM annular vibrodriver (OD 4 1/2″), penetration could theoretically be as much as 20 to 30 m in loose or average consolidated soils.

V.3.3 Offshore support and implementation

The methods of implementation of vibrodrivers differ with the dimensions of the units.

V.3.3.1 Type of vessel

The implementation of a vibrodriver calls for the use of a stable vessel providing considerable free space owing to the manœuvers needed.

The following supports can be used:

– either supply-boats,
– or other types of vessel about 50 m in length (at least).

V.3.3.2 Implementation techniques

Several implementation techniques have been used, depending on the type of device.

The Frabeltra vibropercussion driver has been placed in the water:

– either by means of a crane,
– or by the stern gantry of a supply-boat (see Fig. V.3.4).

The second solution is the most satisfactory.

The TLM annular vibrodriver is designed to be implemented by means of an submerged frame set on the bottom and carrying a slide. Penetration then occurs in successive passes of one to several metres, depending on the consolidation of the formations.

V.3.3.3 Bottom-to-surface link and positioning tolerances

The **semi-flexible** link between bottom and surface consists of:

– the cable or cables carrying the vibrodriver,
– the set of flexible supply and hydraulic control lines.

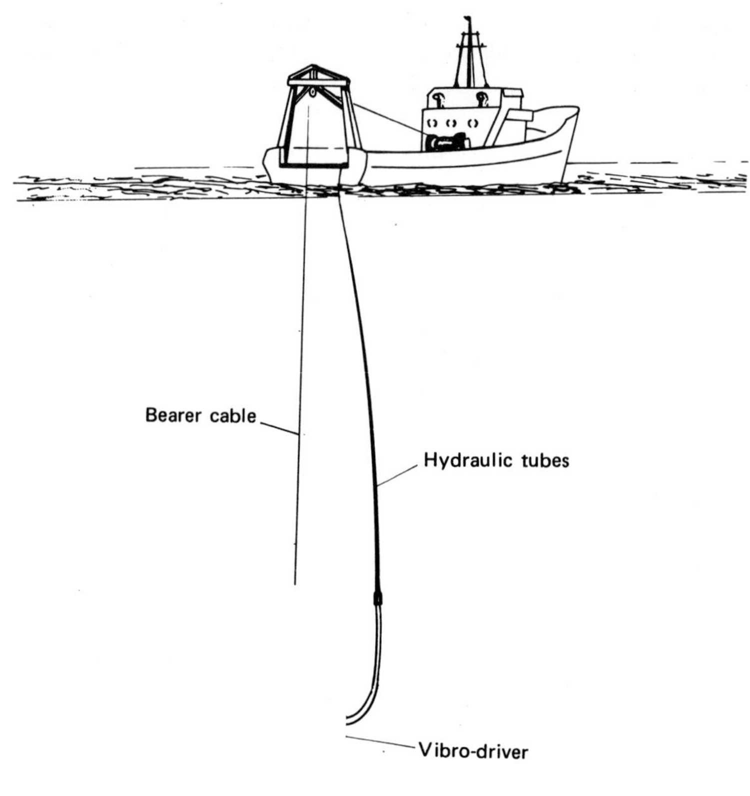

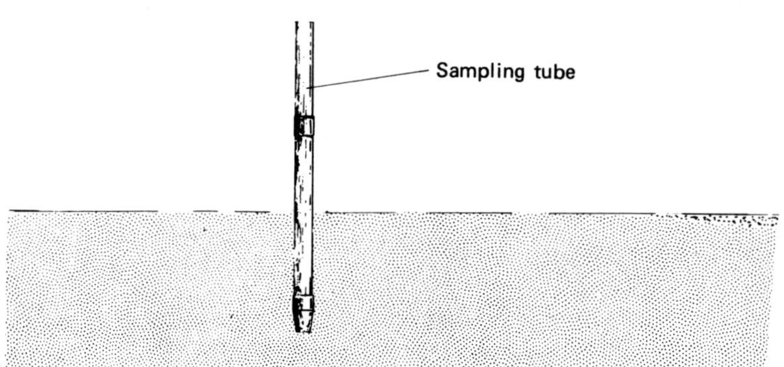

FIG. V.3.4. – Implementation of the Frabeltra Vibrodriver (Institut Français du Pétrole process).

The positioning tolerance depends in particular on the method of implementation:

– if the device is operated from a frame set on the bottom, the semi-flexible bottom-to-surface link allows the vessel to move 10 to 20 m,

– if the device is operated by the "plumb line" method, the vessel cannot be allowed to move by more than 5 to 10 m. In addition, pullout of the barrel requires the vessel to be positioned exactly vertically above the vibropercussion driver (since otherwise the barrel will be twisted and the pullout effort increased).

V.3.3.4 Handling equipment, space and manpower required

A 4-ton (type Frabeltra) vibropercussion driver can be used from the stern of a supply-boat by means of:

– a gantry frame and a 10 to 15 ton constant-tension winch,
– a winch for paying out the set of flexible hydraulic pipes.

Implementation of the TLM type annular vibrodriver is possible using the loading boom of a vessel, provided it has a force of 4 to 5 t and a clearance of about 10 m beneath the hook.

There must also be clear deck space on the vessel:

– first, to position the vibrodriver, handling gantry, winch, etc.,
– second, for storage, assembly and disassembly of the barrels,
– amounting in all to about 80 to 100 m^2.

Implementation of a vibrodriver requires the use of three to four technicians for the various operations of assembly and disassembly of the barrels, testing the hydraulic controls, etc.

V.3.4 Oceanographic conditions

V.3.4.1 Water depth

The maximum water depth in which most of today's vibrodrivers can operate is about 100 to 200 m.

The common depth of water in which the Frabeltra vibropercussion driver operates varies from 30 to 200 m.

V.3.4.2 Waves and current

The 4-ton vibropercussion driver can be lowered into the water only when the wave amplitude is 1 to 1.5 m owing to the considerable pendular movement incurred.

Launching of a lighter weight unit will also be limited by the handling of the submerged frame.

The supply boat can only be anchored provided the current does not exceed 2 to 3 knots.

Furthermore, the stability of the submerged frame set on the bottom for using the annular vibrodriver, for instance, also implies that the current must be weak (no more than 2 knots).

V.3.5 Performance of vibrocoring

We shall confine ourselves to limited data on the performance that can be achieved by the Frabeltra vibropercussion driver (developed by *Institut Français du Pétrole*).

V.3.5.1 Average duration of an operation

A vibrocoring operation consists of:

– assembly of the barrel,
– lowering the unit into the water,
– driving the tube into the soil at a speed dependent on the nature of the terrain, in any case about 0.5 to 1 m/min in loose soils,
– breakout of the tube by pulling or counterdriving,
– rehoisting of the unit and disassembly of the core barrel.

The average duration of a coring operation for a penetration of about 10 m is in the region of 2 to 3 h.

The following examples may be mentioned:

– execution of a reconnaissance campaign of 8 corings per day on the average with a penetration of about 7 m into sands and gravels,
– achievement of a coring operation down to a penetration of 31 m in very soft sediments in 1 h 30 min.

V.3.5.2 Recovery rate

The recovery rate is generally high thanks to the check-valve:

– 80 to 90 % in muds and soft clays,
– up to 80 % in heterogeneous sediments: sands and gravels,
– 100 % in coral sands, with debris of coral,
– 100 % in consolidated clays and boulder clays (Labrador Sea).

If counterdriving has to be used to pull out the coring tube, the recovery rate can be considerably diminished, particularly with sands.

V.3.6 Geotechnical characteristics of the cores sampled

Vibrocoring generally represents a geological reconnaissance method rather than a geotechnical reconnaissance method, owing to the disturbance of the cores obtained, particularly by vibropercussion driving. But large diameter vibrocoring gives excellent results in heterogeneous soils.

V.3.6.1 Types of soil

Vibrodrivers are appropriate in particular for the reconnaissance of loose terrains: sands of low density, soft clays or clays of average consolidation, heterogeneous soils, etc. Penetration is obtained:

– by vibration, in soft soils,
– by vibropercussion, in soils of average consolidation, sands or gravels.

When the vibrodriver has a rotating function (for instance in the case of TLM annular vibrodriver), it is generally possible to sample cores in highly consolidated soils.

V.3.6.2 Limitations and possibilities of use of vibrodrivers

Depending on the nature of the soil and the operating conditions of the device, the amount of disturbance of the cores, and hence their use, will differ widely.

In the case of low consolidated clays, use of the device as a "vibrator" makes it possible to sample cores which have been disturbed relatively little and are hence acceptable for geotechnical studies.

In the case of loose or dense sands, or heterogeneous soils calling for the use of the device by "percussion", the cores invariably undergo disturbance and can be used only for purposes of lithological identification (in the case of little diameter).

Vibrodrivers have numerous applications in addition to coring, namely:

– driving of tubes for anchoring systems,
– application of a pressuremeter (TLM vibrodrivers).

V.3.6.3 Geotechnical qualities of the cores

The "vibration" coring method is all the more appropriate:

– the greater the diameter of the cores,
– the greater the cohesion of the soils, since use of vibrodrivers on "vibration" is limited to terrains with a cohesion < 5 to $10\ t/m^2$ (< 50 to 100 kPa).

The geotechnical coring method by "percussion" is acceptable only:

– provided the diameter of the cores is sufficiently large (at least 15 to 20 cm),

– provided the formations are of high cohesion. In actual fact, with the Frabeltra vibropercussion driver, which has a maximum driving force of ≈ 100 t, penetration exceeds 8 to 10 m if the cohesion is 10 to 15 t/m^2 (100 to 150 kPa).

X-ray laboratory examination of the cores shows that vibropercussion driving represents the coring method involving the greatest disturbance.

V.3.7 Inventory of vibrodriving equipment

The great number of hydraulically, pneumatically or electrically actuated vibrocoring machines makes it impossible to give a detailed description together with the various conditions of use for each one. However, a non-exclusive inventory of the main units grouped together by operating mode has nonetheless been established.

These various units differ in:

– their operating mode: hydraulic, pneumatic, electric,
– their dimensions, powers and hence performance,
– their use: coring and/or recovery of sediments by circulation.

V.3.7.1 Hydraulically-powered vibrocoring units (Table V.3.2)

The vibropercussion drivers developed by *Institut Français du Pétrole* and built by *Geomecanique* are of considerable power and make it possible to extract cores of large diameter down to penetrations of 10 to 20 m, depending on the nature of the sediment. Thanks to the large diameters of the cores, good representivity of the soils can be expected, particularly with clays.

The vibrocorers developed by *TLM* are more compact in size and relatively easy to operate. They have been designed equally well for core extraction, insertion of pressure measuring probes, anchoring pipelines, etc. The TLM annular vibrodriver version is equipped with a rotary capability.

The Geodoff 1 and Geodoff 1 MK2 units (Fig. V.3.5) built by *Conrad Stork* would seem highly robust and very effective, regardless of the nature of the formation encountered. The new MK2 version of the unit is:

– equipped with a rotary capability,
– designed for application of a static penetrometer (reaction ≈ 5 t).

Ocean Science Engineering (OSE) has developed several vibrocorers the penetration of which varies from 6 to 12 m (in theory), depending on the particular model. These units appear relatively easy to operate, but their use is limited in particular to low consolidation soils. It should be noted that the "12 m" penetration version has in all probability been used very little.

TABLE V.3.2

HYDRAULICALLY POWERED VIBRATION OR VIBROPERCUSSION CORERS

Maker and/or contractor	Designation of unit	Type of unit	Type of link	Water depth (m)	Penetration (m)	Samples recovered	Observations
Institut Français du Pétrole Geomecanique (maker)	Undersea vibro-percussion driver	Hydraulic vibropercussion driver	Semi-flexible	200	10 to 20	Cores with with diameter 200 to 300 mm	4-t unit developing a force of 100 t on vibropercussion
Geomecanique (maker)	Undersea vibro-percussion driver	Hydraulic vibropercussion driver	Semi-flexible	200	10 to 15	Cores with diameter 175 mm	3-t unit developing a force of 60 t on vibropercussion
Techniques Louis Menard (TLM)	Annular vibro-driver	Vibrodriver (hydraulic vibropercussion driver)	Semi-flexible	200 to 300	A few dozen metres	Cores with diameter of 80 to 100 mm	1-t unit developing a force of 15 to 18 t on vibropercussion One model has a rotary capability
Techniques Louis Menard	D 700 vibrodriver	Hydraulic vibrohammer	Rigid semi-flexible	200	3 to 30	Cores with diameter 60 or 44 mm	Easy to implement Has little power for core sampling in sands
Ocean Science and Engineering (OSE)	M 1 000 vibrocorer	Hydraulic vibrocorer	Semi-flexible	90	6 or 12	Cores only slightly disturbed (unconsolidated formations)	"12 m penetration" version would not appear to have been used
Conrad Stork	Geodoff 1	Electro-hydraulic vibrocorer	Flexible	150	3 to 14 (depending on variant)	70 mm cores or sediments brought to surface by reverse circulation (water plus air)	Independent multipurpose unit Highly robust
Conrad Stork	Geodoff 1 MK 2	Hydraulic vibrocorer + rotary corer	Flexible	200	Vibrocoring = 7 m Rotary coring = 5 m Reverse circulation = 20 m	Cores (vibrocoring, rotary coring) Sediments (reverse circulation)	Recent unit (improved variant of Geodoff 1) Adaptation of penetrometer (reaction force 5 t)
Ocean Science and Engineering (OSE)	Horton Sampler	Hydraulic vibrocorer	Rigid	30	30	100 mm corers (cable coring) or recovery of sediments by water and/or air circulation	Utilized almost exclusively from the *OSE* vessel. Difficult to implement

V.3.7.2 Pneumatically powered vibrocoring devices (Table V.3.3)

Pneumatically powered corers do not generally allow a penetration of more than a few metres. The small diameter of the cores (5 to 8 cm) would doubtless imply considerable disturbance.

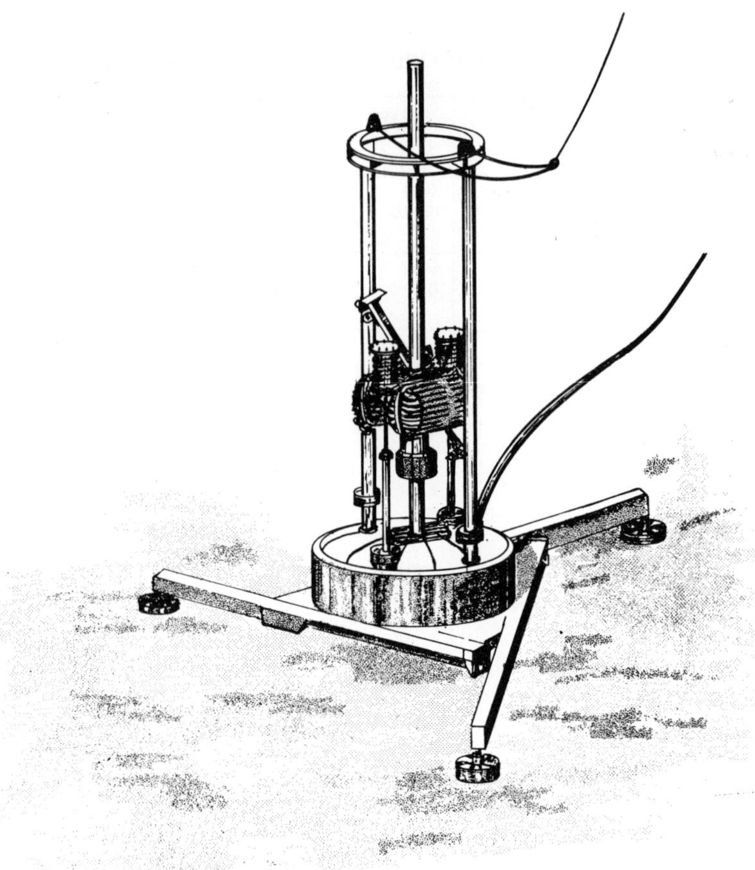

FIG. V.3.5. – Sketch of the Geodoff I MK2 vibrating core sample (*Conrad Stork*).

Alpine's Vibracore device mounted on a heavy, bulky frame would seem difficult to handle and of fairly limited effectiveness (penetration of only a few metres), considering particularly its dimensions.

The "Amdril" method of *Alluvial Mining,* well adapted to the prospection of placers, makes it possible to recover sediment solely by circulation (the airlift method).

V.3.7.3 Electrically powered vibrocoring devices (Table V.3.4)

The high frequency electrical vibropercussion driver of *Hydro Werkstatten* enables cores to be recovered, disturbed only slightly and with a large cross-section (15 × 15 cm). However, the unit, which can be used at great depths of water (theoretically up to 400 m), cannot obtain penetrations of more than a few meters.

TABLE V.3.3
PNEUMATICALLY POWERED CORERS

Maker and/ or contractor	Designation of unit	Type of unit	Type of bottom to surface link	Depth of water (m)	Penetration (m)	Samples recovered	Observations
Terresearch	Pneumatic Seabed Sampler	Pneumatic "Sonnette"	Flexible	120	3 to 6	50 mm diameter cores	Robust but rustic device Low average penetration (3 m)
Alpine Geophysical Associates Inc.	Alpine's Vibracore	Pneumatic vibrocorer	Flexible	100	6 to 12	87 mm diameter cores	Bulky unit Difficult to handle Results not very satisfactory.
Marine Advisors		Pneumatic vibrocorer	Semi-flexible	180	3 to 6	50 mm diameter (approx) cores only slightly disturbed	Unit adapted for coarse sediments
Alluvial Mining	Amdrill	Pneumatic hammer	Semi-flexible	50 to 100	15 to 20	Sediments raised by air-lift	Perfectly adapted to the prospection of placers Samples impossible to recover intact

The Zenkovitch electrical vibrocorer (Fig. V.3.6), compact and easy to handle, is at present very widely used in the North Sea for surface reconnaissance of dense sands and overconsolidated clays. However, the results obtained are not generally very satisfactory: penetration is often no more than 2 to 3 m and there would appear to be considerable disturbance.

The *Terresearch* electrical percussion device (Self-controlled Seabed Tool) offers the advantage of being autonomous (no connection with the vessel). However, it only allows a maximum penetration of 2 m. Other devices of the same type are now in the development stage, but their low penetration will always limit their use.

The electrical vibrocorer with a rotary capability built by *Redlac* and commonly known as the "Scottish corer", would appear to be primarily adapted for coring rock and highly consolidated soils for the reconnaissance of outcrops at great water depths (up to 900 m).

TABLE V.3.4
ELECTRICAL VIBRATION OR PERCUSSION CORERS

Maker and/or contractor	Designation of unit	Type of unit	Type of bottom to surface link	Depth of water (m)	Penetration (m)	Samples recovered	Observations
Hydro Werkstatten	Kergerat vibrohammer	High frequency percussion corer	Flexible	400	1 to 6	Cores with a square section of 150 x 150 mm	Recovery of only slightly disturbed cores
Maker: subcontractor of the *Geological Institute of the Netherlands*	Zenkowitch	Electrical vibrocorer	Flexible	100	Max. 5	65 mm diameter cores only slightly disturbed	Low cost, compact unit Used by the Dutch Geological Service and by *Fugro-Gesco*
Ocean Science and Engineering (OSE)	Diver Vibracorer 1136 Operated	Electrical vibrocorer	Flexible	30-40	3 to 20	Small diameter cores (diameter unspecified)	Light unit implemented by divers
Sonico Inc. (subsidiary of Shell Oil Co.)	Sonico	Electrical vibrator	Rigid	30	30 to 70	52 mm diameter cores or sediment recovered by counter-flow.	Applied in 1967 by the *US Bureau of Mines* Would not appear to have been used since.
Terresearch	Self controlled Seabed tool	Electrical percussion sounding device	None needed (safety cable optional)	300	2	50 mm diameter cores	Fully autonomous unit Very limited penetrating capacity
Maker: *Redlac Ltd.* Inventor: *Institute of Technological Sciences (Scottish Continental Shelf Unit, Edinburgh)*	Combined Vibrocore and Rock Drill (scottish corer)	Electrical vibrocorer (and rotary corer)	Flexible	900	6 (as vibrocorer)	62 to 200 mm diameter cores (by vibrocoring)	Unit in the development stage
Bureau de Recherches Géologiques et Minières (BRGM)	Light vibrosounder (models VS 5 and VS 8)	Electrical vibrocorer	Flexible	180	5 or 8	10 mm diameter cores	Weight: 1 to 6 t Good results obtained in loose soils

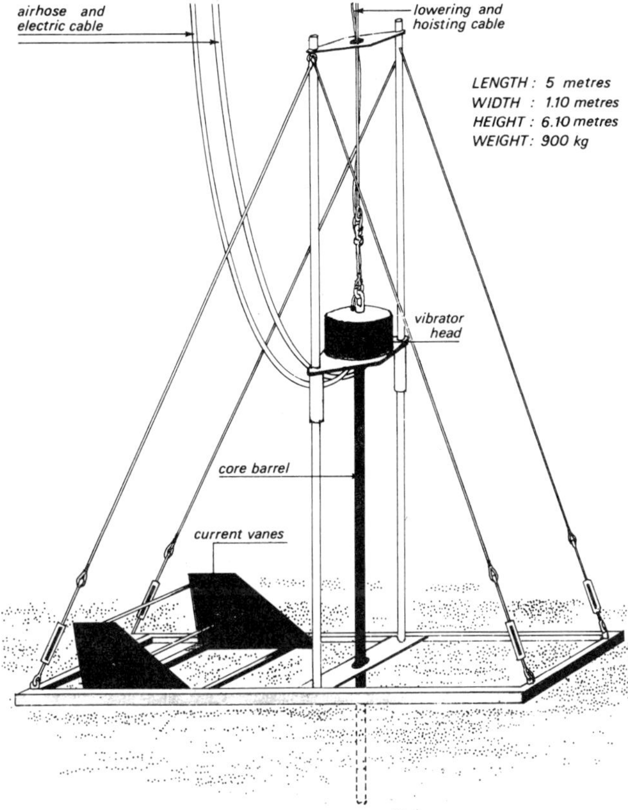

FIG. V.3.6. – Zenkovitch vibrocorer.

V.4 CORING BY MEANS OF TECHNIQUES DERIVED FROM FLEXODRILLING

The *Institut Français du Pétrole* has developed two methods for coring the sea bottom from the flexodrilling method [11] [12] [13]:

– first, the submerged device set on the bottom known as the "ocean-floor electro-corer" (ECSM), that can be used particularly in depths of less than 100 m,

– second, the "plumb line" coring method, usable only from a specialized vessel, but applicable in depths of water of from 200 to 300 m. For the record, only the principle and characteristics of this method will be recalled here, calling for the equipment of a vessel fitted out for flexodrilling (see Section V.4.7).

V.4.1 Principle and description of the ocean-floor electrocorer (ECSM2)

V.4.1.1 Operating principle and corers used

The ocean-floor electrocorer (ECSM2) is a submerged and remote-controlled device for drilling and coring (see Fig. V.4.1):

FIG. V.4.1. – ECSM2 (underwater electric core sampler).

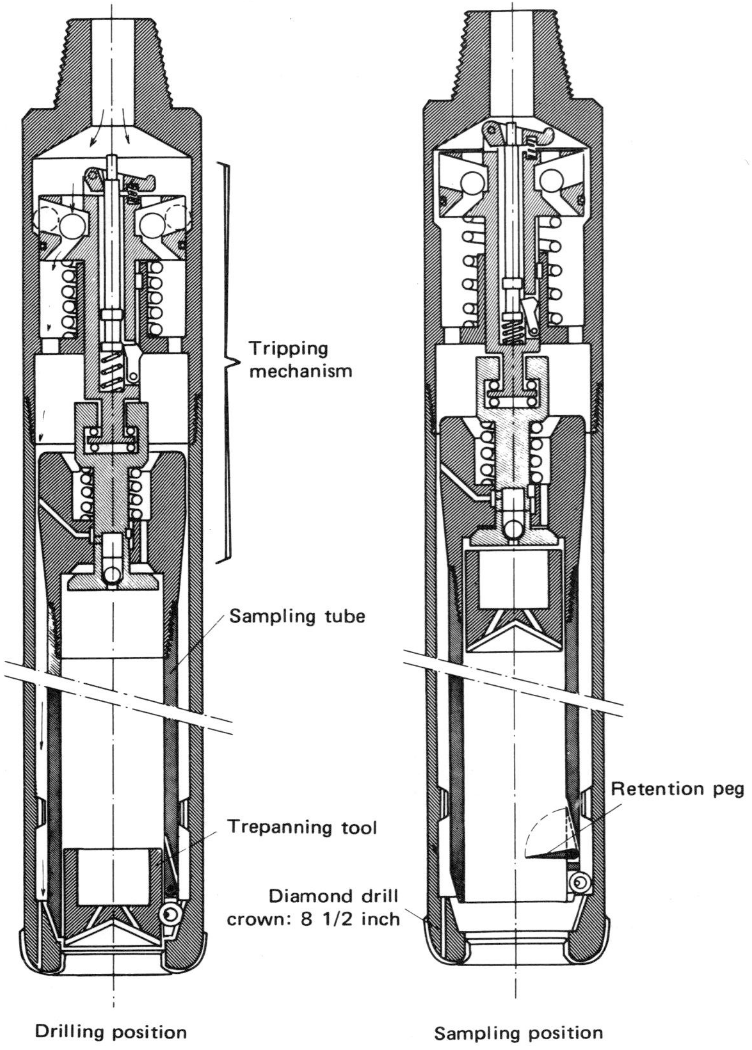

FIG. V.4.2. – Remote-controlled release corer.

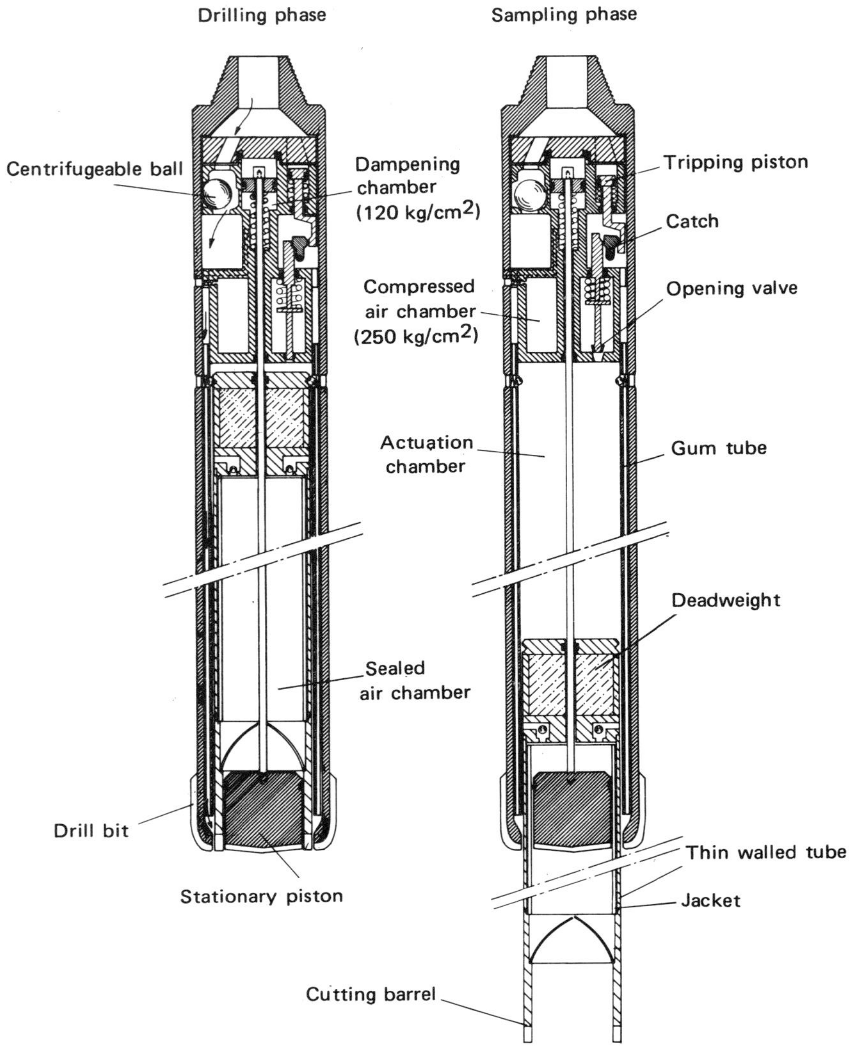

FIG. V.4.3. – Pneumatic propulsion corer.

– employing the flexodrilling method,
– resting on the bottom of the sea mounted on a metal tripod.

The drilling gear consists:

– of a rigid section comprising the corer and electrodrill,
– of a flexible hose wound onto a storage drum.

The ECSM2 is normally used with a remote-controlled release **rotary corer** applicable particularly to consolidated soils (Fig. V.4.2).

It can also be fitted with a pneumatic release **system corer**, which is particularly advantageous in soft soils (2) (Fig. V.4.3).

These two types of corer ensure remote-control changeover from the **drilling** phase to the **coring** phase at the sampling depths desired.

V.4.1.2 Description of the ECSM2 unit

The ECSM2 essentially consists of (Fig. V.4.1):

– a metal tripod about 9 m in height resting on the sea bottom on two 1 m^2 plates and one 2 m^2 plate,
– a drum for storing the flexible rod with an inside diameter of 1.70 m and an outside diameter of 3.10 m,
– a hydraulic drive motor for the storage drum, enabling:
  – the payoff speed of the hose to be adjusted to suit the drilling speed (slow operating conditions),
  – the hose to be rewound on the drum after coring (fast operating conditions),
– a transmission drum for the flexible hose at the top of the tripod,
– a rigid drill stem about 8 m in length consisting of a corer and a 40 hp electrodrill,
– a flexible stem 40 m long (the maximum length that the storage drum can take is about 50 m), 124 mm in diameter and enclosing an electric cable,
– a centrifugal pump for injecting the drilling fluid (seawater).

The surface equipment comprises the following particularly:

– a hydraulic power plant for driving the hydraulic motor,
– a 150 kVA electrogenerator set,
– a drum for storing a maximum length of 60 m of electric and hydraulic cable,
– a remote-control and telemetering cabin.

(2) The existing prototype is not operational and requires extensive modifications to be adapted to the ECSM2.

V.4.2 Characteristics of use of the ECSM2

The ECSM2 ocean-floor electrocorer makes it possible:

– in the initial stage, **to drill down** to the level selected for extracting the sample,
– in the second stage, to trigger the **coring** process.

V.4.2.1 Drilling characteristics of the ECSM2

The maximum load on the bit is limited to the weight of the rigid drill stem on the bottom (electrodrill and corer), i.e. about 1t.

The rotating speed when drilling, as also during rotary coring, using the controlled release corer, is set by:

– the speed of the electrodrill motor (900 rpm),
– the down-gearing of 1/3, i.e. 300 rpm.

The available **torque** when drilling (or coring) at 300 rpm is about 95 m × kg. Without the reduction gear, the electrodrill develops a torque of 30 m × kg.

The flow-rate of the drilling fluid corresponding to operation of the controlled release rotary corer is 450 l/min at 13 kg/cm^2 (1.3 MPa).
The drilling fluid flows through an open circuit, discharging the material on the sea bottom.

The drilling fluid used is always seawater.

V.4.2.2 Coring characteristics of the ECSM2

The remote-controlled release corer, by the rotation of the bit and the circulation of the drilling fluid, makes it possible to change over from the drilling to the coring stage. The release principle of the unlocking mechanism is based on the possibility of separately controlling:

– on the one hand, the rotation of the bit,
– on the other, the circulation of fluid in the corer.

The corer (Fig. V.4.2) consists of three main sections:

– an 8 1/2" core bit with its centre closed off during the drilling stage by a movable central drilling bit,
– a "double corer" section with fingers and wedges for holding the core,
– a release mechanism for unlocking and then raising up the central bit part by a sequence of actions on the circulation of the fluid and the rotation of the bit carried out in a well-determined order.

The following are **the dimensions of the rotary corer**:

- total length about 1.50 m (restricted by the dimensions of the ECSM2),
- diameter of drilling bit (diamond core bit = 8 1/2" (216 mm)).

The dimensions of the core obtained are:

- maximum length = 1 m,
- diameter = 106 to 108 mm.

Only one core is sampled at the end of the drilling stage.

Several cores therefore have to be drilled to reconnoitre at different depths.

The penetration is limited by the storage capacity of the flexible hose on the drum:

- current depth of investigation = 35 m,
- possible maximum depth of investigation = about 45 m.

V.4.3 Offshore support and implementation

V.4.3.1 Type of vessel

Utilization of the ECSM2 from a support vessel or other, about 50 m in length and equipped with a handling crane (Fig. V.4.4), implies highly favourable oceanographic conditions owing to the considerable pendular motion incurred.

Launching of the ECSM2 from the stern of a supply ship 50 to 55 m long from a suitably adapted gantry frame avoids the aforegoing disadvantages. A special frame known as the DMEC (French acronym denoting electrocorer handling device) has been specifically designed for this purpose (Fig. V.4.4).

V.4.3.2 Bottom-to-surface links and positioning tolerance

The flexible bottom-to-surface links consist of:

- a carrying cable,
- a set of lines comprising:
  - electric cables (power supply, remote-control, telemetering),
  - hydraulic pipes (supply to the hydraulic motor).

A tolerance in location of about 10 m can be allowed for the bottom-to-surface links.

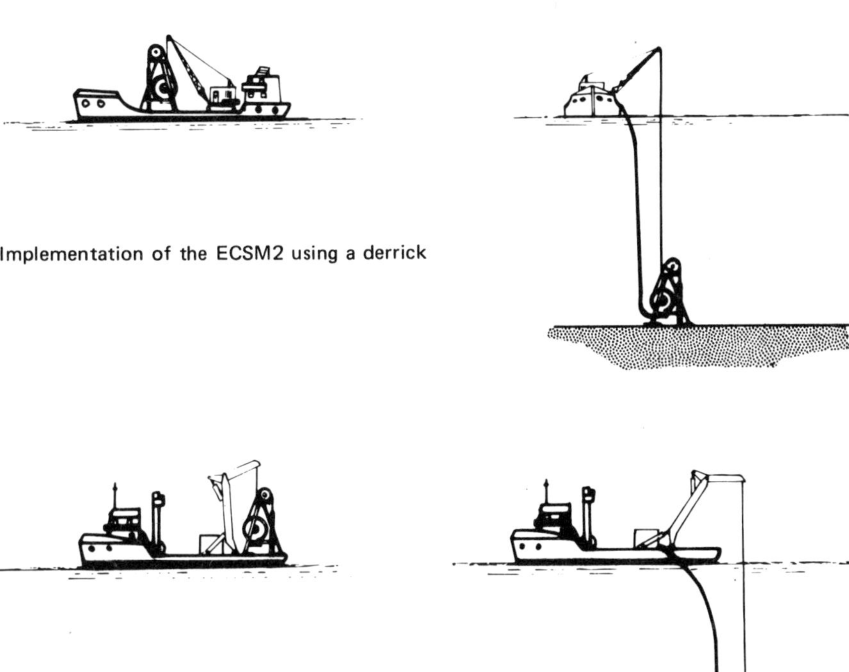

FIG. V.4.4. – Implementation of the ECSM2.

V.4.3.3 Handling equipment, space, power and manpower required

The ECSM2 weights about:

- 13 t in air,
- 11 t in the water.

It has the following dimensions:

- overall height of tripod = 9 m,
- base of tripod lies within a circle with a diameter of 6.80 m.

The crane for handling the ECSM2 must have the following characteristics:

- hoisting capacity ⩾ 20 t,
- minimum clearance beneath hook = 7 m,
- minimum outboard overhang of crane = 1.70 m,
- average overhand desirable = 4.20 m.

The DMEC special gantry frame can be used only on the stern of a supply-boat.

The equipment needed to operate the ECSM2 comprises:

– the storage drum for the electric and hydraulic cables,
– the remote-control cabins,
– the central hydraulic power plant,
– the electrogenerator set.

The necessary deck space on the vessel for this handling and other equipment is about 150 m^2.

The power needed to operate the ECSM2 comprises:

– the electric power provided by a 150 kVA, 380 V, 3-phase electrogenerator set,
– the hydraulic power developed by a variable output plant = 0-50 l/min with a maximum pressure of 200 to 250 kg/cm^2 (20 to 25 MPa).

Operating the ECSM2 requires two technicians, one to keep a check on the recorder of the drilling parameters (weight on bit, penetration rate, etc.).

V.4.3.4 Conditions of use of the ECSM2: stability on the bottom and pullout force of the rigid drill stem

The soil supporting load constraint beneath the base plates of the ECSM2 (total area 4 m^2) is about 3 t/m^2 (30 kPa).

From this it can be deduced that the settlement of the unit into the sea bottom will be small provided the shear strength of the surface sediments is greater than 0.5-1 t/m^2 (5-10 kPa) (soft clays or very loose sands).

On a flat sea bottom, the ECSM2 is stable provided the speed of the currents is less than 4 to 5 knots.

On an inclined sea bottom, the ECSM2 is less stable, since there is no way to compensate for the gradient of the soil (estimated by means of an inclinometer). In this case, drilling first starts inclined to the vertical and then is gradually corrected, thus increasing the risk of jamming the rigid drill stem.

The considerable difference in diameter between the corer (216 mm) and the flexible drill stem (124 mm) involves a certain risk of jamming the rigid drill stem in loose sediments (in particular sands) as a result of caving in.

The pullout force that can be applied to the flexible stem (up to 10 to 12 t) generally makes it possible to free the rigid drill stem without any major difficulties.

V.4.4 Oceanographic conditions

V.4.4.1 Water depth

The depth of use of the ECSM2 depends in particular on the type of anchor system of the vessel and the length of the cables available:

– the unit is commonly used in depths in the range of 30 to 60 m,
– operational depths of up to 100 to 150 m could be possible by adapting a subsea hydraulic power plant to the unit.

For greater depths, flexocoring by the "plumb line" method from a specialized vessel would be preferable (see Section V.4.7).

V.4.4.2 Waves and current

In the event of the ECSM2 being implemented by means of a crane, the wave amplitude must not be more than 1 to 1.50 m owing to the considerable pendular motion involved.

Should the unit be implemented from a gantry frame, this maximum could be increased to 1.5 to 2 m.

The vessel is generally anchored in slack water.

A current of below 3 knots does not disturb the core extraction operations. However, on certain coastal sites, where tidal currents are considerable, it will be preferable if possible to operate at slack water.

The effect of the tide amplitude is negligible thanks to the flexible links between the bottom and the surface.

V.4.5 Performances of the ECSM2 ocean-floor electrocorer

V.4.5.1 Time needed to drill a borehole

After anchoring the vessel, the operation comprises:

– handling of the ECSM2 (lowering into and below the water),
– drilling down to the sampling level (at a rate of about 1 m/min in loose sediments),
– taking a core 1 m long,
– rehoisting the ECSM2 aboard.

The duration of an operation varies with the depth of penetration sought after.

In reconnaissance for the Antifer site at Le Havre, the average duration, consisting of drilling down to about 15/20 m and coring was less than 1 h.

After coring, the following repreparation operations on the corer are required:

– disconnection of the electrodrill,
– extraction of the core,
– cleaning of the core triggering mechanism,
– resetting of the corer and replacement of the corebarrel,
– reconnection of the electrodrill.

All these operations take about 2 h.

V.4.5.2 Actual performances of the unit

Implementation of the ECSM2 on coastal sites generally takes place:

– in slack water, to facilitate anchorage of the vessel,
– in wave amplitudes of $\leqslant$ 1 m.

Under these conditions, the following are possible:

– a maximum of 2 to 4 boreholes, per day without coring (no resetting of the corer),
– in general, only 1 to 2 boreholes with coring, whenever weather conditions permit.

For example, during the Antifer reconnaissance campaign at Le Havre (in a maximum of 25 m of water) from February to May 1972, the following work was carried out:

– 84 boreholes in 31 working days (out of a total of 89 days):
  – total depth drilled = 776 m,
  – average penetration = 9.25 m,
  – maximum penetration = 20 m,

– 20 coring operations were triggered, of which 9 were effective, the low recovery rate being the result of the sandy nature of the sediments.

V.4.6 Geotechnical characteristics of the cores sampled

V.4.6.1 Types of soils and recovery

The controlled-release rotary corer lends itself to the geotechnical and geological reconnaissance of:

– rock which is not hard,
– clays of medium or high consolidation (cohesion > 5 to 10 t/m^2) (> 50 to 100 kPa),
– highly dense sands. The rate of recovery is generally low in sands which are not partially cemented.

The pneumatically powered piston corer could advantageously supplement the range of application of ECSM2 by making it possible to extract cores in:

– muds or soft or low consolidation clays (cohesion $c_u < 5$ t/m^2) ($c_u < 50$ kPa),
– sands.

V.4.6.2 Limits and possibilities of use of the controlled-release rotary corer

In the case of loose sediments, i.e. muds, soft clays, sands, etc., the controlled-release rotary corer is poorly adapted to **geotechnical** reconnaissance, owing to:

– the high rotating speed (300 rpm); it is generally recognized that disturbance of cores taken from soft soils (cohesion < 5 to 10 t/m^2) (< 50 to 100 kPa) is invariably considerable if the rate of rotation is more than 15 to 20 rpm (see Section V.2.6),

– the break-up of cores brought about by high flows of water (450 l/min) required to keep the electrodrill cooled; the arrangements adopted for the flow of water at the end of the corer do however limit the disturbances induced in the core.

In the case of consolidated clays (cohesion > 10 to 20 t/m²) (> 100 to 200 kPa), the rotary corer enables cores which are acceptable for geotechnical reconnaissance to be extracted.

In the case of hard rocks, the penetration speed is low owing to the low bit load (≈ 1 t).

V.4.6.3 Representativity of cores sampled by the rotary corer

In soils of low consolidation (muds, soft clays) or loose structure (sands, shells, etc.), the cores sampled using the controlled-release rotary corer are often only slightly representative:

– from the **geological** standpoint, the low rate of recovery (15 to 20% is commonly observed) will not allow a continuous stratigraphic profile to be made,
– from the **geotechnical** standpoint, the disturbance brought about by the rotation and circulation of water results in considerable modification of the mechanical properties. The shear strength calculated from the unconfined compression tests is often below that measured by triaxial "undrained-consolidated" tests.

In consolidated soils (silts and clays), the cores sampled by the rotary corer are generally representative:

– the recovery rates commonly reach 80% or more,
– the mechanical characteristics measured on samples cut in the centre of the core (diameter 10.6 cm) are significant.

V.4.6.4 Drilling and coring logs

The ECSM2 enables only one core 1 m long to be sampled per borehole drilled whatever the corer used.

This limitation is partly offset by the possibility of continuous recording of the drilling parameters:

– weight on bit,
– rate of penetration,

in terms of the depth (Fig. V.4.5).

This continuous recording goes to make a veritable instantaneous log of the drilling properties of the formation, sometimes making it possible to choose the correct depth at which to start coring.

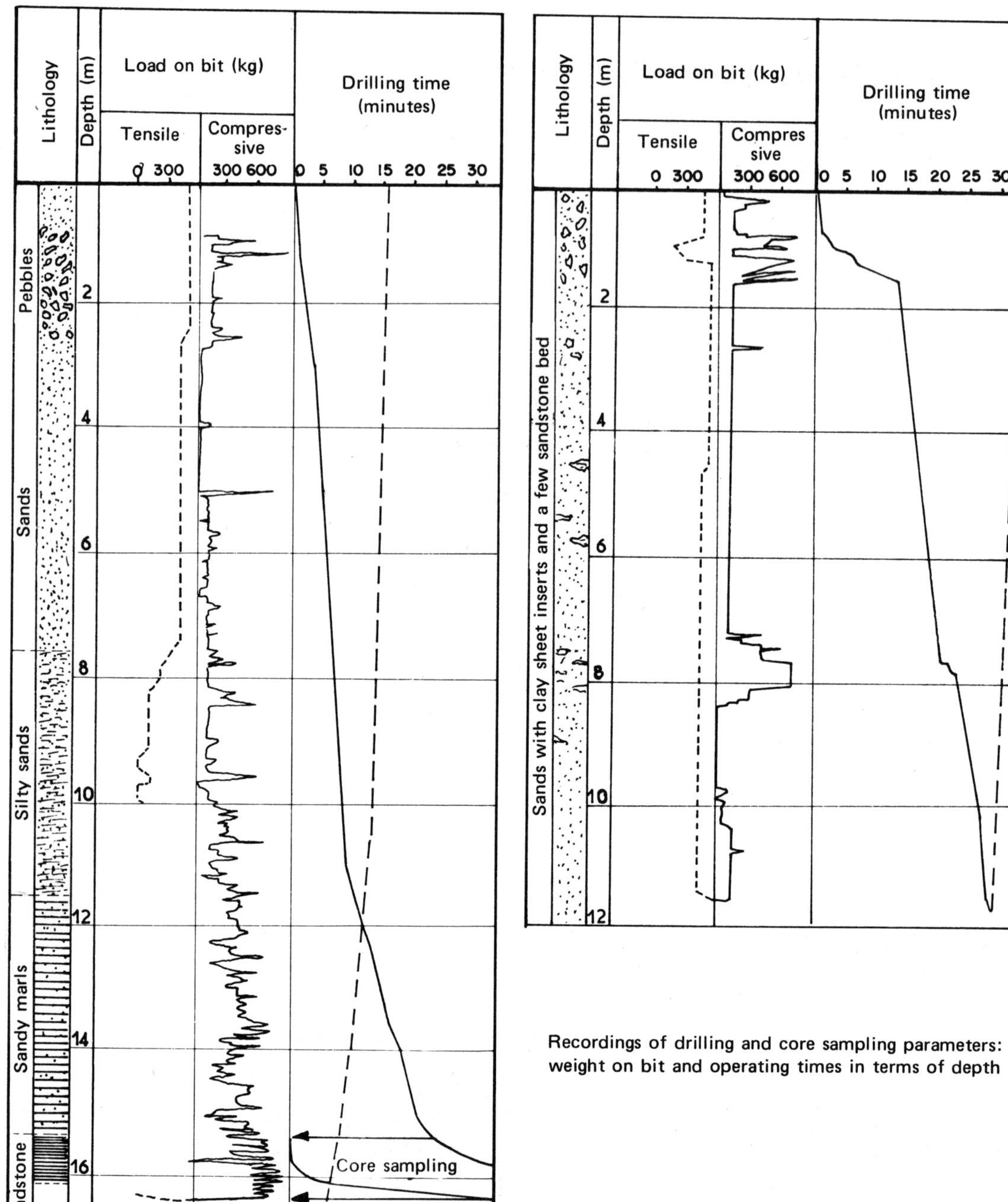

FIG. V.4.5. – Drill logs recorded using the ECSM2.

V.4.7 Flexocoring by the "plumb line method"

Flexocoring by the "plumb line method" requires a vessel fitted out for flexodrilling (Fig. V.4.6). The method has been implemented from the "Terebel", the early experimental vessel of the *Institut Français du Pétrole* (Fig. V.4.7).

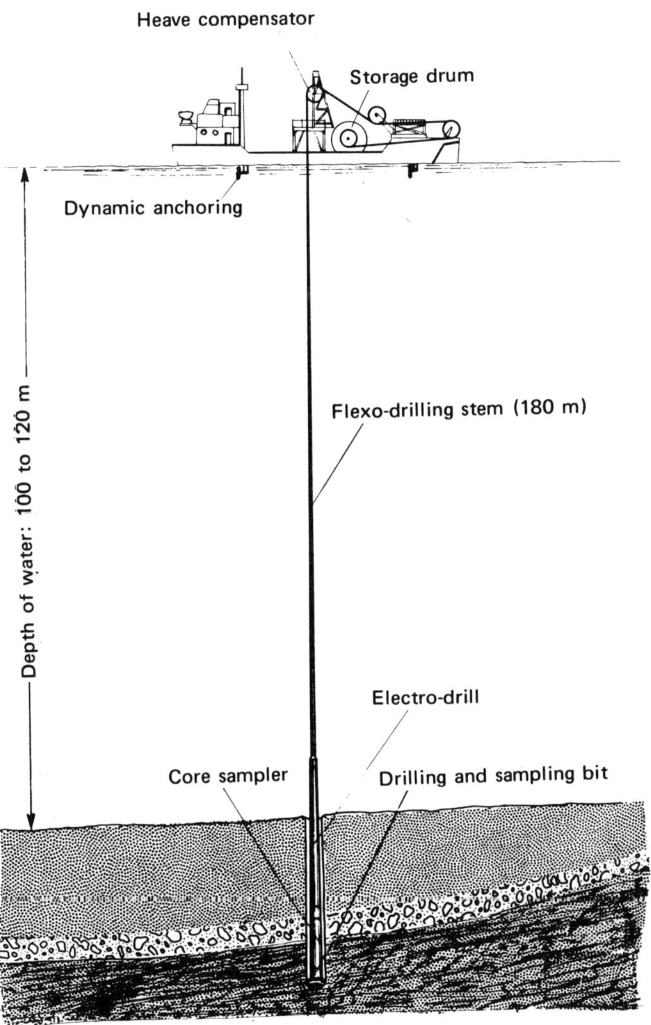

FIG. V.4.6. – Principle of flexo-core drilling by the plumb line method.

FIG. V.4.7. – The Terebel, the early experimental vessel of the *Institut Français du Pétrole.*

V.4.7.1 Principle of flexocoring

The flexocoring method is characterized by the fact that bottom gear consisting of instruments needed for drilling and coring is simply hung from the end of a flexible conductor of electric energy.

A heave compensator on the vessel enables the flexible line and the bottom gear to be rendered independent of the heave of the ship.

The flexible line is held taut by means of a controlled-tension winch or a traction gripper device.

The equipment used ensures:

– in the initial stage, full diameter drilling,
– in the second stage, drilling to the depth desired, remote controlled from the surface.

V.4.7.2 Characteristics of use of flexocoring

The drill stem 13.5 m in length and weighing (in air) 1.7 t, consists of:

– a controlled release corer with an outside diameter of 8 1/2″,
– a 150 hp electrodrill 9 m long,
– a flexodrilling hose the length of which is adjusted to the water depth, with an outside diameter of 124 mm and weighing 17 kg/m in water.

The electric power is transmitted to the electrodrill via conductors set into the walls of the flexible stem.

Water is circulated through the drill stem by means of a pump on the ship.

The following are the drilling characteristics:

- weight on bit load = about 1.5 t,
- rotating speed (loaded) = 300 rpm,
- maximum torque about 350 m x kg.

Just as with the ECSM2, with the "plumb line" method, only one core can be taken per borehole.

The following are the dimensions of the core:

- length 4 m,
- diameter 106 to 108 mm.

V.4.7.3 Equipment needed to implement the flexocorer

The necessary handling equipment comprises:

- the storage drum for the flexodrilling hose,
- the traction gripper for the hose capable of exerting an effort of 10 to12 t,
- the heave compensator capable of withstanding a load on the drill stem varying between about 1 to 15 t and absorbing amplitudes of about 4 m with variations in the tension of about 0.5 t for a mean value of 8 t.

The remaining surface equipment is necessary for using the ECSM2:

- 150 kVA, 380 V, 3-phase electrogenerator set,
- circulating pump,
- control and monitoring console.

Remark. A variant of the "plumb line" flexocoring method has been developed for coring in great water depths (down to 2,600 m) by substituting an electric cable for the flexodrilling hose. In this case, the bottom gear comprises an electric pump in addition to the equipment listed above.

V.5 CORING USING SUBMERGED REMOTE-CONTROLLED ROTARY CORERS

During recent years, several subsea remote-controlled rotary coring devices, whether of the barrel type or not, have been either planned, partially built, or made and tested, with a view to geological or geotechnical reconnaissance of marine soils in varying water depths (200 to a maximum of 1,800 m).

We shall merely confine ourselves:

- to a description and listing of the problems of implementation of the "Maricor" (of *Wimpey*) which at present is the main example of this type of equipment,
- to giving a few indications on the *NCEL* and *Terresearch* automatic corers now being tested, or information concerning other equipment in the project stage.

V.5.1 The "Maricor" barrel type remote-control rotary corer of Wimpey

The "Maricor" was built in about 1970 by *Atlas Copco* for *Wimpey Laboratories Ltd.* [14] [15] [16].

V.5.1.1 Operating principle of the Maricor

The Maricor (Fig. V.5.1) is a subsea, remote-controlled rotary coring unit.

The device is hydraulically powered and renders possible continuous coring down to penetrations of 60 m and in maximum water depths of 200 m.

The drill pipes, which are added in sequence by remote-control, are stored in a rotary magazine (the "barrel").

The wireline-operated core barrel is re-raised to the surface by means of a lift system after each coring operation.

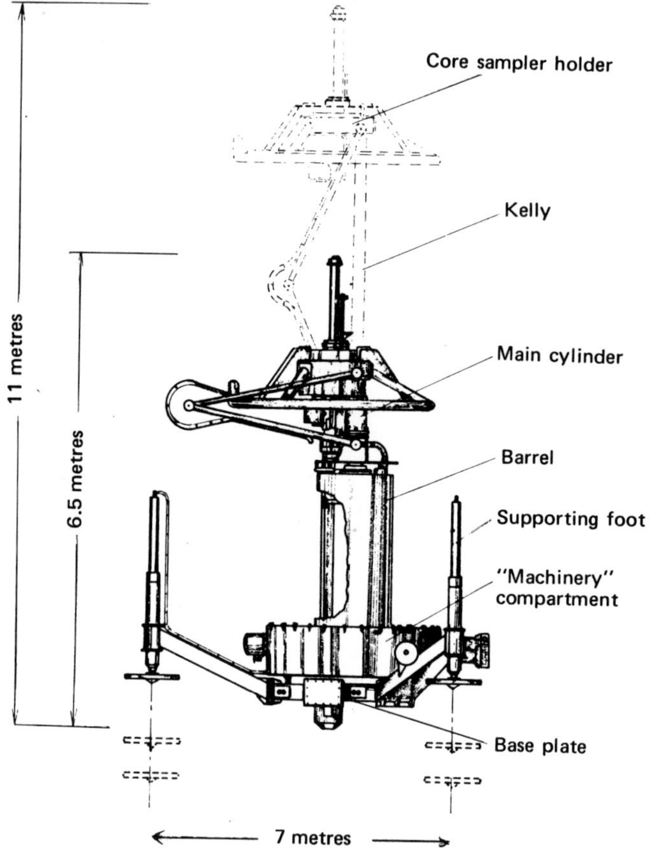

FIG. V.5.1. – General sketch of the Maricor.

V.5.1.2 Description and remote-control of the Maricor

The submerged device consists of the following parts (Fig. V.5.1):

– a rigid supporting base carries the system as a whole and rests on 3 feet –equipped with hydraulic jacks for stabilization– on the sea bottom,

– the "machinery" compartment, connected to the flexible supply line, contains all the hydraulic pumps, valves and electric motors, immersed in an oil bath,

– the revolving magazine built around a master cylinder connected to the supporting base contains 20 drill pipes (each with a unit length of 3 m) and two core barrels.

– the central section connected to the top of the master cylinder constitutes the top of the fixed part of the system; it contains the set enabling the string of drill pipes to be run in and out, the coupling flange to the torsion joint and the upper locking clamp,

– the torsion joint consisting of stainless tubes contains the flexible hydraulic conduits and electric cables ; its purpose is to prevent the rotary drive box from shifting out of centre while drilling,

– the rotary drive box enables the drill string to be rotated; it also contains a cylinder for uncoupling the drill pipes,

– the core holder (lifting device) raises the core barrel to the surface and then relowers it. The operations are accomplished by means of a winch along two guide cables which are kept taut.

The submerged equipment is controlled from the surface from a console on the vessel, The link runs via a 78 conductor main cable to ensure:

– power supply,
– control,
– monitoring and measurements.

The controls cover:

– settling on the sea bottom,
– implementation of the various monitoring systems,
– start-up of the rotary drive box,
– control of the rotation of the magazine (barrel) to supply the string of drilling rods,
– start-up of the core barrel hoisting system.

The parameters recorded on the surface are:

– the rotating speed of the drill string,
– the torque,
– the weight on the bit,
– the water pressure (drilling fluid),
– the rate of penetration into the formation.

V.5.2 Drilling and coring characteristics of the Maricor

V.5.2.1 Drilling characteristics

The weight on bit applied by jacks can vary from 0 to 4 t.

The rotating speed of the drill bit varies from 0 to 800 rpm (driven by a hydraulic motor).

The torque developed is:

- 500 m x kg at 50 rpm,
- 50 m x kg at 500 rpm.

The flow-rate of the circulating pump must be preset (it has 5 operating settings). The maximum flow-rate is 110 l/min at a pressure of 50 kg/cm² (5 MPa).

The drilling fluid is sea water.

V.5.2.2 Coring characteristics

The rotary coring bit has a diamond crown (Craelius).

The outside diameter of the drilling bit is 96 mm.
The dimensions of the core are:

- diameter 57 mm (2 1/4"),
- the maximum length = 2.20 m.

Coring is continuous owing to the automatic assembly of the 3 m drill pipes as coring proceeds.

The theoretical penetration of the unit can reach 60 m, since the magazine contains 20 three-metre pipe lengths.

The theoretical penetration could be increased to 120 m with a magazine containing 40 three-metre lengths.

V.5.3 Offshore support and implementation of the Maricor

V.5.3.1 Offshore support

The Maricor can be implemented from the central well of a vessel, namely:

- either the "Sealab", the 99 m long *Wimpey* vessel,
- or another stable vessel with a central well with a diameter of more than 7 m which has available the necessary handling equipment.

The unit can alike be operated from the stern of a supply vessel by means of a davit and a platform hung outboard.

V.5.3.2 Bottom-to-surface links and positioning tolerances

The Maricor is connected to the vessel by flexible and semi-flexible links consisting of (Fig. V.5.2):

- the electric power, remote-control and telemetering cable (78 conductors),
- the guide and hoisting cables,
- the corer operating wireline (semi-flexible links).

The guide cables must be kept taut (constant tension) for the core carrier operations. Because of this, any movements of the vessel around its initial position exert an overturning moment on the submerged unit through the guide cables. The unit as a whole would, however, appear to be stable thanks to the considerable ballast formed by the "machinery" compartment (at least 6 t).

In actual fact, it would seem that the vessel should not move by more than 2 m.

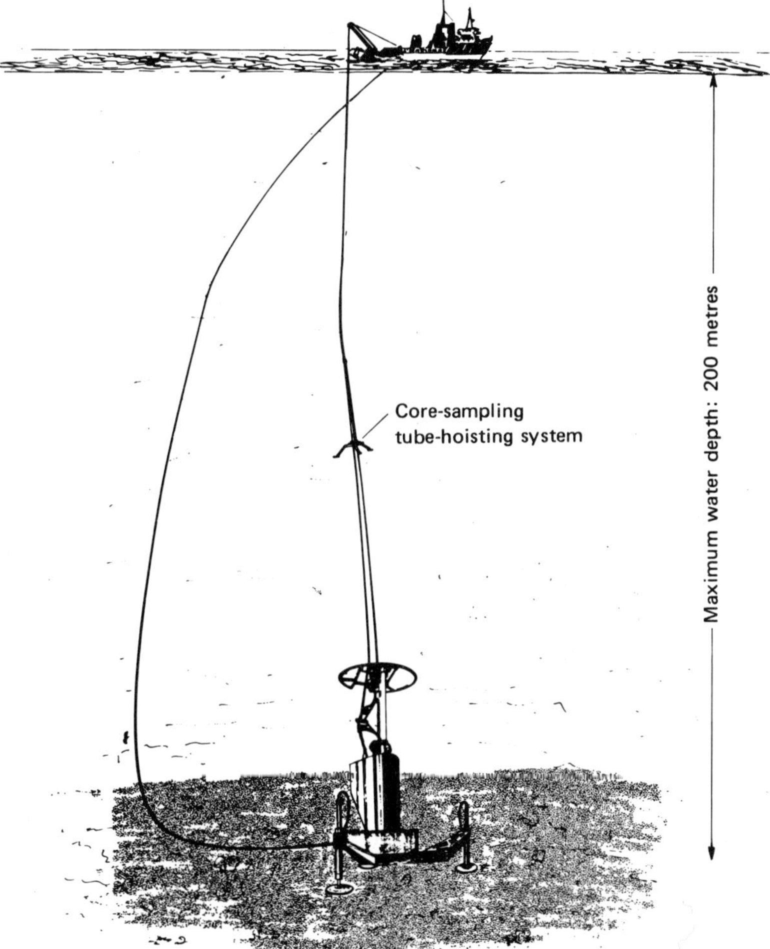

FIG. V.5.2. – Implementation of the Maricor.

V.5.3.3 Equipment, space, power and manpower necessary on the vessel

The weight of the Maricor is approximately:

– 14 t in air,
– 11 t in water.

It has the following outside dimensions:

– overall diameter of unit resting on bottom = 7 m,
– diameter of machinery compartment = about 3 m,
– minimum height of unit (not operating) = 6.50 m,
– maximum height of unit (operating) = about 11 m.

Implementation of the Maricor requires:

– for the case of a vessel with a central well with a diameter of 7 m, a gantry frame with a capacity of over 15 t and a minimum clearance beneath the hook of 7.50 m,
– for the case of a supply vessel, a davit with a capacity of 15 t (and an outboard overhang of 4.50 m) and a height of 11 m.

In either case, the hoisting or guide cable tension winch must have:

– a tensile effort capability of over 10 t (2 strand pulley-block),
– the ability to store a length of twice the depth of water, plus a margin (i.e. 450 m for a depth of 200 m).

Implementation by means of a crane would appear out of the question because of:

– the weight of the unit (14 t) in view of the considerable pendular motion,
– the guide cables and the wireline device.

A deck area of about 150 m² is needed on the vessel in order to use to Maricor.

The electric power is supplied from a 75 kVA, 380 V, 3-phase electrogenerator set.

Use of the Maricor requires at least 4 technicians.

V.5.3.4 Remarks concerning the use of the Maricor

At the present stage of utilization of the unit, no judgement can be expressed concerning its reliability and performance.

The system for recovering the cores, which are raised to the surface after each sampling, require a complicated "lift" device together with constant tension on the guide cables (limiting the possibilities of movement of the vessel).

The unit is set level on the sea bottom by means of jacks provided the slope of the soil is less than about 10 to 15°.

This coring device is suitable in particular for the reconnaissance of relatively consolidated soils and rock.

V.5.4 The NCEL automatic sampler

Ocean Science Engineering (OSE) on behalf of the *US Navy Civil Engineering Laboratory (NCEL)* began in 1970-1971 to build a subsea remote-controlled rotary coring device for use at great water depths (down to 1,800 m) [2] [17].

This unit, the construction of which was temporarily postponed from 1972 to 1973, is now undergoing trials.

V.5.4.1 Principle and description of the NCEL device

The *NCEL* subsea remote-controlled corer is a device featuring:

- electro-hydraulic powering,
- drill pipes and core barrels supplied from 2 separate magazines.

The frame of the unit rests on retractable plates on the sea bottom (see Fig. V.5.3), endowing it with great stability.

The frame has the following dimensions:

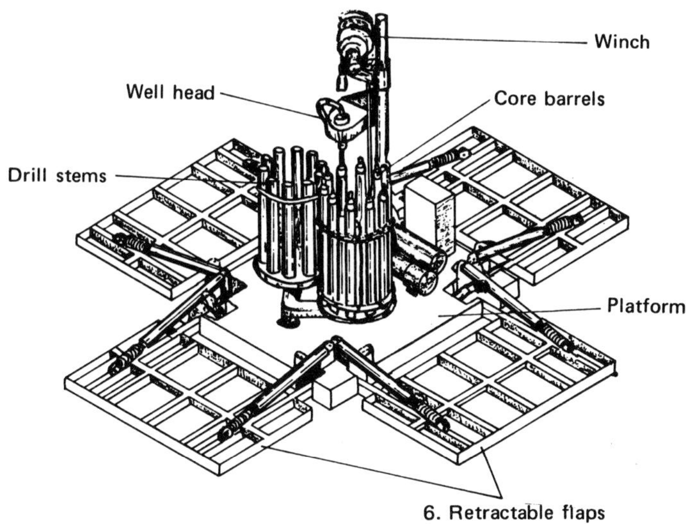

Characteristics:

Approximate height:		5 m
Flaps down	length:	7 m
	width:	7.8 m
Flaps retracted	length:	4 m
	width:	3 m
Approximate weight in air:		9 t
Approximate weight submerged:		6.8 t

FIG. V.5.3. – The NCEL barrel type automatic core sampler.

– height = 5 m,
– width = 3 m (6.90 m with the plates unfolded),
– length = 2.90 m (7.80 m with the plates unfolded).

The weight of the unit is:

– about 9 t in air,
– about 7 t in water.

The hydraulic motor of the unit is driven by a central hydraulic power plant set in the frame and supplied with electric power from the vessel. A total power of about 80 hp is developed.

The two magazines contain:

– the first, ten 1.50 m drill pipes,
– the second, ten 1.50 m core barrels.

The sequence of operations consists of the following in turn:

– automatic lowering and locking of a core barrel,
– automatic addition and locking of a drill pipe length,
– coring,
– extraction of the core barrel by means of a winch with a capacity of 5.4 t mounted at the top of the magazine,
– storage of the core barrel (filled) in the magazine.

V.5.4.2 Characteristics of the NCEL unit

The maximum penetration of the corer is 15 m (ten 1.50 m drill pipes).
Coring occurs in successive steps of 1.50 m.
The cores have the following dimensions:

– diameter = 76 mm (3″),
– maximum length = 1.50 m.

The rotary drive head is capable of developing a torque of:

– 270 m x kg at 100 rpm,
– 800 m x kg at 30 rpm.

This unit is designed so as to be capable equally well of extracting cores from all types of formations i.e. loose sediments or rock.

V.5.5 The T.S.O. seabed sampler

V.5.5.1 Principle of the Terresearch device

Terresearch (the English *Taylor Woodrow* group) have produced a subsea rotary corer of the barrel magazine type which is fully remote-controlled (Fig. V.5.4).

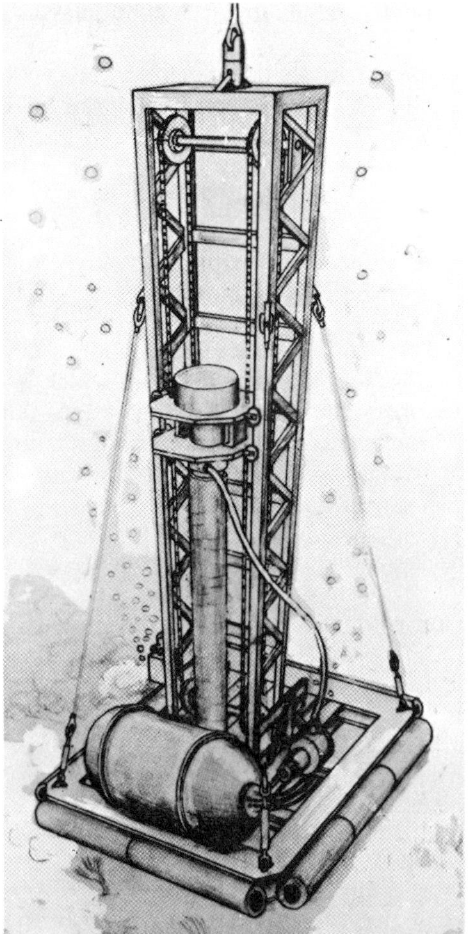

FIG. V.5.4. – The *Terresearch* barrel type automatic core sampler.

The unit is to ensure:

– either the taking of cores,
– or in situ measurements (penetrometry),

in a borehole without having to withdraw the drill pipe.

V.5.5.2 Description of the Terresearch device

The following are the dimensions and weight of the unit:

– total height varies from 7.50 to 10.50 m,
– length of side of unit resting on ground = 3 m,
– weight = 8 t.

The present core extraction depth is limited to 9 m but may be improved.

This device, which is at present in the experimental stage, appears to offer considerable advantages owing to its ability to combine the functions of core drilling and in situ measurements.

However, implementation will be relatively difficult owing to the considerable height (pendular motion).

V.5.6 Inventory of subsea remote-controlled rotary coring devices either in service or in the project stage

In addition to magazine units, there are several rotary coring devices which are electrically or electro-hydraulically powered, the penetration of which is limited to a few metres. Without entering into the operating details of these various units, the following tables summarize their leading characteristics of use. A certain amount of information is also given on a number of units which are in the project stage, but whose development would appear to have been postponed or abandoned.

V.5.6.1 Rotary coring devices without a magazine with low penetration

Various rotary corers, not equipped with a magazine system, have a penetration of no more than a few metres (3 to 6 m depending on the case). Table V.5.1 provides a non-exhaustive inventory of devices of this type.

TABLE V.5.1

LOW PENETRATION ROTARY CORING UNITS

Maker and/or contractor	Designation of unit	Type of unit	Type of connection	Water depth (m)	Penetration (m)	Samples recovered	Remarks
Conrad Stork	Geodoff 1 (variant 2)	Electro-hydraulic rotary corer	Flexible	150	3.50	70 mm cores	Robust device Possibility of coring rock Variant of vibropercussion device
Conrad Stork	Geodoff 1 MK 2	Electro-hydraulic rotary corer + vibrocorer	Flexible	200	Rotary coring : 5 m Vibrocoring : 7 m	Cores	Recently developed (improved version of Geodoff 1) Adaptation of penetrometer (reactive force 5 t)
Redlac Ltd. (Inventor: *Institute of Geological Sciences in Edinburgh)*	Combined Vibrocore and Rock Drill	Electro-hydraulic rotary corer vibrocorer	Flexible	900	Rotary coring : 4.5 m Vibrocoring : 6 m	62 mm cores	Recent corer Rotary function is suitable for coring rock
Maker: *Hill Offshore Inc.* Contractor: *Hyco*	Hill Sub-Sea Corer	Electro-hydraulic rotary corer	Flexible	180	3 to 6	44 or 54 mm cores	Lightweight unit Usable for coring rock or loose sediment
Geomecanique	ESCM3	Sounding frame with electric rotary head (electrodrill)	Flexible	200	5	102 mm cores	Unit particularly well adapted for reconnaissance of consolidated formations

These corers, which are generally electro-hydraulically powered, often ensure several functions, i.e. rotation, vibro-percussion, etc.

Rotary coring devices are designed particularly for the reconnaissance of highly consolidated soils and rocks.

V.5.6.2 Coring devices in the project stage

Table V.5.2 recapitulates the characteristics of three subsea remote-controlled rotary coring devices which were in the project stage in about 1970, the development of which has been interrupted.

TABLE V.5.2

ROTARY CORING UNITS IN THE PROJECT STAGE

Maker	Designation of unit	Principle of unit	Characteristics of unit		Remarks
			Maximum depth of water (m)	Maximum penetration (m)	
Texas A and M University (USA)	Seacore 50	Electro-hydraulic power Drill pipes and core barrels supplied in lengths of 1.35 m by magazines (barrels)	300	15	Projected in about 1970 Construction interrupted
Otto and Sichling (USA)	Submarine Coaxial Coring unit	Telescopic deployment coring unit		90	In the project stage since 1970-1971
Conrad Stork (Holland)	Geodoff II	Electro-hydraulic drive Drillpipes and core barrels supplied in lengths of 4 m by magazines (barrels) Corer with retractable cutting tool	200	50	Unit was partially built around 1971-1972 Projected weight 17 t

The Seacore 50, under study by *Texas A & M University,* [18] is a twin-barrel (magazine) unit (Fig. V.5.5), the characteristics of which closely approach those of the *NCEL* corer:

- penetration = 15 m,
- diameter of cores = 76 mm.

On the other hand, its dimensions (height 3.70 m) and weight (4.5 t) are very modest and should facilitate its handling.

The Geodoff II, now planned by *Conrad Stork,* is also a twin-barrel (magazine) device enabling cores to be taken:

- diameter 180 mm,
- penetration up to 50 m.

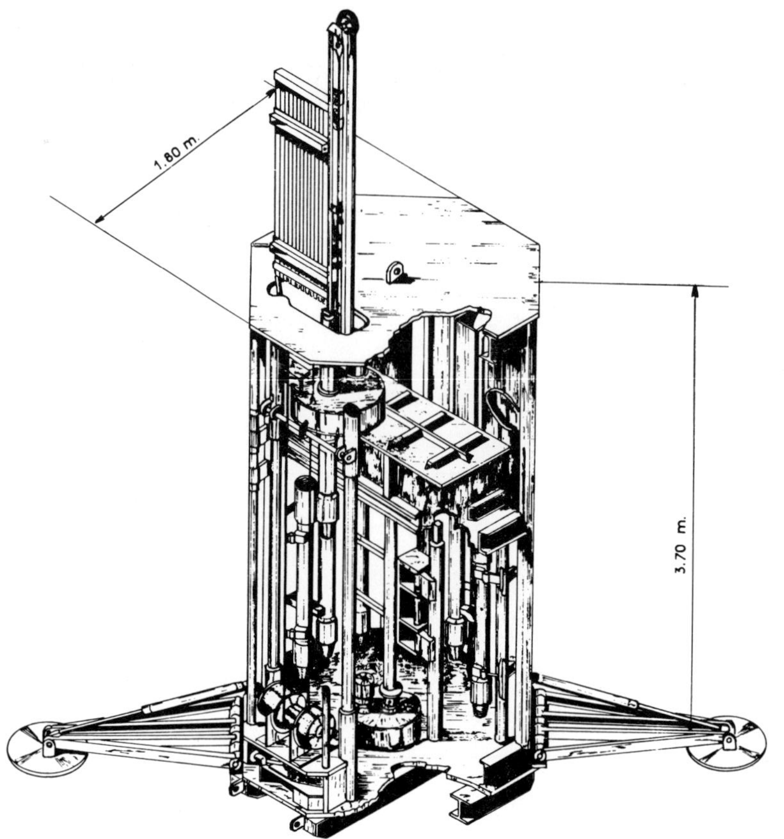

FIG. V.5.5. – The Seacore 50 underwater automatic core sampler *(Texas A & M University)*.

The intended system, consisting of the differential advance of the coring tube, should make it possible equally well to take cores in loose formations or rocks.

Furthermore, water is to be circulated at the bottom of the corer to reduce the deterioration of cores taken in loose soils.

The dimensions (height 10 m) and weight (17 t) of the unit were very comparable to those of the Maricor.

The *Otto and Sichling* telescopic deployment unit does not appear to have gone beyond the project stage.

V.6 SURFACE CORING BY MEANS OF GRAVITY OR STATIONARY PISTON CORERS

There can be no question of describing the great many variants of the different free-fall corers (gravity, stationary piston, etc.), used for surface reconnaissance of soils at sea [19] [20] [21] [22] [23].

We shall merely confine ourselves:

– first, to describing the operating principle and leading characteristics of gravity and stationary piston corers,
– second, their implementation and results and performances obtained.

V.6.1 Principle and description of gravity and stationary piston corers

The stationary piston corers (or Kullenberg corer (3)) is an improvement on the gravity corer.

V.6.1.1 Principle of penetration of gravity and stationary piston corers

The gravity corer, which drops in free-fall from a limited height (Fig. V.6.1) penetrates into the soil merely under gravity.

The stationary piston corer is a gravity corer which also drops in free-fall from a limited height (Fig. V.6.2) but has a lower end enclosed by a piston until penetration starts into the soil.

The piston, connected to the main (bearing) cable by a wire which becomes taut when the coring tube comes into contact with the bottom, remains approximately stationary as the tube penetrates into the terrain.

The presence of the piston creates a negative pressure in the coring tube as it penetrates into the soil, thus enabling the frictional forces of the core on the walls of the tube to be overcome. This generally results in recovery rates which are better than those of corers which penetrate under gravity alone.

V.6.1.2 Description and operation of the stationary piston corer

The description and operation of the stationary piston corer are shown in sketch V.6.2. The device essentially comprises:

– a main power cable,
– a release system,

(3) The first corer of this type was developed in 1947 by *Kullenberg*.

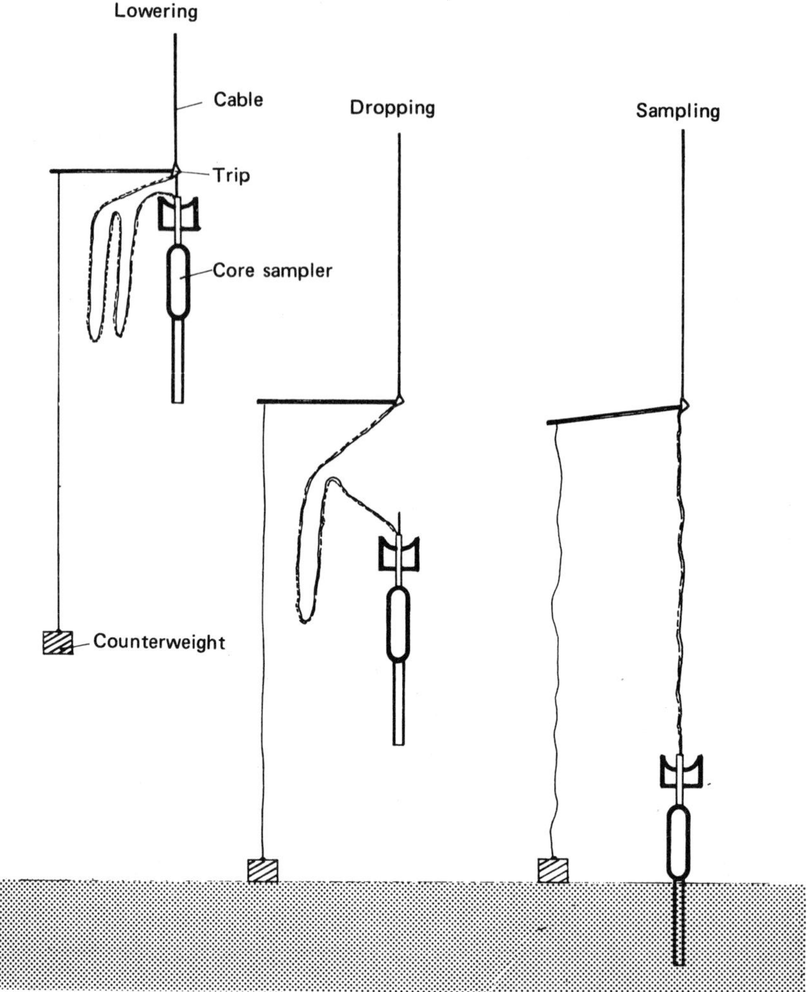

FIG. V.6.1. – Operating principle of the gravity core sampler.

– a weight (ballast),
– a counterweight hung by cables from the release system,
– a core barrel:
  – the length of which varies considerably from a few metres to over 20 m,
  – the diameter of which is generally between 40 and 120 mm, depending on the type of unit,
– a piston connected to the main cable by a wire which becomes taut when the coring tube comes into contact with the soil,
– a PVC internal tube enabling the core to be extracted without additional disturbance.

The weight of the unit can vary from about 300 to 1,500 kg, depending on the amount of ballast.

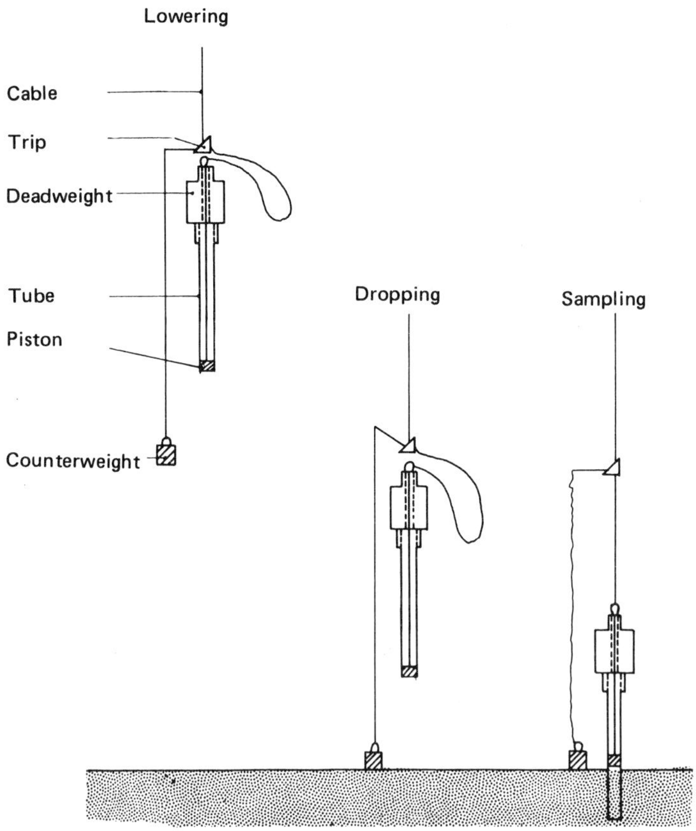

FIG. V.6.2. – Operating principle of the stationary piston core sampler (Kullenberg type).

A considerable number of improvements have been made since 1947 on the Kullenberg corer:

– larger core barrel diameter (up to at least 120 mm),

– addition of a device for holding back the sediment (indispensable in the case of sand),

– attempt to render the piston as immobile as possible when in contact with the sediment, from contact with the bottom until the core is recovered on the vessel,

– limitation of inadvertent release of the device during lowering,

– monitoring of the loads exerted on the cable during the various phases of the operation by means of dynanometers and recorders,

– increase in the weight of the corer and improvement to the overall hydrodynamic profile.

The diameter of the core barrel varies with the model from about 40 to 120 mm. The length of the core barrels is chosen in the light of the presumptive nature of the terrains:

– a few metres in sands,
– 10 or 20 m in muds and very soft clays.

V.6.2 Characteristics of stationary piston corers

The penetration of the core barrel into the soil is essentially a function of:

– the kinetic energy impact, i.e. the impact velocity and the weight of the ballast,
– the nature of the formations.

V.6.2.1 Falling velocity of the corer

The falling velocity V of the corer in the water at time t depends:

– on the mass M and mean density d of the corer,
– on a drag coefficient (approximately $0.12\ V^2$ dyn/cm²),
– on the distance x between the point of release of the corer at time t_0 (at velocity V_0) and the position of the corer at time t.

The velocity $V_0 \approx 1$ m/s is generally negligible compared to the velocity V at the end of free fall.

Figure V.6.3 shows the variations in velocity V in terms of the distance x travelled and the mass M of the device.

As one can see, the kinetic energy on impact $1/2\ MV^2$ is a non-linear function of the mass.

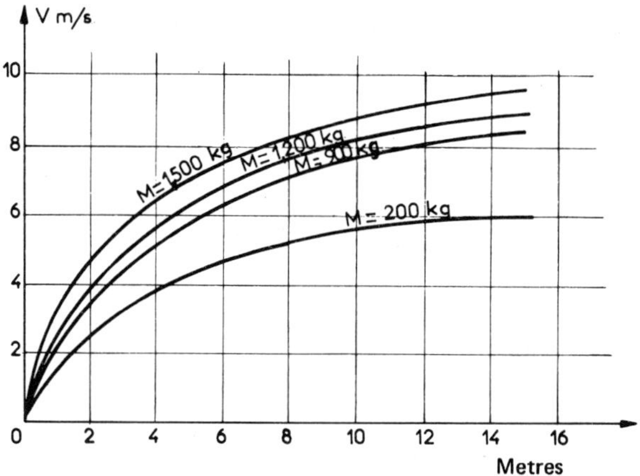

FIG. V.6.3. – Freefall speed of the sampler in water, depending on the dropped height x and the weight M.

V.6.2.2 Water depth

The pressure p beneath the piston (above the core) decreases as the penetration increases. The difference between the hydrostatic pressure and pressure p tends to offset the resistance exerted by friction between the barrel and core.

The water depth is a factor (though difficult to evaluate in figures) which is favourable to the penetration of the core, by substituting a suction effect by the top for a "stuffing" effect by the bottom.

V.6.2.3 Influence of the resistance of the formation to penetration of the corer

The absorption of the initial kinetic energy on impact $1/2\, MV^2$ depends:

– on the resistance to penetration of the tip of the corer,
– on the lateral – internal and external – friction resistances between the core barrel and the formation.

This absorption of the potential energy as the coring tube penetrates is very much dependent on the nature of the formation. A few general indications, though highly approximate, can be given:

– **for muds:**
  – proportional to penetration,
  – proportional to $D_o + D_i$ outside and inside diameters of core barrels,
  – a linear function of the difference $D_o - D_i$,
– **for sands:**
  – proportional to the square of penetration,
  – proportional to $D_o + D_i$,
  – function of the difference $D_o - D_i$.

The penetration achieved varies commonly:

– between 1 and 3 m in relatively dense sands,
– to about 10 m in soft sediments.

In certain highly consolidated soils (dense sands or clays), the penetration is practically zero.

On the other hand, in very soft muds, the penetration may reach about 20 m.

V.6.3 Offshore support and implementation

V.6.3.1 Types of vessel and hoisting facilities

Kullenberg coring is generally carried out during the geophysical reconnaissance of sites.

The tonnage of the vessel to be used therefore depends mostly on the oceanographic conditions rather than the requirements of the coring to be carried out:

– in certain calm seas, reconnaissance can be achieved by means of a vessel of light tonnage about 30 m long,

– in rough seas like the North Sea, geophysical reconnaissance and Kullenberg coring must be conducted from a vessel about 50 m long.

The pullout effort for extracting the core barrel depends on penetration, i.e. the nature of the formation:

– for penetrations of 1 to 3 m (3 to 4 m coring tube) in sands in the North Sea for instance, the pullout force is relatively low,

– for penetrations of 8 to 10 m (10 m core barrel) in relatively consolidated soils, the vessel must be equipped with sufficient hoisting facilities such as a 5 loading boom with a clearance of at least 4 to 5 m above the deck.

V.6.3.2 Bottom-to-surface link and positioning tolerances

The bottom-to-surface link is provided simply by the corer releasing cable, unwound from a winch on the vessel.

The great rapidity of execution obviates the need to moor the vessel (which is furthermore impossible in great water depths).

The flexible bottom-to-surface link does not require the vessel to be positioned to within a few metres.

V.6.3.3 Rate of operations

The theoretical rate of taking cores 2 to 3 m in length at water depths of 200 m can reach 2 per hour.

The actual sampling rate reaches the following on the average:

– 15 to 25 cores of 2 to 3.5 m per day in the North Sea at water depths of 100-150 m,

– 20 to 25 cores of 5-10 m per day in the soft soils of the Mediterranean at water depths of 60-70 m.

V.6.4 Geotechnical characteristics of cores sampled by Kullenberg coring

The geotechnical quality of the cores sampled depends at one and the same time on the shape of the corer and how it penetrates into the soil. The best results are obtained with:

– large diameter corers with well adapted form coefficients of the penetration "nose",

– high constant penetration velocities.

V.6.4.1 Form coefficients of coring tubes and quality of the cores

The form coefficients of the penetration "nose" of a coring tube, or Hvorslev coefficients, are defined as follows (see Fig. V.6.4):

$$C_i = \frac{d_s - d_e}{d_e} \times 100$$

$$C_o = \frac{d_w - d_t}{d_t} \times 100$$

$$C_a = \frac{d_w^2 - d_e^2}{d_e^2} \times 100 \text{ (coefficient of area)}$$

$$C_l = \frac{L}{d_s}$$

where:

d_e = inside diameter of cutting edge,
d_w = outside diameter of cutting edge,
d_t = outside diameter of core-barrel,
d_s = inside diameter of sleeve,
L = length of core barrel.

Table V.6.1 provides a comparison between:

– the extreme values of these coefficients, for an acceptable geotechnical quality for **mud** cores,

– and the values of these coefficients for the *Saclant CEN* corer considered as one of the best Kullenberg type units that can be used in muds.

TABLE V.6.1

Hvorslev's coefficients and inside diameter of corer	Acceptable maximum values for muds	Values achieved with the Saclant CEN corer
Inside backlash C_i (%)	< 1	0.8
Outside backlash C_o (%)	< 3	5.1
Coefficient of area C_a (%)	< 10	44
Coefficient of length C_l	< 20 in muds and clays < 10 in sand	< 100
Angle of cutting edge (°)	5	5
Inside diameter d_s (mm)	> 40	120

For reasons of manufacture and in order to reach high penetrations, the coefficients C_a and C_l of Kullenberg corers are invariably much higher than the limits compatible with recovery of cores disturbed only slightly. The result is that the geotechnical quality of the cores is often mediocre, particularly when the diameter is small ($<$ 8 to 10 cm).

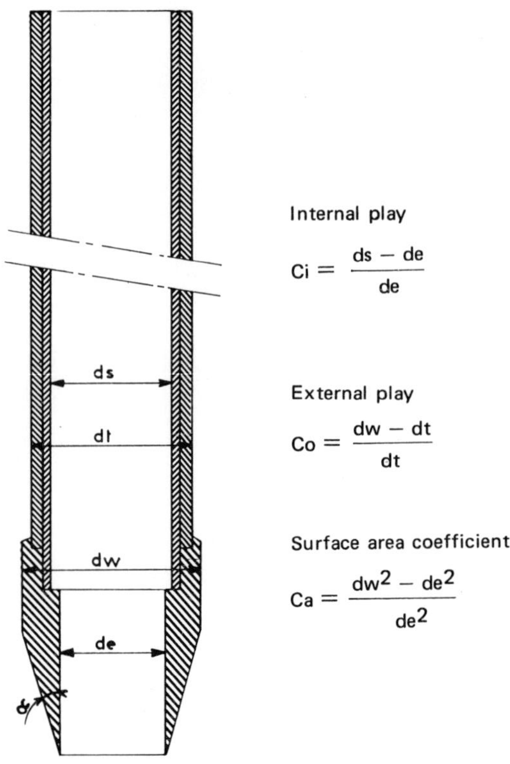

FIG. V.6.4. – Dimensions and coefficients of core sampler nose pieces.

V.6.4.2 Penetration velocity of the corer and qualities of the cores

The equation for the motion of the corer (penetration velocity V in terms of penetration depth z in the formation) depends:

– on the mass M and the mean density d of the corer,
– on the inside and outside diameters D_i and D_o of the corer,
– on the force of friction f,
– on a coefficient K.

Figure V.6.5 gives the curves of $V(z)$ for:

– a mass M of 1,200 kg,
– an inside diameter D_i of 80 mm,
– a force of friction of 1 kgf/cm (per centimetre length of the core barrel),
– various values of the initial impact velocity V_i.

A high initial impact velocity V_i hardly increases the penetration z at all but does on the other hand offer the advantage of maintaining a high penetration velocity which

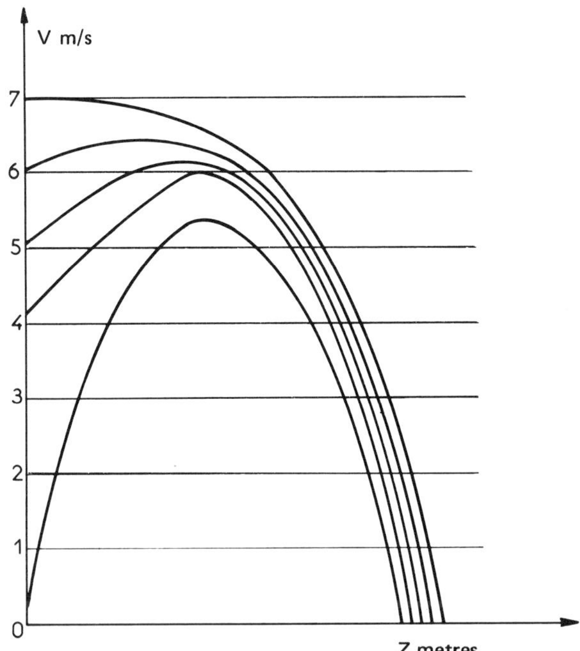

FIG. V.6.5. – Penetration velocity of the sampling tube into the terrain in terms of the initial velocity and the depth z (Kullenberg type sampler).

is less variable during the initial insertion stage, thus improving the quality of the core sampled.

V.6.4.3 Disturbance of cores caused by immobilization of the piston

The direct connection between the piston and the operational cable is an important factor in the disturbance of the sediments. At the moment of impact of the corer against the soil, the sudden release of tension on the operating cable causes a shock wave which propagates through the cable. In shallow water depths this shock wave can bounce several times between the piston and the winch, resulting in disturbing the sediments beneath the piston.

"Split piston" devices (i.e. pistons with 2 chambers) or "recoilless piston" devices in the corers developed by Kermabon [21] alleviate this drawback.

V.6.4.4 Qualities and utilization of the core taken

Owing to the considerable disturbance, the cores taken are generally not very representative of the geotechnical qualities of formations. Measurements of shear strength and settlement made on these cores are probably of only little significance. Figure V.6.6

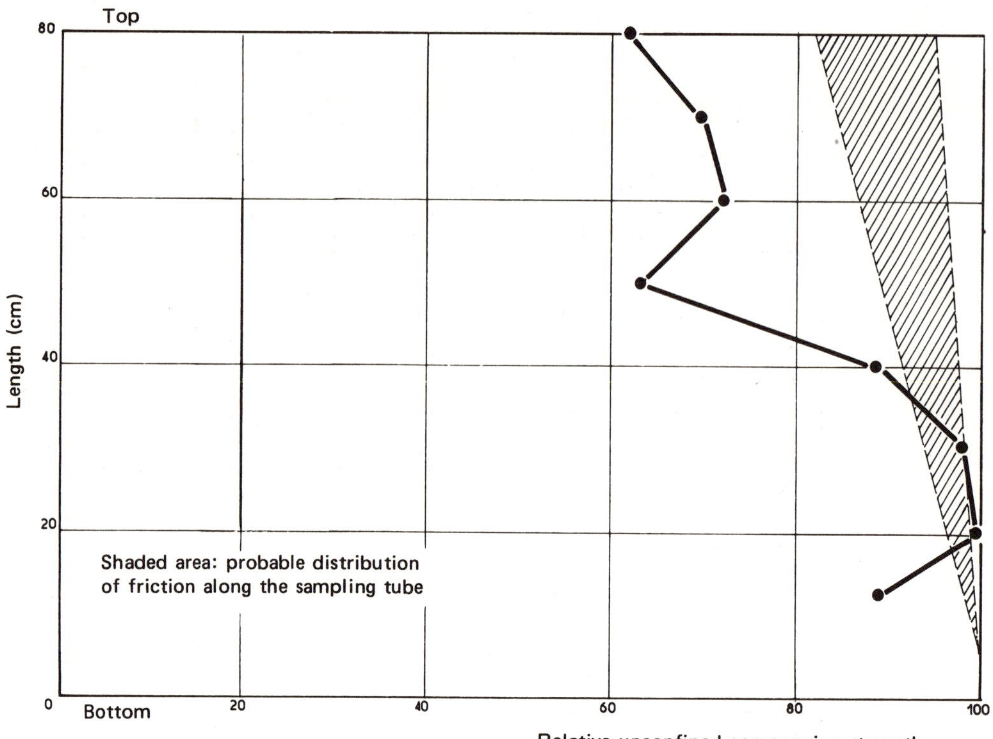

FIG. V.6.6. – Disturbance caused by penetration of the core sampling tube, characterized by the unconfined compression strength.

shows the extent of disturbance at the ends, characterized by the relative value of the direct compressive strength. Furthermore, the depths of penetration, generally confined to 10 to 15 m at the most, are notoriously insufficient for the reconnaissance of soils with a view to siting structures on deep foundations (pile-mounted platforms) or of the gravity base type.

On the other hand, this method of coring is particularly well suited to the reconnaissance of alignments for pipelines in particular. The cores taken at depths of 3 to 4 m at the most allow conventional soil identification tests to be carried out: Atterberg limits, size distributions, densities, water contents, etc.

V.6.5 Inventory of gravity, stationary piston and free-fall corers

The list of gravity, stationary piston or simply free-fall corers given in the following tables is certainly not exhaustive, and its purpose is simply to provide the following infor-

TABLE V.6.2

GRAVITY OR STATIONARY PISTON CORERS

Maker and/or Contractor	Designation of unit	Type of unit	Type of link	Water depth (m)	Maximum penetration (m)	Samples recovered	Remarks
Inventor: Kermabon	Saclant CEN	Improved Kullenberg type	Flexible	Unlimited	About 10-15	Cores about 120 mm in diameter (soils of low consolidation)	Unit has been improved to reduce disturbance Large diameter cores (samples are representative)
Benthos	Model 2,450 Gravity piston	Gravity and piston corer (Kullenberg type)	Flexible	Unlimited	15	65 mm diameter cores	Unit can be used in sediments of low consolidation
Alpine Geophysical Associates	Inventor: Ewing	Piston corer	Flexible	Unlimited	3-7.5	58 mm diameter cores	Unit can be used in soft soils
Comex	Casting corer	Gravity corer	Flexible	Unlimited	3 or 5	86 mm diameter cores	Unit can be used in muds and soft clays
Norvegian Geotechnical Institute (NGI)	NGI Gas-Operated Sea-Floor Sampler	Stationary piston corer: – lowered by gravity – penetration by deflagration	Flexible	350	⩽ 10	Cores: length = 1.65 m diameter = 54 mm	Disturbance is limited owing to the short length of the core
Techniques Louis Menard (TLM)		Kullenberg corer	Flexible	Unlimited	5 to 10	63 and 85 mm diameter cores	Can be adapted to pressuremeter
Station de Géodynamique sous-marine Villefranche-sur-mer	Corer	Kullenberg corer	Flexible	Unlimited	20	54 mm diameter cores	Unit can be used in soft soils
	"Giant Corer"	Kullenberg corer	Flexible	Unlimited	> 20	140 mm	Unit used in soft soils

mation for certain units:

– operating principles,
– maximum penetration that can be expected,
– dimensions of core recovered,
– optimum conditions of use (nature of formations, etc.).

V.6.5.1 Stationary piston corers (see Table V.6.2)

As indicated earlier (see Section V.6.1), stationary piston corers penetrate into the soil under the combined effect of:

– gravity,
– the negative pressure created by the presence of the piston.

The penetration varies from a few metres to about 20 m at the most in very soft sediments.

Mention should be made of the particularity of the *Norvegian Geotechnical Institute (NGI)* corer where the coring tube penetrates into the soil by deflagration. This device enables a short (1.65 m) core to be extracted at a depth of ⩽ 10 m in the soil.

V.6.5.2 Gravity and free-fall corers (see Tables V.6.2 and V.6.3)

Among the many gravity type corers, the "Bouma Box Sampler" should be pointed out, enabling large dimension (200 x 75 mm) rectangular cores to be sampled.

Various autonomous corers generally offer very restricted performance with penetrations of only about 1 m. Their main advantage lies in the facility with which they can be implemented.

TABLE V.6.3

GRAVITY OR FREE-FALL CORERS

Maker or Contractor	Designation of unit	Type of unit	Type of link	Water depth (m)	Maximum penetration (m)	Samples recovered	Remarks
Hydro-Products	Bouma Box Sampler	Gravity corer	Flexible	Unlimited	0.45	Rectangular cores measuring 200 x 75 mm	Device usable for sampling fine sands
Hydro-Products	Moore Free Corer	Free corer	Autonomous	Unlimited	0.60 1.50 3	62 mm diameter cores	Autonomous unit that can be used in soft soils
Benthos	Boomerang Free Fall		Autonomous	6,000	1.20	65 mm diameter cores	Unit can be used in soft soils
Reinecke	Long Reinecke	Gravity corer	Flexible	Unlimited	< 10	150 x 150 mm cores	Collapsible Sheath

BIBLIOGRAPHY

[1] EMRICH (W.J.). – "Performance Study of Soil Sampler for Deep Penetration Marine Borings". *Symposium ASTM-STP 501,* 1972.

[2] McCLELLAND (B.). – "Techniques Used in Soil Sampling at Sea". *Offshore,* March 1972.

[3] EIDE (O.). – "Marine Soil Mechanics. Applications to North Sea Offshore Structures". *Offshore North Sea Technology Conference,* Stavanger, 3-6 Sept., 1974.

[4] de RUITER (J.). – "Reconnaissance des sols marins pour plates-formes pétrolières en mer du Nord". *Annales de l'ITBTP* No. 332, octobre 1975.

[5] de RUITER (J.). – "The Use of In Situ Testing for North Sea Soil Studies". *Offshore Europe Conference,* Aberdeen, 1975, paper No. OE-75-219.

[6] de RUITER (J.). and FOX (D.A.). – "Site Investigations for North Sea Forties Field. *OTC,* Houston, 1975, Paper OTC No. 2246.

[7] *Data sheet of COMEX* on "Rotary sounders".

[8] *Data sheets of Vriens Diving Co* on "Underwater Working Chamber".

[9] *Data sheet of IFP* on "Subsea Vibrodriver". Ref. 17928 A, March 1970.

[10] MENARD (L.). – "Intérêt technique et économique du vibromarteau hydraulique annulaire pour le prélèvement d'échantillons en mer et la réalisation d'essais géotechniques in situ". *Oceanexpo,* Bordeaux, 1975.

[11] CASTELA (A.). – "Emploi d'appareils télécommandés pour la reconnaissance des sols marins". *Colloque de l'ASTEO,* Paris, 18-20 oct. 1972.

[12] DELACOUR (J.) et MORIN (P.). – "Application du flexoforage à la reconnaissance des grands fonds marins". *Oceanexpo,* Bordeaux, 9-14 Mars 1971.

[13] TIREY (G.B.). – "Recent Trends in Underwater Soil Sampling Methods". *Symposium ASTM-STP 501,* 1972, p. 42-54.

[14] "Diamond Core Drill for Seabed Sampling", Technical Instructions, *ATLAS COPCO.*

[15] "Diamond Core Drill for Seabed Sampling", *Gas World,* 7 July 1973.

[16] "Maricor, carottier sous-marin télécommandé", *Pétrole Informations,* 22-28 Juin 1973.

[17] NOORANY (J.). – "Underwater Soil Sampling and Testing: a State of the Art Review". *Symposium ASTM-STP 501,* 1972, p. 3-41.

[18] BAILEY (E.I.), DAVIS (G.L.) and HENDERSON (H.O.). – "Design of an Automatic Corer", *OTC,* Houston, 19-21 April 1971. Paper N° OTC 1365.

[19] KULLENBERG (B.). – "The Piston Core Sampler Svenska Hydrografisk", *Biologiska Kommissionffens Shrifter,* 1947.

[20] KERMABON (A.), BLAVIER (P.), CORTIS (V.) and DELAUZE (H.). – "The "Sphincter Corer" a Wide Diameter Corer with Watertight Core-catcher", *Marine Geol.,* 4, 1966.

[21] KERMABON (A.) and CORTIS (V.). – "A New "Sphincter Corer" with a Recoilless Piston", *Marine Geol.,* 7, 1969.

[22] RICHARDS (A.F.). – *Marine Geotechnique.* University of Illinois Press, 1967.

[23] INGHAM (A.E.). – *See Surveying* (Vol. I and II). John Wiley and Sons, London-New York, 1975.

CHAPTER VI

Seabed Exploration by in situ Measurements

CONTENTS

INTRODUCTION

1. The increasing difficulties in obtaining undisturbed core samples as depths of water increase are resulting more and more in the recourse to in situ measurements for the reconnaissance of soils at sea. However, one should not lose sight of the fact that whilst in situ measurements, even when carried out under optimum conditions, make it possible to obtain representative and comparative figures, the sampling of cores remains indispensable to the geological knowledge and identification of the soils. **Cores and in situ measurements are therefore perfectly complementary.**

2. The in situ measurement techniques and devices (penetrometers, pressuremeters, vanes, drilling logs) used for the reconnaissance of soils at sea are directly derived from devices used on land, only differing in their implementation.

3. The various techniques used in the laboratory or in situ to determine the shear strength of soils in fact take place under widely differing conditions. Experience has shown that the measurement of the shear strength or settlement of sands or clays by means of various in situ or laboratory techniques sometimes lead to very different results .

Furthermore, there is no obvious reason why equipment as different as laboratory or in situ measurement techniques (such as the penetrometer, pressuremeter, etc.) should yield magnitudes identical to those actually brought to bear to ensure the stability of structures.

Lastly, the parameters measured by this various equipment are more or less directly relatable to the intrinsic characteristics (cohesion and friction) of the soil.

It is therefore of importance to ensure the signification and representivity of the magnitudes measured with the various devices and if possible to compare the results obtained.

4. It is proposed in the present chapter to examine in turn the principle, operation, implementation at sea and the interpretation of the results:

- of modular penetrometers,
- of static and dynamic cable operated penetrometers,
- of the Menard pressuremeter,
- of the wireline remote vane,
- of log probes (in particular nuclear) used in the borehole.

VI.1 MODULAR PENETROMETERS INTERPRETATION OF PENETROMETER RECORDINGS

Several modular penetrometers set on the sea bottom are used in deep water:

– either by divers, on several coastal or harbour sites or again in the Persian Gulf,

– or from jack-up platforms, like the device used by *NGI* for reconnaissance of the Ekofisk site in a depth of 70 m.

Otherwise, *Shell* have developed a telescopic rod penetrometer with a bell frame anchored to the bottom by negative pressure.

However, the development by *Fugro* several years ago of the Seacalf penetrometers weighing about 10 to 30t has led to a veritable boom in penetrometer measurements for soil reconnaissance in the North Sea.

For some years several modular divices have been built or have been being developed.

In the present section:

– it is first of all proposed to describe the principle, characteristics and implementation of modular Seacalf penetrometers with an electrical cone,

– and then to give a number of indications as to other modular penetrometers used less widely or now being developed.

VI.1.1 Principle and description of Seacalf modular penetrometers

Seacalf penetrometers practically only differ in the dimensions and weight of the frame (about 5, 10 or 30t) (see Fig. VI.1.1) [1] [2] [3].

VI.1.1.1 Operating principle

The Seacalf is a penetrometer frame lowered to the sea bottom by means of bearing cables and remote-controlled from the surface.

A penetrometer cone is inserted into the soil in successive sequences by means of a hydraulic cylinder. The necessary reactive force is absorbed by the weight of the frame resting on the bottom.

The cone of the penetrometer carries two strain gauges (Fig. VI.1.2) providing a continuous and simultaneous indication of the following:

– tip resistance R_p,

– lateral friction resistance f,

read on a recorder on the vessel.

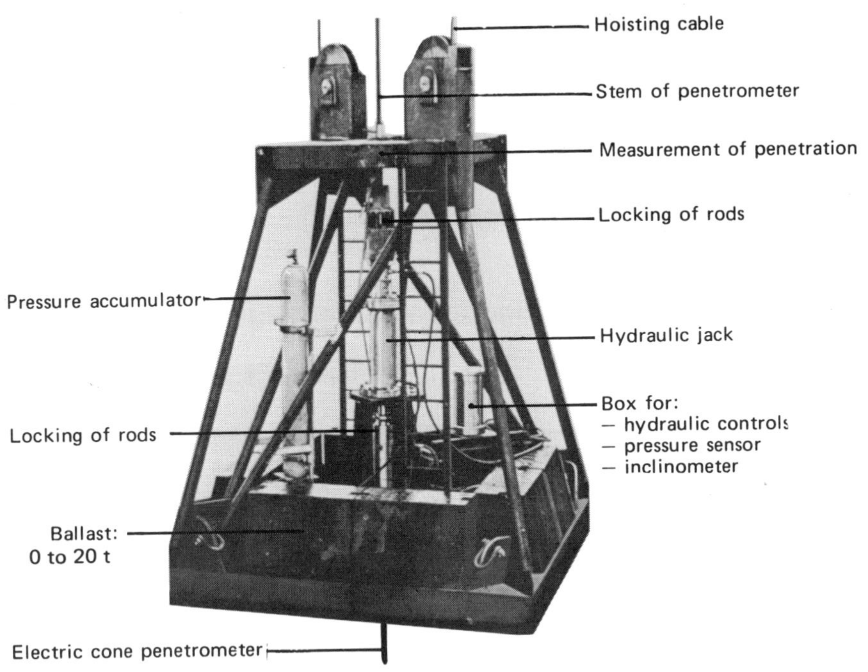

FIG. VI.1.1. – Functional sketch of the Seacalf (*Fugro*)

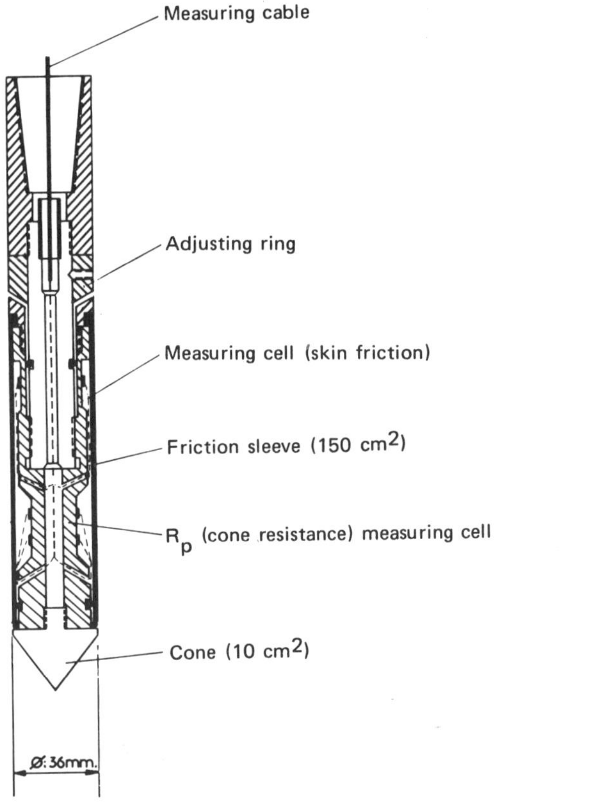

FIG. VI.1.2. – Electric cone penetrometer equipped with strain gauges (*Fugro*).

VI.1.1.2 Description of unit

The subsea **unit** consists essentially of (Fig. VI.1.1):

– a ballasted frame weighing 10 to 30 t resting on the sea bottom on a square base with size about 2.5 to 3.0 m long,

– a hydraulic cylinder controlled on the surface to push in the drilling string thanks to a special clamping device; the piston returns under the pressure from a hydraulic accumulator,

– a string of drill pipes 36 mm diameter and about 30 m in length carrying the 36 mm diameter penetrometer cone,

– an inclinometer to monitor the deviation of the drilling string.

The surface equipment comprises:

– an operating winch for raising and lowering the subsea unit by means of two bearing cables,

– a winch for extracting the drilling string from the terrain by means of a carrying cable,

– a central hydraulic power plant for supplying the cylinder,

– a recording instrument connected to the cone of the penetrometer by an electric cable (8 conductors) running through the drilling string.

VI.1.2 Characteristics of the Seacalf penetrometer

VI.1.2.1 Equipment characteristics

The cone dimensions are the same as those of several conventional penetrometers used ashore (Delft Laboratory, Begemann, etc.) [4] [5]:

– cone summit angle = 60°,
– cone diameter = 36 mm (cross-section 10 cm^2),
– lateral surface area of lateral friction measuring sleeve (diameter 36 mm) = 150 cm^2.

These dimensions, which are practically standard, permit:

– rapid comparison of the results of soil reconnaissance at different sites,

– characterization of the soils in terms of conventional magnitudes (R_p and f), which are functions of the cone dimensions.

The maximum reaction for penetration of the rods depends on the weight of the frame and the characteristics of the hydraulic equipment:

– about 8 t in the case of a 10-ton frame,
– about 20 t in the case of a 25-ton frame.

VI.1.2.2 Functional characteristics

The cone penetrates into the soil discontinuously in successive sequences of about 0.60 m.

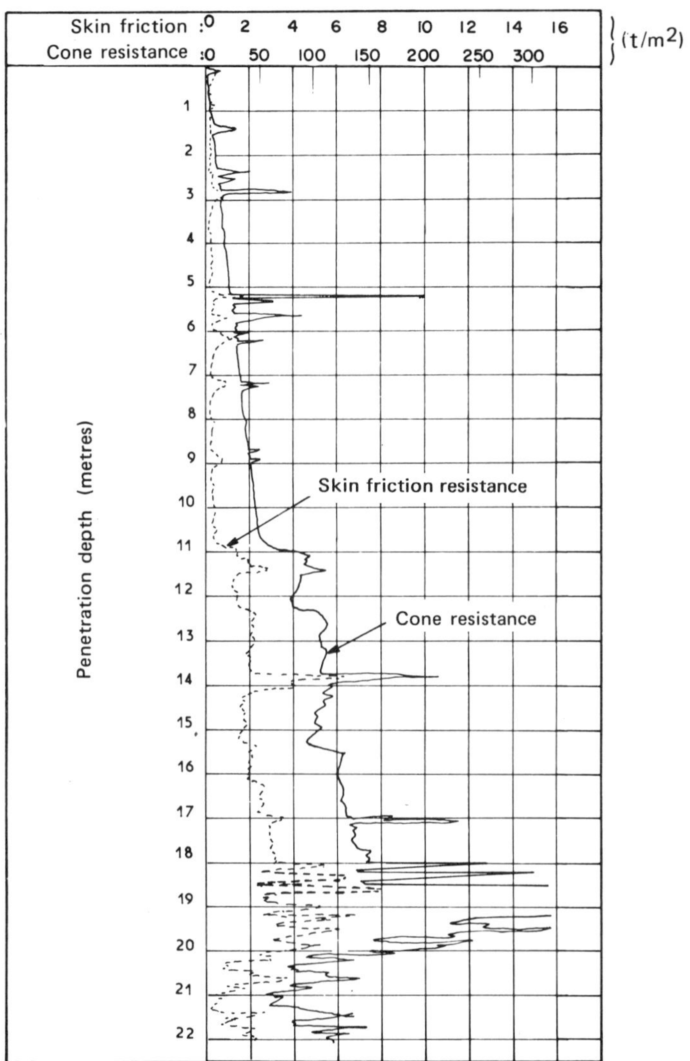

FIG. VI.1.3. – Example of logs plotted with the Seacalf penetrometer.

The penetrometry measurements R_p and f recorded during each sequence are combined to produce continuous profile (1) of the mechanical characteristics of the strata crossed through (Fig. VI.1.3).

The rate of penetration of the cone theoretically lies between 1.5 and 2 cm/s, i.e. about 1 m/min.

Allowing for the penetration discontinuities, the average "driving rate" of the rods is about 20 m/h (i.e. 0.3 m/min).

Pullout of the rods by a winch is done at about the same speed.

The load applied by the jack must partially offset the lateral frictional forces of the formation on the string of rods:

– in clays, friction falls off quickly owing to the movement of the string of rods,

– in sand, on the other hand, friction may be considerable and thus limit the penetration of the cone.

The measuring cell of cone resistance R_p is designed:

– in the case of sands overconsolidated clays, for a maximum R_p value of 5,000 to 6,000 t/m² (50 to 60 MPa),

– in the case of clays or muds, for a maximum resistance R_p of 750 t/m² (7.5 MPa).

The lateral friction measuring cell is generally designed to withstand a load of 750 kg, i.e. a maximum stress f of 50 t/m² (500 kPa).

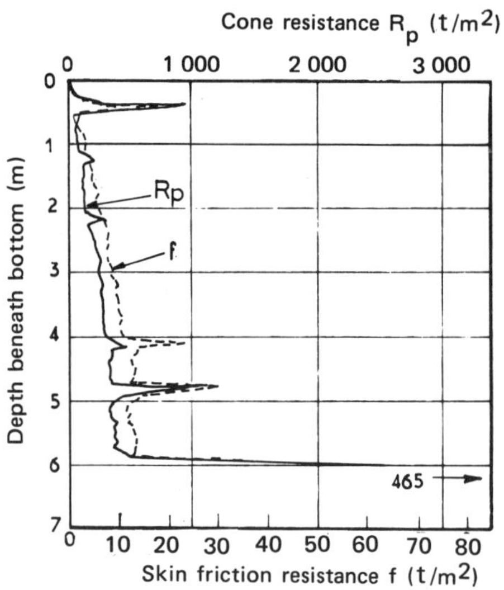

FIG. VI.1.4. – Example of Seacalf penetrometer recording with refusal on encountering a layer of dense sand.

(1) In contrast to mechanical penetrometers for instance, which measure R_p and f in alternation at depths that differ by 20 to 30 cm.

The penetration depth depends on the frame used and the formations encountered:

– merely a few metres in the dense sands of the North Sea, using a 10-ton unit (Fig. VI.1.4).

– about 15 m in the consolidated soils of the North Sea, using a 25-ton unit,

– a maximum of 25 to 30 m in only slightly consolidated soils (length of string of rods = 30 m).

VI.1.3 Offshore support and implementation of Seacalf penetrometers

VI.1.3.1 Type of vessel

So far, Seacalf modular penetrometers have been implemented in the North Sea from the Heerema drilling vessels and other vessels (with a central well), anchored at 4 or 6 points (see Section V.1).

– The Explorer:

length = 56 m,
displacement = 500 t,
derrick = 10 t.

– The Surveyor:

length = 75 m,
displacement = 800 t,
derrick = 30 t.

– The Mariner:

length = 82 m,
displacement ≈ 1,000 t,
derrick = 40 t, 20 m.

Implementation of a modular penetrometer from a supply vessel would call for a very calm sea owing to the pendular motion caused by the considerable weight of the equipment.

By hauling the vessel up to its anchors, several penetrometer soundings can be made quickly over an area measuring about 100 to 200 m along the sides (or in diameter).

VI.1.3.2 Bottom-to-surface links and positioning tolerance

The bottom-to-surface link is flexible and consists of:

– the carrying cables,
– the hydraulic cables connecting the central hydraulic power plant to the cylinder,
– the electrical measuring and control cables.

This type of flexible link enables the vessel to move a few metres.

VI.1.3.3 Space, implementation equipment and manpower required. Duration of an operation

The space needed on the vessel:

– for the frame of the penetrometer,
– for the handling, supply, control and measuring equipment,
– for assembly and disassembly of the drilling string,

is about 80 to 100 m².

The equipment for implementing the modular penetrometer has already been described in brief (see Section VI.1.1.2).

Implementation of the modular penetrometer requires at least three technicians.

After anchoring the vessel, the operation of soil reconnaissance using the Seacalf consists of:

– lowering the equipment (frame + drilling string) into the water through the central well of the vessel,
– lowering it to the bottom on the carrying cables,
– insertion of the drilling string (at a speed of about 1 m/min) while recording R_p and f,
– withdrawal of the drilling string from the terrain,
– raising of the equipment and rehoisting inboard.

A reconnaissance operation down to a penetration of 15 to 20 m would normally take 4 to 5 h.

VI.1.4 Oceanographic conditions

VI.1.4.1 Water depth

Seacalf penetrometers are commonly used in depths which at present vary from 100 to 150 m in the North Sea.

Use at a depth of a few hundred metres would call for a submerged central hydraulic power plant mounted on the frame of the unit.

VI.1.4.2 Waves and current

The maximum wave amplitude compatible with use of the 10 or 25 t modules is 1.50 to 2 m owing to the considerable pendular motion.

The vessels used can be anchored provided the currents are below 3 to 4 knots.

VI.1.5 Hyson penetrometer operated by divers

VI.1.5.1 Principle of the Hyson penetrometer

The Hyson penetrometer (Fig. VI.1.5) weighs 1,500 kg and is held by ballast or anchored to the sea bottom and operated by divers. It consists essentially of :

– a 13-ton thrust device formed by two hydraulic jacks,
– a subsea central hydraulic power plant driven by an electric motor power supplied from the surface,

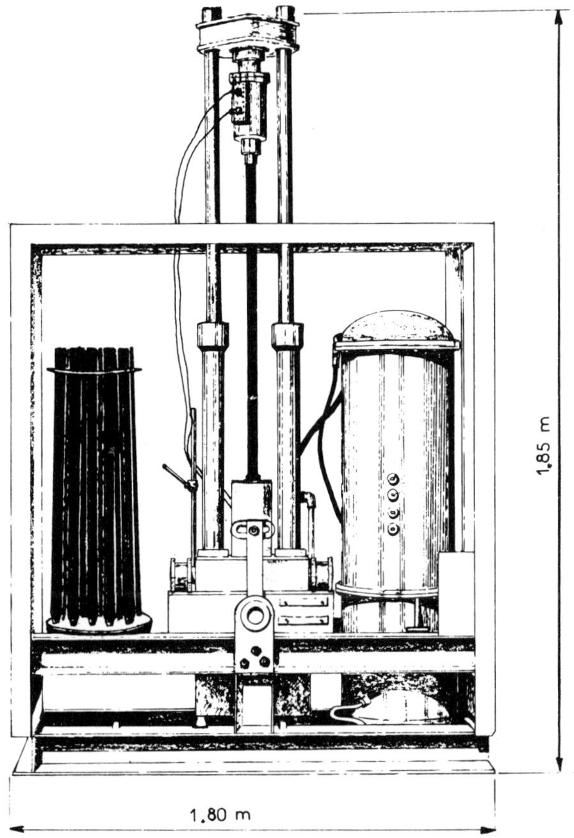

FIG. VI.1.5. – The Hyson static penetrometer.

– a frame carrying the jacks, power plant and magazine of drill pipes.

The penetrometer cone measures the following continuously and simultaneously:

– the tip effort,
– the total effort (tip effort + effort due to lateral friction).

The measuring cells are connected by cable to a recorder on the vessel.

VI.1.5.2 Description and implementation of the Hyson penetrometer

The constituent equipment of the penetrometer comprises the following in particular:

– the complete hydraulic sounding device equipped with its thrust system, sealed container for the electric motor and central hydraulic power plant,

– the barrel type distribution magazine containing 40 drill pipes 1 m in length and 36 mm in diameter,

– the penetrometer cone and strain gauges.

The Hyson penetrometer can be implemented:

– either by ballasting the unit to about 15 t of the apparent weight in water, presupposing ideal sea conditions,

– or by anchoring it to the bottom by screw anchors.

VI.1.6 The NGI penetrometer lowered by means of a drilling string

VI.1.6.1 Principle of the NGI penetrometer

The *Norwegian Geotechnical Institute (NGI)* has developed a modular penetrometer implemented from a jack-up platform by means of the drilling string [6].

The operating principle is shown in Fig. VI.1.6. The penetrometer cone is pushed into the soil in successive steps by the action of the drilling string.

VI.1.6.2 Cursory description of the NGI penetrometer

The unit essentially consists of:

– a 5-ton module, guided and moved by the drilling string,

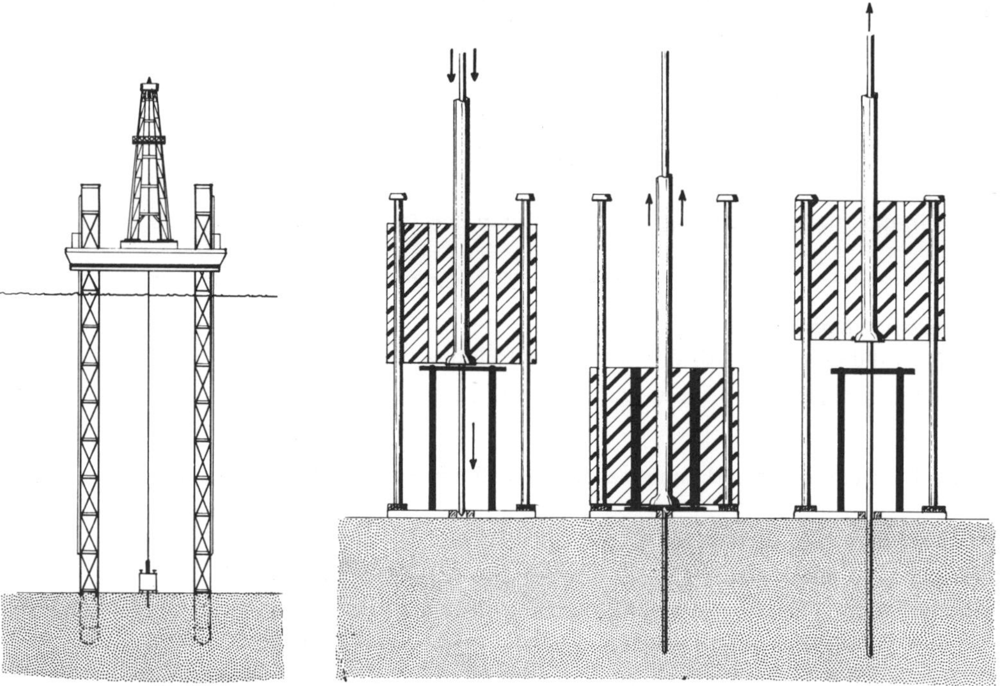

FIG. VI.1.6. – Implementation device for the NGI penetrometer at Ekofisk.

– a small diameter drilling string carrying a penetrometer cone passing through the main drilling string.

The tip resistance is measured by means of a vibrating cord electric cell.

VI.1.7 The telescopic rod penetrometer of Shell

VI.1.7.1 Principle of the Shell penetrometer

The penetrometer built by *Shell* in 1970 is the outcome of adapting (see Fig. VI.1.7) [1] [7]:

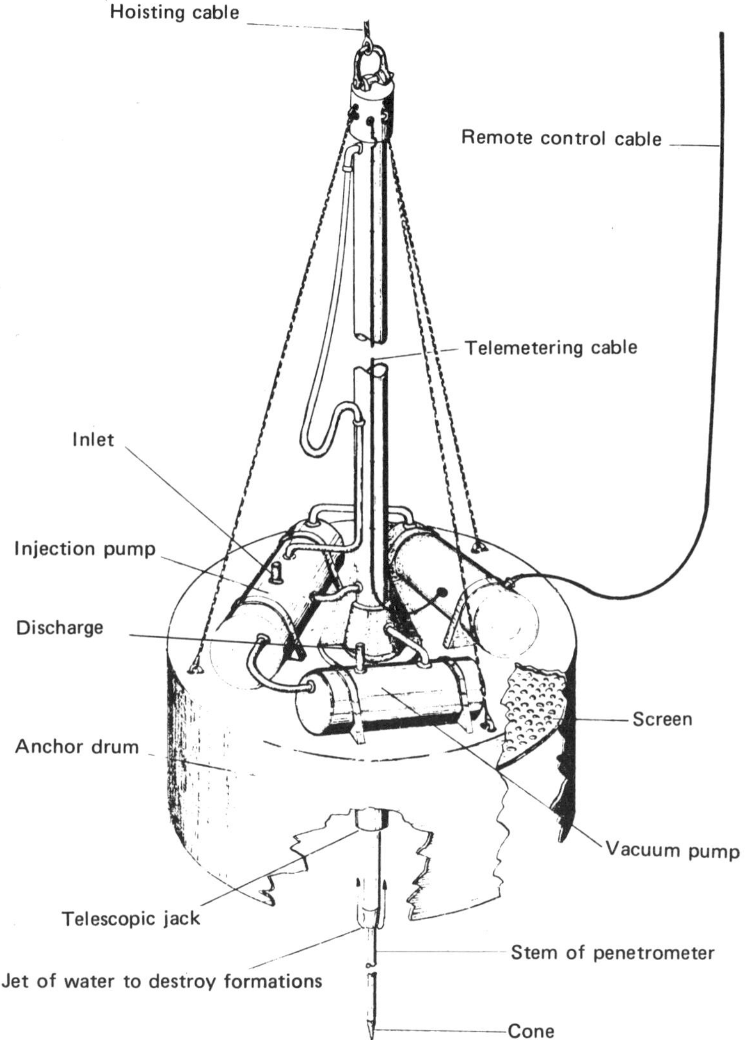

FIG. VI.1.7. – Sketch of the telescopic penetrometer (*Shell*).

– a telescopic development penetrometer measuring device,
– and an undersea anchoring system consisting of a drum held against the soil by negative pressure. The reaction needed to cause the cone to penetrate in the terrain is absorbed by the anchoring force of the drum against the bottom of the sea.

The measuring system consists of two strain gauge cells ensuring continuous and simultaneous measurement of:

– the cone resistance R_p,
– the lateral force of friction f.

VI.1.7.2 Brief description of the Shell penetrometer

The unit consists essentially of:

– a telescopic jack measuring 3 x 9 m (i.e. 27 m when developed), carrying the cone of the penetrometer which is of standard dimensions (cone angle 60°, diameter 36 mm, i.e. a cross-section of 10 cm^2),
– a system for anchoring the unit to the sea bottom consisting of a 3 m diameter drum held against the soil by negative pressure (about 3 t/m^2) (about 30 kPa) and fitted internally with a screen preventing the unit from settling into fine sediments.

The drum carries:

– a pump for the application and constant maintenance of the negative pressure inside the drum,
– an injection pump is order to destroy by jet effect the terrain previously crossed by the cone of the penetrometer, thus reducing the force of friction of the telescopic jack as it advances into the soil. The water is injected with each change in the cross-section of the cylinder.

The unit weighs about 6 to 7 t.
The maximum reaction due to the negative pressure beneath the drum can amount to about 20 t.
The unit does not yet appear to be operational (2).

VI.1.8 The Preussag penetrometer associated to the Syminex nuclear log probe

The *Preussag* modular penetrometer, which is fully automatic, became operational in 1976.

---

(2) A smaller size device weighing about 5 t and endowed with a maximum penetration of 10 m was developed by *Shell* in 1976 for the reconnaissance of pipeline routes.

VI.1.8.1 Principle of Preussag penetrometer

The frame of the penetrometer which weighs 30 t, is lowered to the sea bottom by means of a 5″ drilling string.

The string of penetrometer rods, 30 m long, which carry a penetrometer cone and a nuclear log probe (γ – ray and $\gamma - \gamma$) is:

- introduced through the drilling rods,
- thrust into the soil after clamping by means of hydraulic jacks.

The device enables the following parameters to be measured at one and the same time:

- the resistance R_p of the penetrometer cone,
- the lateral force of friction f,
- the inclination of the rods in the soil,
- the natural radioactivity of the soil (γ – ray),
- the density of the soil in place by $\gamma - \gamma$.

VI.1.8.2 Brief description of the device and the variants envisaged

The pyramid shaped frame of the penetrometer has the following characteristics:

- dimensions = height 4 m, diameter at base 2.85 m,
- weight = 30 t (frame 28 t, drilling string 2 t),
- equipment for insertion of the penetrometer rods: motor, pump, jacks, etc.

The penetrometer rods consist of 30 elements of 1 m (diameter 36 mm).

The dimensions of the probe (complete with cone) are:

- length 3 m,
- diameter 36 mm (cross-section 10 cm^2).

The friction of the probe against the soil can be reduced by jetting of water or air at the top of the probe.

Various arrangements are planned for using the device:

- drilling equipment on a frame to enable hard strata and pebbles to be crossed,
- coring device down to 10 m by hydraulic thrust on the coring tube.

VI.1.9 Principle and description of the Stingray and Seajack units

McClelland Engineers and the *Norvegian Geotechnical Institute,* in 1975-1976, developed two modular units, rather similar to each other, one called Stingray for *McClelland,* [8] and the other Seajack for *NGI* [9].

VI.1.9.1 Principle of Stingray and Seajack

The two units, Stingray (Fig. VI.1.8) and Seajack, essentially consist of a frame fitted with jacks, resting on the seafloor and providing a reaction which makes possible the operation of:

– a penetrometer (or possibly any other measuring device such a vane or pressure-meter),
– a wireline push sampler.

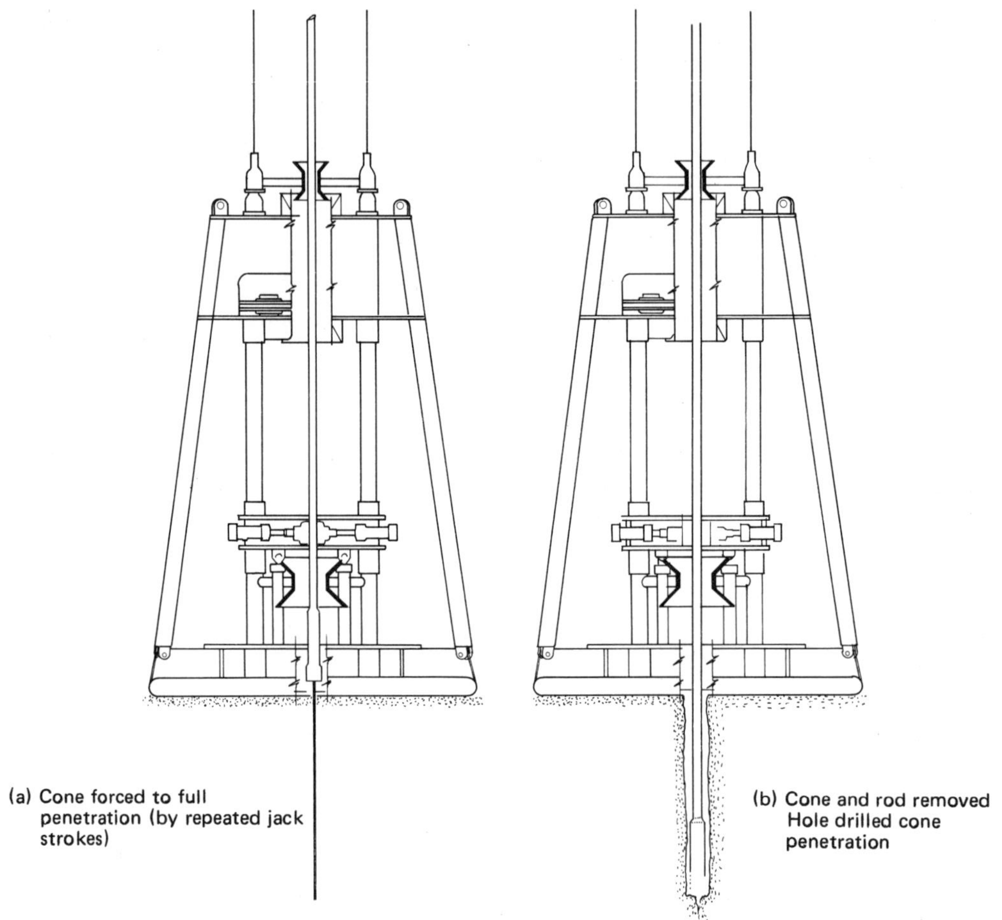

FIG. VI.1.8. – Stingray : cone penetrometer system for marine geotechnical investigations *(McClelland Engineers).*

VI.1.9.2 Characteristics and application of units for the implementation of the penetrometer

The dimensions of the Stingray and Seajack are:

– base = 3 x 3 m,

– height = about 3.5 m
– weight = 16 t.

Implementation of a penetrometer by means of these units comprises the following sequences:

– the jacking frame is set on the bottom, and the drill pipe is lowered with the aid of the bit guide,
– the drill pipe is damped and held off the bottom, the wireline cone penetrometer is passed through the drill pipe and locked into the drill bit with the cone rod attached,
– the cone is forced to full penetration by repeated jack strokes, while the cone point and sleeve-friction resistance are recorded by an electronic readout system at surface,
– the cone and rod are removed, and a hole is drilled down to penetration level of the cone. The unit is ready to repeat cone test cycle.

VI.1.10 Use of modular penetrometers and interpretation of the recordings

The modular penetrometer is at present the best-adapted device for measuring the mechanical characteristics of sands and consolidated clays.

VI.1.10.1 Possibilities and limits of use of modular penetrometers

In the case of loose or dense sands, only the penetrometer enables any significant measurement of the shear strength to be achieved.

The modular penetrometer is probably the in situ measuring device which results in the least disturbance of the soil. However, in sands of medium density ($\gamma_d \approx 1.5$ to 1.6 t/m³), the continuous operation static penetrometer compacts the terrain beneath the cone, thus indicating excessively high values of cone resistance.

The use of modular penetrometers is limited by:

– the excessively loose nature of the sea bottoms; the frame sinks into the soil if the cohesion of the surface sediments is under 2 or 3 t/m² (20 or 30 kPa),
– the presence of banks of pebbles and indurated or highly dense sand zones where $R_p > 5{,}000$ t/m² ($R_p > 50$ MPa).

VI.1.10.2 Interpretation of the penetrometer plots

In sands, the tip resistance R_p increases very rapidly with the dry density γ_d (or the relative density D_r (³)). Figure VI.1.9 gives an example of the mean variations of R_p in terms of γ_d and D_r for a sand.

(³) It is recalled that the relative density is defined by formula:

$$D_r = \frac{e_{\max} - e}{e_{\max} - e_{\min}}$$

where e designates the voids index of the sand (see Section I.1.1.2).

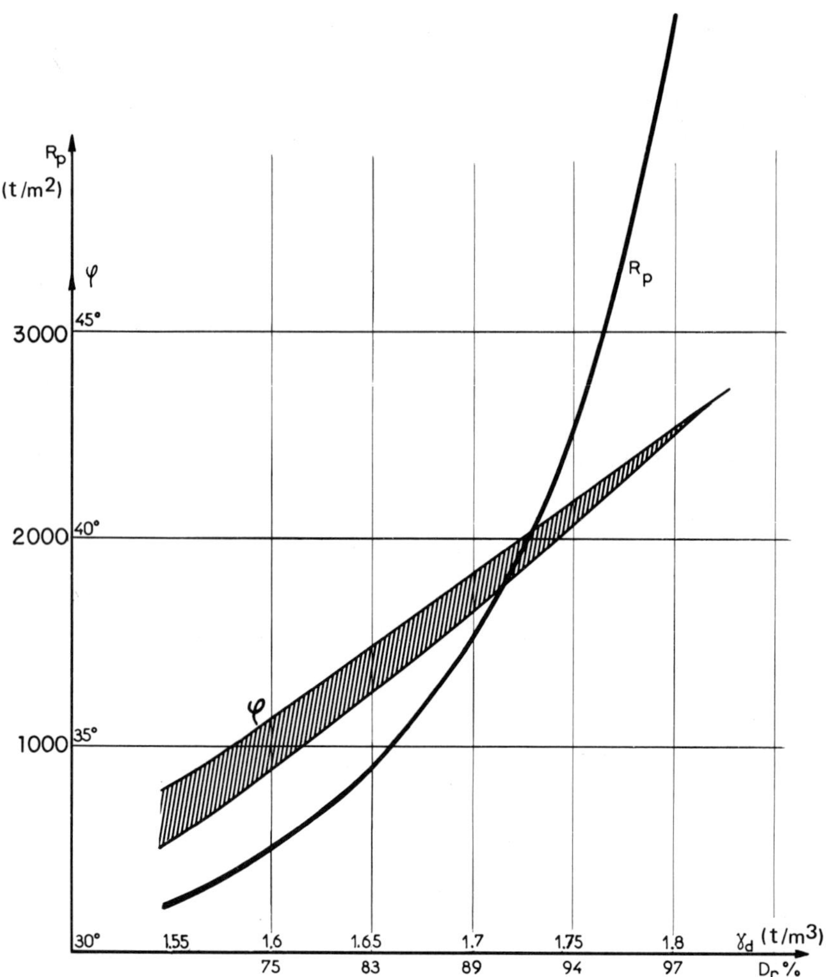

FIG. VI.1.9. – Angle of friction and tip resistance of penetrometer in terms of the density γ_d and relative density D_r of a sand.

The tip resistance for the dense sands of the northern part of the North Sea, which have relative densities approaching 90 to 100%, lies between 700 and 5,000 t/m² (7 and 50 MPa).

In clays, the undrained cohesion c_u lies between:

$$c_u = \frac{R_p}{10} \text{ to } \frac{R_p}{20}$$

depending on the consolidation.

According to experiments made on various Scandinavian sites, the shear strength s_u

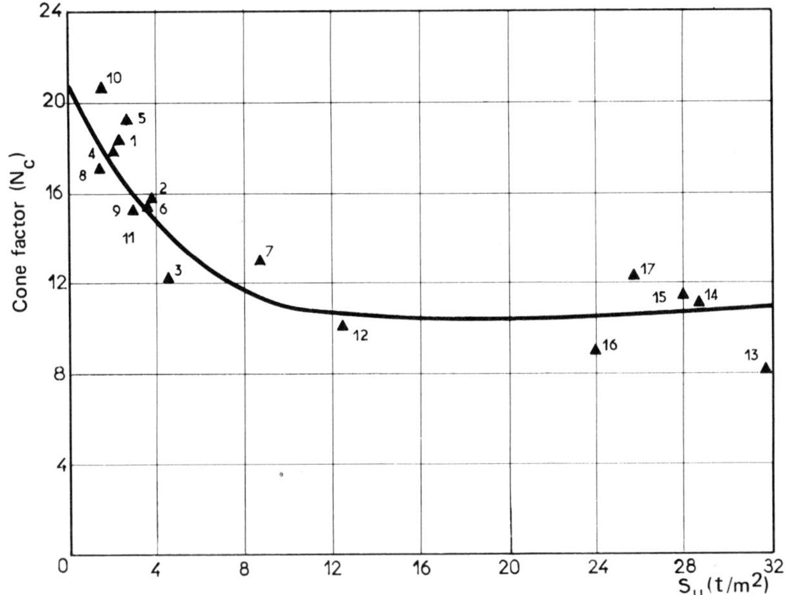

FIG. VI.1.10 – Correlation between the shear strength measured on a vane (scissometer) and the cone factor N_c in various Scandinavian clays (according to Ove Eide).

would be given by (Fig. VI.1.10):

$$s_u = \frac{R_p - \gamma . z}{N_c}$$

where:

$N_c \approx 8$ to 12 in stiff clays where s_u varies from 5 to 30 t/m² (50 to 300 kPa),
$N_c \approx 12$ to 20 in soft clays.

The penetrometer recordings also generally make it possible to identify the thin strata of soils from the values of the ratio $\frac{f}{R_p}$.

If the strength $R_p > 300$ t/m² ($R_p > 3$ MPa), the classification of soils given in Table VI.1.1 is generally accepted (according to Begemann).

TABLE VI.1.1

$\frac{f}{R_p}$	Nature of soils
Between 0.6 and 2%	Sands and gravels
Between 2 and 4%	Mixtures of sands + silts
Between 4 and 8%	Clays

The interpretation of some very marked "peaks" on the plots of R_p often appears a very delicate matter. In some cases, these "peaks" can be explained by the erratic presence of pebbles in front of the cone of the penetrometer and must be neglected in the interpretation.

VI.2 CABLE OPERATED STATIC AND DYNAMIC PENOTROMETERS

Two cable operated penetrometers, the operating principle and implementation of which differ considerably, are at present used or being tested:

– first, the **Wison static penetrometer** developed by *Fugro,* operated by wireline through the drilling string [1]; the operating priciple and implementation of this device, which is used in particular in the North Sea, are set forth in the present section,

– second, the **dynamic penetrometer** now being tested at the University of Newfoundland (Canada), implemented by means of a "Kullenberg type" releasing device [10]. The principle of the device is described briefly at the end of the present section.

VI.2.1 Principle and description of the Wison wireline static penetrometer operated through the drilling string

VI.2.1.1 Operating principle

The Wison [1] [2] is a penetrometer (with an electric cone), operated by wireline, lowered through a 4 or 5″ drilling string (Fig. VI.2.1).

The 8 1/2″ (approximate) insert drilling bit that can be picked up by wireline is similar to that used for wireline coring.

The penetrometer proper consists of:

– a system for locking the device into the lower part of the drilling string; this locking system is operated by wireline,

– a hydraulic jack controlled from the surface, with a capacity of 2 to 3 t, depending on the conditions of use of the device. The piston of the jack has a maximum stroke of 2 m and carries the penetrometer cone. The cone is retracted manually after raising the device to the surface by wireline.

VI.2.1.2 Surface equipment

The surface equipment needed to operate the wireline penetrometer comprises:

– first, the drilling equipment,

– second, the power, control and measuring devices of the penetrometer.

The drilling equipment commonly used consists of:

– a rotating tong for driving the 5″ drilling string,

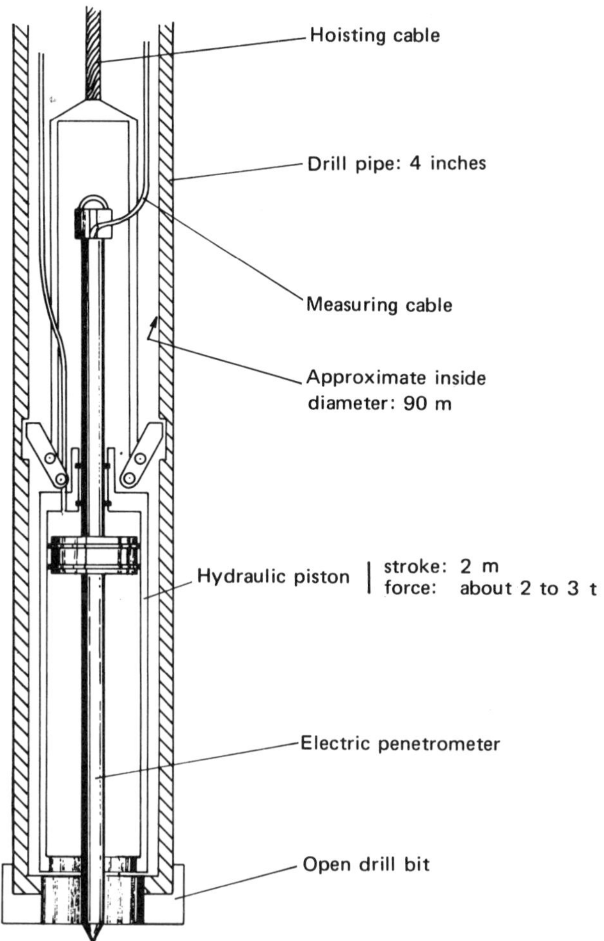

FIG. VI.2.1. – Functional diagram of the Wison cable penetrometer (*Fugro*).

– the drawworks,
– an open circuit mud circulation pump and the mud tanks.

The specific equipment for the penetrometer comprises:

– a power winch for the insert bit and the penetrometer,
– a central hydraulic power plant for actuating the jack,
– a recording device connected by electric cables to the cone of the penetrometer.

VI.2.2 Characteristics of the wireline penetrometer operated through the drilling string

VI.2.2.1 Drilling characteristics

Clearly, the drilling characteristics are identical to those indicated for the wireline drilling and coring method (see Section V.1).

The **weight-on-bit** resulting from the drill-collar and drill pipes is a maximum of approximately 5 to 6 t.

The **rotating speed** of the drilling string is generally 60 to 80 rpm.
The **torque** developed at 60 rpm is about 900 m × kg.

The **circulating pump has a flow-rate** of about 500 l/min.
The density of the drilling fluid is generally from 1.05 to 1.10, depending on the nature of the formations crossed through.

VI.2.2.2 Characteristics of the penetrometer

The penetrometer rod deploys telescopically:

– diameter of rod = 36 or 25 mm,
– penetration of rod into soil below the bottom of the boring = about 2 m.

The normal dimensions of the penetrometer core are:

– cone angle = 60°,
– diameter of cone = 36 mm (cross-section 10 cm^2) or 25 mm (cross-section 5 cm^2),
– lateral friction measuring sleeve = area 150 cm^2.

The maximum reactive force that can be applied may vary from 2 to 3 t.
This reactive force is developed only after a minimum penetration of the drilling rods into the soil of 5 to 10 m.

VI.2.2.3 Characteristics of use of the penetrometer

The combination of wireline drilling and penetrometry makes it possible to obtain a discontinuous profile of the mechanical properties of the formations:

– the penetrometer cone is driven into the soil at a speed of about 1 m/min (2 cm/s) in layers which are only of low and medium consolidation,
– drilling makes it possible to cross through hard layers where it would be impossible for the cone to penetrate.

Drilling prior to penetration of the point of the penetrometer results in considerable disturbance of the terrain over a depth of 0.5 to 1 m ahead of the drilling tool. This disturbance, caused by the circulation of the drilling fluid and relaxation of the stresses at the bottom of the borehole partially modifies the penetrometer profiles recorded.

Maximum penetration is limited only by the drilling possibilities. The procedure thus makes it possible to cross any hard strata encountered: pebbles, dense sands, indurated zones, etc.

Implementation of the wireline penetrometer consists in repetition of the following operations:

– drilling the borehole down to the depth selected for recording the penetrometer profile,
– wireline pick-up of the insert bit,
– lowering the Wison wireline penetrometer,
– penetration of the cone into the terrain (over a depth $\approx$ 2 m) under the action of the remote-control hydraulic jack,
– reraising the penetrometer to the surface and retraction of the rod,
– reboring to a depth of $\geqslant$ 2 m.

VI.2.3 Offshore support and implementation of the Wison wireline penetrometer

VI.2.3.1 Types of vessel

Implementation of the wireline penetrometer is in all respects comparable to that of wireline coring and thus takes place from the same supports.

So far, the wireline penetrometer has been implemented only by *Fugro* from several drilling vessels (Surveyor, Mariner, Ferder, . . .), the principal characteristics of which are indicated in Paragraph VI.1.3.

VI.2.3.2 Bottom-to-surface link and positioning tolerance

The bottom-to-surface link consists of the drilling string.

This type of rigid link requires strict positioning of the vessel (to the nearest 1 or 2 m).

VI.2.3.3 Space, equipment for implementation and manpower necessary

Use of the wireline penetrometer requires the same clear space on the vessel as for drilling and wireline coring i.e. about 150 m^2.

The equipment needed to carry out wireline penetrometry is briefly described in Paragraph VI.2.1.2.

Implementation of the wireline penetrometer calls for:

– two drillers and two workers for preparing the mud,
– two technicians for control, monitoring and operation of the penetrometer.

VI.2.4 Oceanographic conditions

VI.2.4.1 Water depth

The common depth of use of the wireline penetrometer is that of drilling and wireline coring in the North Sea, i.e. 100 to 150 m.

The maximum depth with the present support vessel anchored to the bottom could reach 180 m.

The method would appear applicable in depths of several hundred metres, from vessels equipped for dynamic positioning.

VI.2.4.2 Waves and current

Drilling implies a wave amplitude of less than 2 m.

The penetrometer cannot be used if the amplitude is more than 1 to 1.5 m without disturbing the recording of the penetrometer profile owing to heave.

It would appear possible to anchor the drilling vessels used in currents of up to 3 to 4 knots.

VI.2.5 Use of the Wison wireline penetrometer and obtained results

Paragraph VI.1.5 gives a number of indications on how to interprete the penetrometer obtained results. The inherent characteristics of the wireline penetrometer should however be specified.

VI.2.5.1 Possibilities and limits of use of the Wison wireline penetrometer

The wireline penetrometer:

– makes it possible to attain considerable penetrations, since indurated strata can be crossed by drilling,

– also makes it possible if required to carry out wireline coring at certain levels ([4]).

The use of the Wison wireline penetrometer is however limited by:

– the relatively low reactive force (2 to 3 t) ([5]), thus making penetration impossible when

(4) However, it does not appear that this method is normally used.

(5) The construction of a device fitted with packers will be able to provide a reaction of at least 5 t.

starting drilling or crossing layers where the cone resistance exceeds 1,500 to 2,000 t/m^2 (15 to 20 MPa),

– disturbance caused by drilling, particularly in sandy sediments.

VI.2.5.2 Geotechnical appraisal of the obtained results by the Wison wireline penetrometer

Interpretation of the charts must allow for the possibility of disturbance caused by the prior drilling of the layer. This disturbance is caused:

– first, by circulation of the drilling fluid, particularly in permeable formations (sands or silts); the use of viscous mud as drilling fluid will however alleviate this deterioration,

– second, by the relaxation of the stresses around the borehole.

Yet this disturbance often seems limited to a small depth (< 10 to 20 cm) (Fig. VI.2.2).

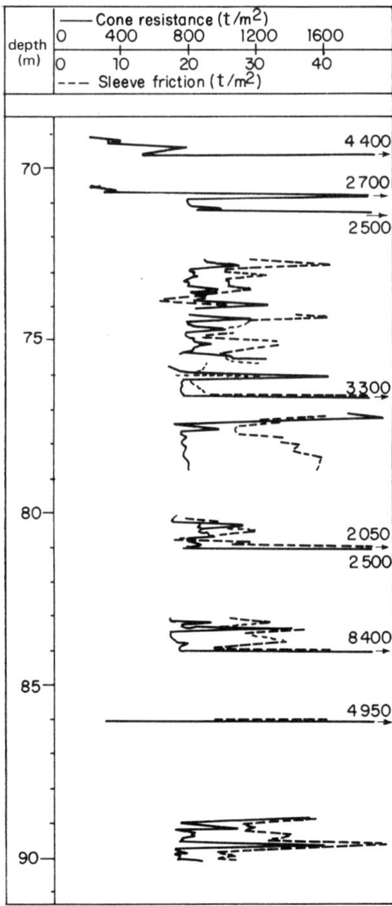

FIG. VI.2.2. – Example of logs plotted with the Wison penetrometer.

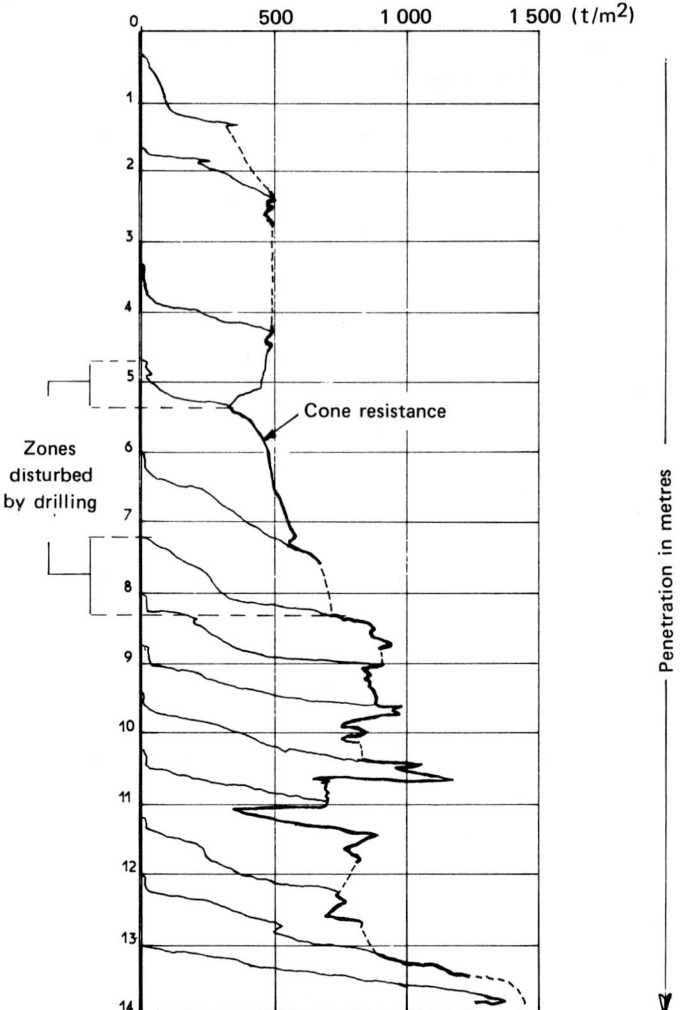

FIG. VI.2.3. – Tip resistance measured using an IFP penetrometer with a telescopic cone (effect of drilling on strength R_p).

However, charts recorded with a different device have shown a disturbance involving a depth of at least 0.50 m (Fig. VI.2.3.).

The penetrometer profiles obtained are of essentially discontinous character:

– first, penetrometry cannot generally be accomplished at all levels, particularly in highly consolidated soils,

– next, disturbance practically eliminates the initial part of the recording of each sequence.

For this reason prudence must be exercised in interpreting the profiles recorded.

VI.2.6 The dynamic penetrometer released by means of a Kullenberg type device

VI.2.6.1 Operating principle and dimensions of the dynamic penetrometer

The principle of the dynamic penetrometer, built and tested in the University of Newfoundland (Canada) is identical to that of a Kullenberg corer with the core barrel replaced by a rod equipped with a penetrometer cone [10].

The dimensions of the cone are those of common static penetrometers, namely:

- cone:
  - cone angle = 60°,
  - diameter of cone 36 mm (cross-section 10 cm^2),
- lateral friction measuring sleeve = area 150 cm^2.

The length of the rod used obviously varies with the nature of the formations encountered :

- 2 to 3 m in sands,
- 10 to 15 m in very soft soils.

VI.2.6.2 Recordings and results obtained

The device enables the following to be recorded simultaneously during penetration:

- acceleration (from which the velocity profile can be deduced),
- the dynamic cone resistance,
- the dynamic lateral friction resistance.

The results obtained by this device:

- are relative in nature, but,
- may be compared, thanks to empirical relationships, to the results derived from static penetrometry.

VI.3 MENARD PRESSUREMETER TECHNIQUES AND INTERPRETATION OF PRESSUREMETER MEASUREMENTS

The Menard pressuremeter is a device which was initially designed for studying the mechanical properties of soils ashore. The utilization techniques of the device, now developed for several years, enable it to be used likewise for reconnaissance of soils at sea [11] [12].

VI.3.1 Principle and description of the Menard pressuremeter

VI.3.1.1 Principle of the pressuremeter and the pressuremeter test

The Menard pressuremeter essentially consists of (Fig. VI.3.1):

– a pressuremeter probe (with cylindrical expansion) lowered into a borehole from the end of a drilling string; the probe follows the radial deformations of the borehole according to the pressure applied against the walls,

– a probe measuring and pressurizing device set up on the deck of the vessel and connected to the probe by means of concentric flexible piping. This device enables the variations in volume of the probe with pressure to be recorded.

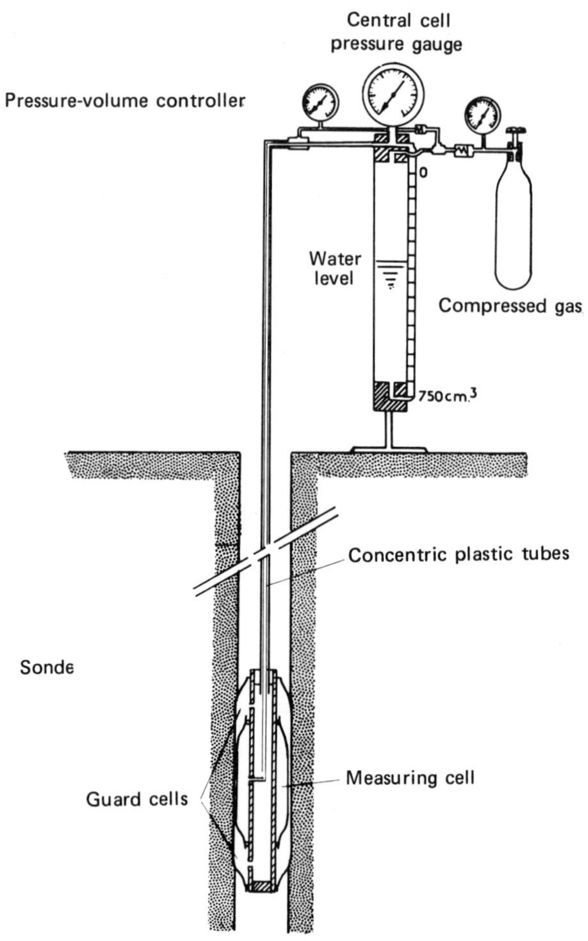

FIG. VI.3.1. – Functional diagram of Menard pressuremeter.

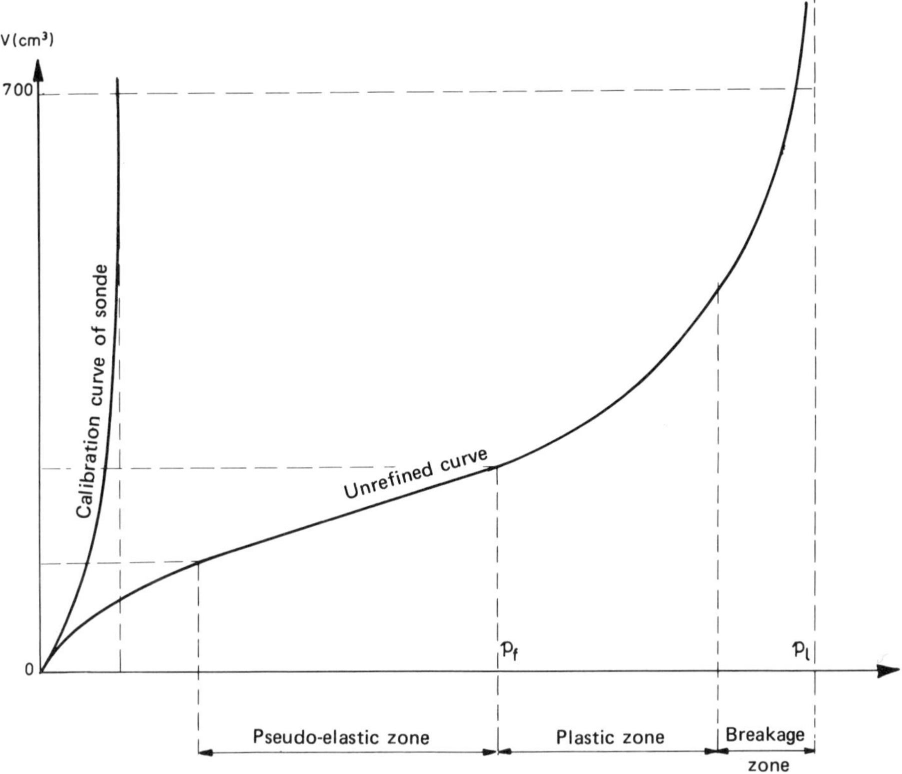

FIG. VI.3.2. – Typical pressuremeter diagram.

The pressuremeter test provides a stress-strain relationship of the soil in place under the conditions of planar deformation.

The essential characteristics needed for designing the foundations are then obtained by analysing the "load-shift" curve of the terrain (see Fig. VI.3.2). These characteristics are:

– the modulus of deformation, upon which settlement depends,
– the maximum pressure related to the bearing capacity of the terrain,
– the creep pressure.

VI.3.1.2 Construction of the pressuremeter probe

The pressuremeter probe comprises three expansion cells (Fig. VI.3.1):

– one central main cell for measurement, inflated with water at a pressure p,

– two guard cells at the end of the probe independent from the main cell and inflated by air at a pressure of 0.1 to 1 kg/cm² (10 to 100 kPa) below p.

This arrangment theoretically makes it possible to ensure that the radial stress field at the main measuring cell is uniform.

Various types of pressuremeter probes (type E, G, GB, etc.) have been built in turn. The "G" metal sheath probe is practically the only one to be used at sea. It consists of a steel tube surrounded:

– by an inner rubber membrane forming the central measuring cell,

– by an outer membrane, also in rubber, forming the two guard cells and covered with metal blades in the form of an umbrella.

The inertia correction for this probe, which is calibrated before being positioned in the borehole, is about 2 kg/cm^2 (200 kPa) for the maximum expansion usually allowed.

VI.3.2 Characteristics of the Menard pressuremeter

VI.3.2.1 Characteristics of pressuremeter probes

The dimensions of the probes, adapted to match diameters commonly used in civil engineering boreholes, are characterized by their nominal diameter. In practice, the two most commonly used probes are the following models:

AX, with a diameter of 44 mm,
BX, with a diameter of 58 mm.

The total length of the probe is about 60 to 70 cm.
The length of the central measuring cell is about 20 cm.

The maximum **pressure and expansion** of the probes depends on the model used. With the metal sheath probes which are generally used at sea:

– the maximum pressure applied is about 25 kg/cm^2 (2,5 MPa),
– the maximum expansion corresponds to a volume of about 700 cm^3.

VI.3.2.2 Characteristics of use of the pressuremeter

The utilization techniques of the pressuremeter probe depend on the nature of the soils, the depth of water, etc. (see Section VI.3.3). At sea, the probe is generally positioned:

– either by vibrodriving from the end of a small diameter drilling string,
– or by lowering it by cable through the drilling string,
– or by ordinary driving.

The pressuremeter makes discontinuous measurements of the mechanical characteristics of soils, generally at intervals of 1 to 2 m (whereas the mean length of the measuring cell is about 20 cm).

The values measured by the pressuremeter are partly dependent on the conditions of location of the probe, owing to the varying degree of disturbance of the soil (see Section VI.3.5).

The normal pressuremeter test has to be made in a series of ten successive pressure steps of about the same value until the maximum pressure is reached (a tolerance of 6 to 14 steps is accepted).

The readings of volume with time are made respectively at intervals of 15s, 30s and 1 min after pressurizing the probe at each step.

The normal duration of the pressuremeter test at each sounding level is therefore about 15 min.

The pressuremeter test is **an undrained shear** of the soil. The duration of the standard test (1 min per pressure level) obviates any dissipation of the interstitial pressure resulting from insertion of the probe.

The depth of penetration obtained is largely dependent on the conditions of implementation:

- usual depth = 20 to 50 m,
- maximum depth = 60 to 80 m.

VI.3.3 Offshore support for implementing the pressuremeter

While the operating mode for the pressuremeter test is perfectly defined, implementation techniques have considerably changed with the increasing depths of water, the development of the vibrodrivers used to cause the probe to penetrate, etc.

The implementation techniques of the pressuremeter probe will be described on the basis of:

- a guide tube,
- the plumb line method,
- the annular vibrodriver,
- the Kullenberg system.

VI.3.3.1 Types of vessel

In a calm sea and depths of water of a few tens of metres, the pressuremeter can be used from a vessel 40 to 50 m in length (supply-boat or other type vessel) providing it offers the following possibilities:

- 4-point anchoring, with each anchor moored by an independent winch,
- loading mast capable of hoisting 2t,
- clearance beneath hook of 10 m,
- clear deck space of 30 to 50 m^2 depending on the implementation technique used.

For operations well out to sea, in depths of over 60 to 70 m, it is indispensable to have a highly stable vessel at least 50 to 60 m long and equipped with the necessary hoisting facilities.

The implementation of the pressuremeter during exploratory drilling from jack-up platforms or semi-submersible platforms should be the aim as far as possible. Indeed, the stability of the support facilitates the operations and enables the quality of the measurements made to be improved.

VI.3.3.2 Implementation of the pressuremeter probe by means of a guide tube

In this method, the bottom-to-surface link is **rigid** and consists of a guide tube (pipe conductor) 85 mm, 4 1/2" or 7" in diameter, applied from a "lateral" platform (i.e. secured to the side of the ship) and inserted into the terrain to an average depth of 7 to 8 m (Fig. VI.3.3).

The pressuremeter probe insertion rods running through the guide tube are added as the instrument penetrates into the formation.

Various methods of penetration (vibrodriving, preliminary drilling or ordinary driving) are used, depending on the consolidation of the formations. Prior drilling with a rotary vibrohammer often requires bentonite grouting to enable the walls of the borehole to hold.

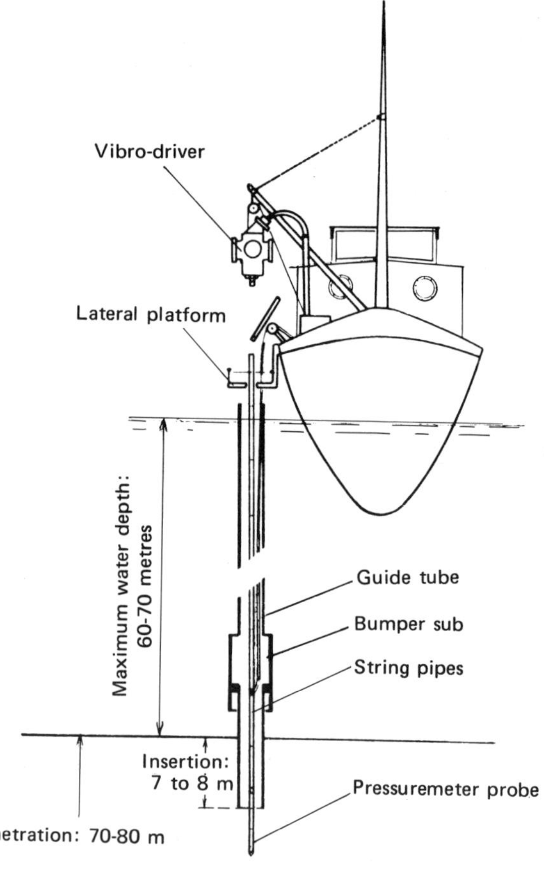

FIG. VI.3.3. – Application of the presssuremeter probe using a guide tube.

A bumper sub is needed to offset the vertical oscillations of the vessel.

This method, which is difficult to apply when the depth of water exceeds 60 to 70 m, calls for perfect anchoring of the vessel, any movement exceeding 2 to 3 m resulting in bending of the rods, breakage of the probe-to-measuring cell flexible connecting lines and the impossibility of recovering the equipment.

The maximum penetration that can be obtained by this method is about 70 to 80 m.

VI.3.3.3 Implementation of the pressuremeter probe by the "plumb line" method

This method, which is becoming less and less commonly used, can only be applied by vibrodriving the pressuremeter probe into the soil.

The bottom-to-surface link is **semi-flexible** and consists of (see Fig. VI.3.4):

– first, the flexible connection between the vessel and the vibrodriver attached to the top of the drilling string,
– second, the drilling string as it penetrates into the soil.

The vessel-to-vibrodrivers links consist of:

– the bearing cable of the vibrodriver (weighing about 450 kg) and the drilling string (weighing 4 kg/m for a diameter of 44 mm),
– the flexible tubes between the central hydraulic power plant (on the vessel) and the vibrodriver,
– the flexible tubes linking the pressuremeter probe to the measuring cell on the surface.

The application of this method is limited by the danger of the drilling string buckling, and penetration is generally no more than 40 to 50 m.

The motion of the vessel must be restricted to a few metres at the most since otherwide the drilling string would penetrate at an angle.

Correct vertical penetration of the drilling rods can be ensured by means of an inclinometer.

VI.3.3.4 Implementation of the pressuremeter probe by means of the annular vibrodriver

At present, development work is now going on to replace the vibrodriver mounted on the end of the drilling string by an annular vibrodriver, for operations at sea ([6]).

(6) The developement of the annular vibrodriver, which has a rotary capability, should also make it possible to replace to advantage certain methods used previously, such as that of the self-stable frame resting on the bottom (Fig. VI.3.5) to implement the low penetration pressuremeter in depths of water of 100 to 200 m or more.

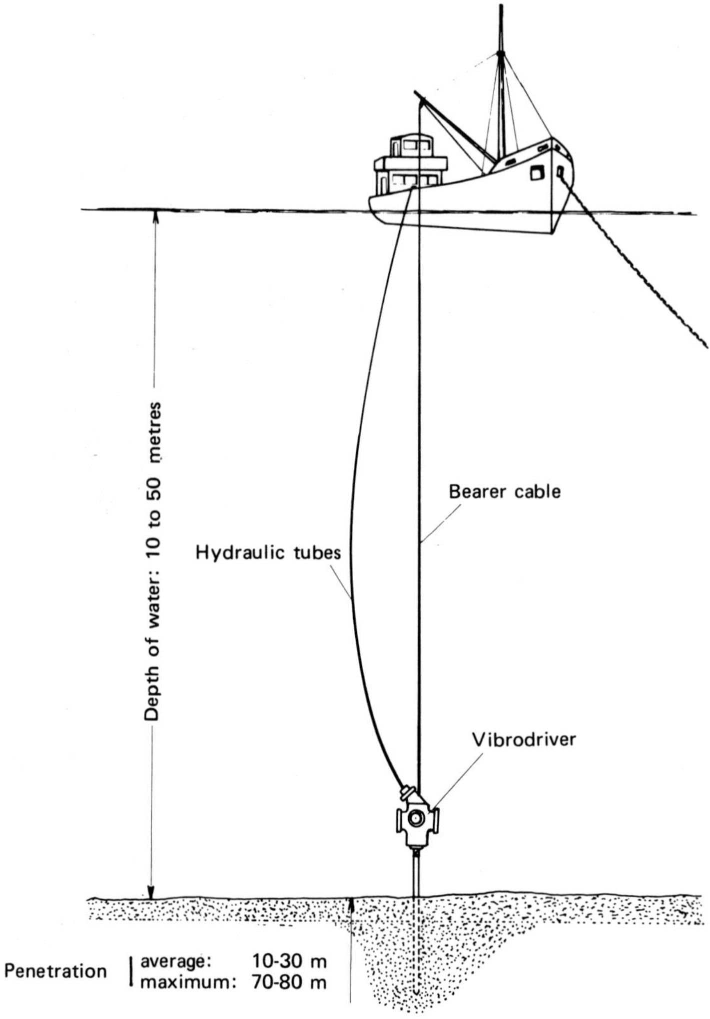

FIG. VI.3.4. – Implementation of the pressuremeter probe using the plumb line technique.

The bottom-to-surface link is **flexible** and consists (see Fig. VI.3.6) of:

– the carrying cable of the annular vibrodriver (weighing about 1 t) and the drill pipe,

– the flexible hoses between the central hydraulic power plant and the annular vibrodriver,

– the flexible hoses (running through the drilling string) connecting the pressuremeter probe to the measuring cell on the surface.

A float would theoretically enable the drilling string to be held approximately vertically provided the speed of the currents is low.

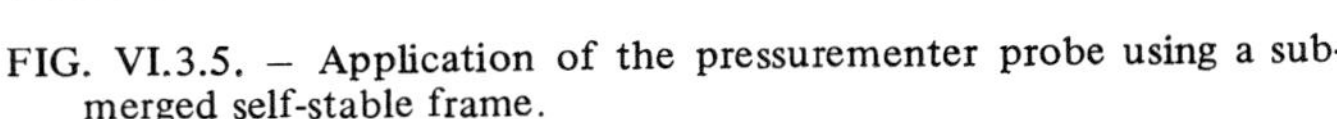

FIG. VI.3.5. – Application of the pressurementer probe using a submerged self-stable frame.

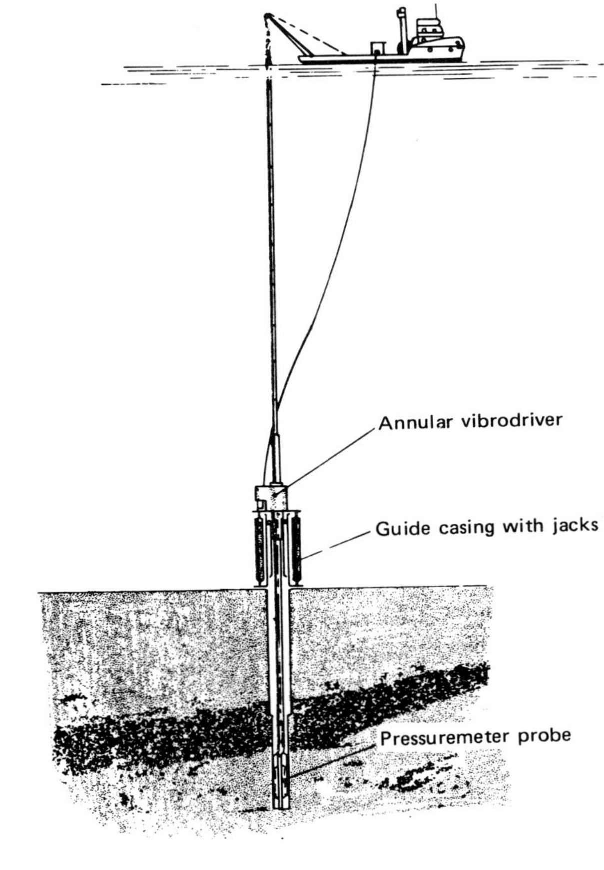

FIG. VI.3.6. – Application of the pressuremeter probe using the annular vibrodriver.

The annular vibrodriver is held about 2 to 3 m above the sea floor by means of a guide tube equipped with 2 cylinders. Penetration occurs in successive stages.

Thanks to the flexible "bottom-to-surface" links, the vessel can move a few metres.

VI.3.3.5 Implementation of the pressuremeter probe by the Kullenberg coring technique

The Kullenberg technique (release from a predetermined height above the sea floor of a tube which penetrates into the ground by gravity) is applicable to the pressuremeter (as for the dynamic penetrometer, see Section VI.2.6).

The Kullenberg corer is replaced by a pressuremeter probe surmounted by a 2 to 6 m long tube and connected to the ballast and the ring of the trigger.

The pressuremeter test is made with or without anchoring the vessel, which can move slightly without influencing the results of the operation.

VI.3.3.6 Power and manpower required

The central hydraulic power plant for supplying the vibrodriver is driven by a Diesel engine.

The use of pressuremeter techniques therefore calls for practically no electrical power.

The use of these techniques does however require:

- at least two to three technicians for assembly and disassembly of the drilling string,
- one technician for making the measurements.

VI.3.3.7 Time needed to carry out a reconnaissance by a pressuremeter

It is assumed that the vessel has previously been anchored and that favourable weather conditions prevail.

Reconnaissance of the soil by means of a pressuremeter involves the following operations :

- lowering the drilling string (if necessary after insertion of a guide tube),
- penetration of the probes,
- execution of the pressuremeter test at intervals of 1 to 2 m,
- raising the probe and rehoisting the equipment inboard.

The average duration of a pressuremeter test at any given sounding level is about 15 min.

The approximate duration of reconnaissance at a depth of 50 m at intervals of 1 m is about 24 h.

VI.3.4 Oceanographic conditions

VI.3.4.1 Water depth

The maximum depths compatible with application of pressuremeter techniques depend on the method of implementation used, namely the type of bottom-to-surface link.

In the case of a rigid link (guide tube method), reconnaissances have been made:

- commonly in depths of 60 to 70 m,
- more rarely from 80 to 100 m.

In the case of a semi-flexible link (plumb line method), the depths have varied:

- commonly from 30 to 40 m,
- fairly rarely from 70 to 80 m.

In the case of a flexible link as for:

- the self-stable frame or "Capsub" set on the bottom and carrying a vibrodriver,
- the annular vibrodriver held near the bottom,

the depth of water can attain 200 to 300 m.

VI.3.4.2 Waves and current

Both with respect to taking measurements and implementing pressuremeter techniques, a support of good stability is needed, generally implying wave ampitudes of under 2 m.

The anchoring of the vessels and the stability of the equipment near the bottom imply a current speed of less than 2 to 3 knots.

VI.3.5 Use of the pressuremeter and interpretation of its measurements

VI.3.5.1 Possibilities and limits of use of the pressuremeter

The conditions of location of the pressuremeter probe are a factor of prime importance as regards the quality of the measurements made.

In silts and consolidated clays, execution of a borehole prior to introduction of a pressuremeter probe results in minimum disturbance of the soil, and the test can be conducted under satisfactory conditions.

In muds or soft clays, the penetration of the probe by vibrodriving or hammer-driving is accompanied by a "punching" effect of the terrain which may result in reduction of the shear strength. However, in view of the difficulties of sampling representative cores, the pressuremeter test still makes it possible to attain a reliable indication as to the properties of the soil.

In sands (saturated), the vibrodriving of the probe is accompanied by disorganization of the structure of the terrain near the walls, generally resulting in more or less accentuated packing, depending on the density in place. It should be noted that the penetration of the probe by vibrodriving into sand will be highly limited.

VI.3.5.2 Characteristics deducible from the pressuremeter diagram

The following can be deduced from the pressure-volume pressuremeter diagram (see Fig. VI.3.2):

– the pressure modulus of the soil E,
– the limit pressure p_l,
– the natural pressure of the soil at rest p_0,
– the creep pressure or elastic limit p_f,

The first two magnitudes constitute the fundamental characteristics of the soil. The last two are not involved in the common run of stability studies.

The pressure modulus E measured in a deviatoric stress field characterizes the pseudo-elastic phase of the test and is defined by the relationship:

$$E = K \cdot \frac{\Delta p}{\Delta v}$$

where:

K is a geometric constant of the pressuremeter probe.
Δp and Δv are concommitant variations of pressure and volume in the pseudo-elastic phase of the test.
It can be shown that:

$$K = 2.66\,(v_0 + v_m)$$

where:

v_0 is the volume of the measuring cell at rest ($v_0 = 535$ cm^3 for the AX and BX type probes,
v_m is the volume of water introduced to the measuring cell for the mean pressure, p_m applied.

To establish this relationship, the Poisson coefficient ν of the soil is assumed arbitrarily to be 0.33.
The modulus E plays a fundamental part in calculation of the settlement of foundations and the stability of piles under the effect of horizontal loads.

The limit pressure p_l by definition corresponds to the extreme state of rupture of the terrain subjected to a uniform and increasing stress flux against the wall of a cylindrical cavity. The value of the limit pressure p_l equals the abscissa of the asymptote of the pressuremeter curve.

This fundamental mechanical characteristic is involved in all calculations of the stability of foundations made using pressuremeter methods.

No simple theoretical relationship exists between the limit pressure p_l and the conventional mechanical characteristics of soils. However :

– in clays of normal consolidation, the following formula would serve an initial approximation:

$$p_l - p_0 = 5.5\, c_u$$

where c_u designates the undrained cohesion of the soil,

– in various formations, the following relationships would exist between p_l and the penetrometer cone resistance R_p :

$$\text{clays: } R_p/p_l = 2.5 \text{ to } 4$$

$$\text{silts: } R_p/p_l = 5 \text{ to } 6$$

$$\text{sands: } R_p/p_l = 7 \text{ to } 9$$

VI.3.5.3 Interpretation of pressuremeter diagrams

The form of the uncorrected pressuremeter curve (p, v) depends on the consolidation of the formation and the rigidity of the probe used.

It is not always possible to adapt the pressuremeter probe to suit the nature of the soils, particularly in the case of highly heterogeneous strata.

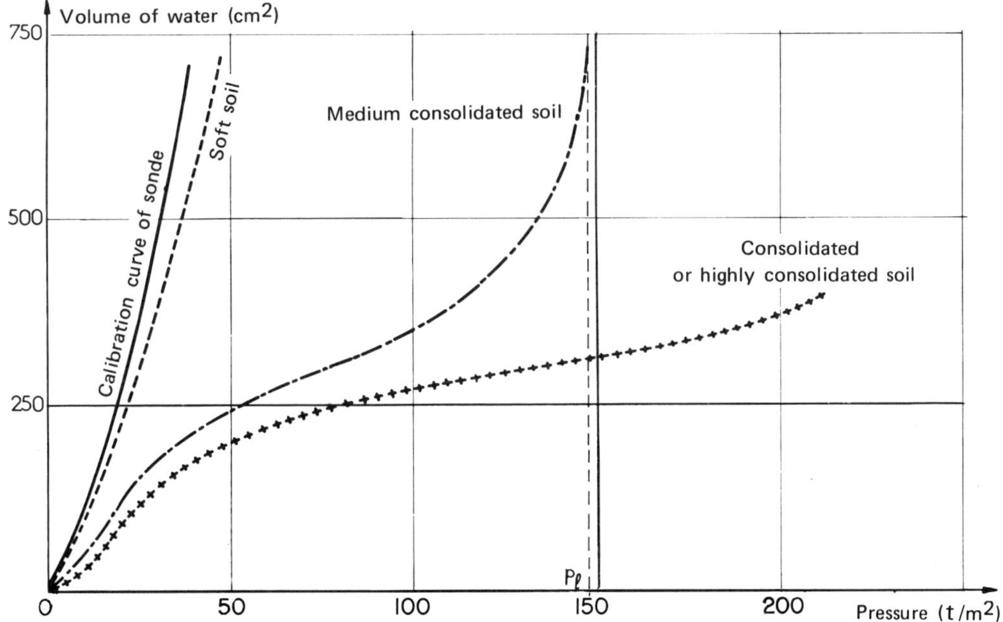

FIG. VI.3.7. – Different types of pressuremeter diagrams depending on the degree of consolidation of the soil.

In the case of the "metal sheath" probe (a typical calibrating curve of which is shown in Fig. VI.3.7), and used in a sounding through strata with widely differing mechanical characteristics, one obtains:

– in **soft soils**, a gross curve comparable to the calibration curve of the probe; this results in inaccuracy both in E and p_l (see Fig. VI.3.7),

– in **soils of medium consolidation**, a curve of conventional form; determination of E and p_l will be satisfactory,

– in **consolidated or highly consolidated soils**, a curve without any well defined asymptote; the limit pressure is poorly determined.

VI.3.5.4 Applications for pressuremeter results

The results of the pressuremeter test (limit pressure p_l and modulus E) are directly involved in the design of foundations by the *TLM* method. Figure VI.3.8 gives an example of diagrams of E and p_l as a function of depth, through strata of widely differing kinds.

The pressuremeter test determines only the mechanical properties of the soil in the horizontal direction:

– for calculating the bearing capacity and settlement of structures, it is therefore

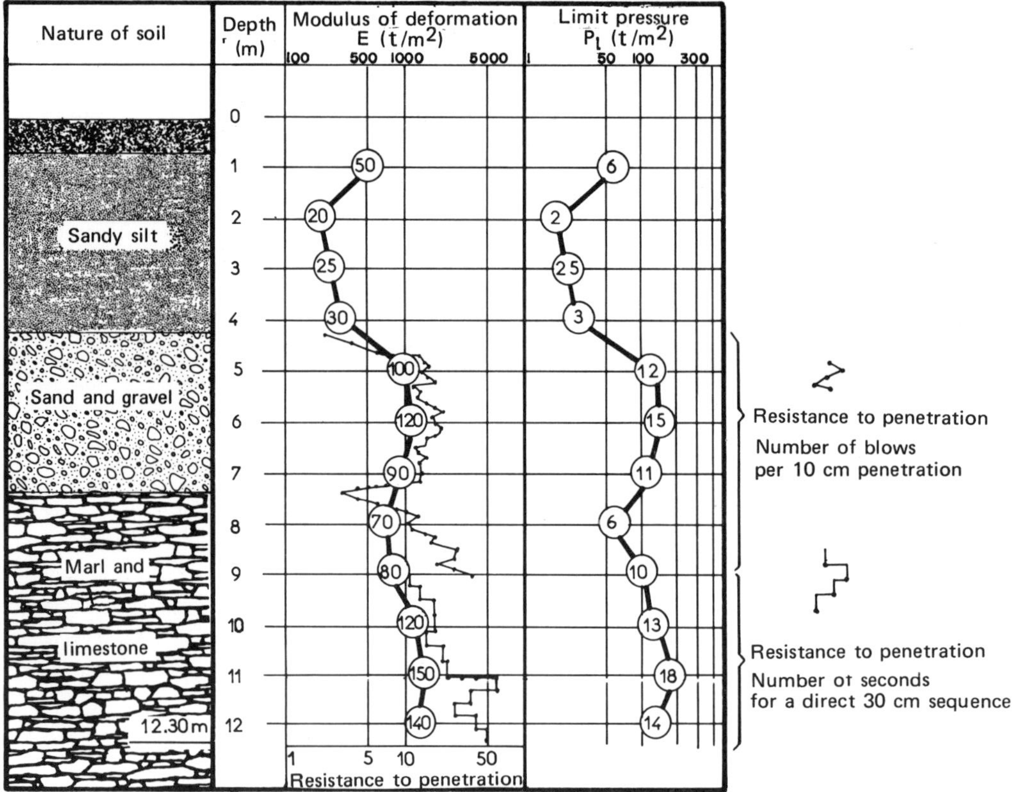

FIG. VI.3.8. – Examples of "limit pressure and modulus of deformation" diagrams.

necessary to allow for the thickness of the strata, their anisotropy and their heterogeneity,
– for calculating piles subjected to horizontal loads, the pressuremeter is probably the ideal method.

The pressuremeter is at present the only in situ measuring instrument capable of directly measuring the deformability characteristics of soils.

The pressure modulus (not directly linked to the oedometric or triaxial moduli) enables the settlement of the soil to be calculated.

VI.3.5.5 "Self-boring" pressuremeter

The characteristics of the "self-boring" pressuremeter of *Laboratoires des Ponts et Chaussées* (France), which is not yet usable at sea, have not been described in the present section [12 bis].

However, it should be pointed out that the implementation of the "self-boring" pressuremeter probe by means of a cutting tool, the diameter of which is equal to the initial diameter of the probe, enables the disturbance of the soil to be reduced to the minimum and the pressuremeter measurements thus to be made under ideal conditions.

The *Institut Français du Pétrole* is developing an applicable "self boring" device under three hundred metres (then one thousand metres) water depth.

VI.4 REMOTE VANE OPERATED BY WIRELINE THROUGH THE DRILLING STRING. IN SITU MEASUREMENT OF SHEAR STRENGTH

Of the various instruments for in situ measurement of the mechanical characteristics of the soils, only the vane directly yields the value of the undrained shear strength, Although widely used for reconnaissance of soils ashore, the vane is applied very little at sea owing to the difficulty in implementing it in depths of any magnitude. At present, the only units for reconnaissance of soils at sea are :

– the wireline vane (Remote Vane) developed by *McClelland Engineers Inc.*, implemented through the drilling string,
– the *CEL* vane designed for great depths and implemented by means of a frame set on the bottom, but with a penetration of no more than 3 m.

The present section describes only the operating principle and characteristics of use of the wireline vane, at the moment the only device of this kind offering any interest for the reconnaissance of soils to depths of a few tens of metres. The implementation, offshore support needed and oceanographic conditions are obviously identical to those for wireline penetrometry and coring (see Section VI.2).

Next, a number of indications are given on the comparison of the values of the shear strength deduced :

– first, from the vane test, considered as a reference,
– second, from other penetrometer and pressuremeter in situ tests.

VI.4.1 Principle and description of the wireline vane

VI.4.1.1 Principle of the wireline vane

The wireline vane, developed a few years ago by *McClelland Eng. Inc.* [13] [14] can be operated through the drilling string (Fig. VI.4.1a) (either alone, or in alternation with the wireline corer), from vessels fitted out for the purpose (7) (8).

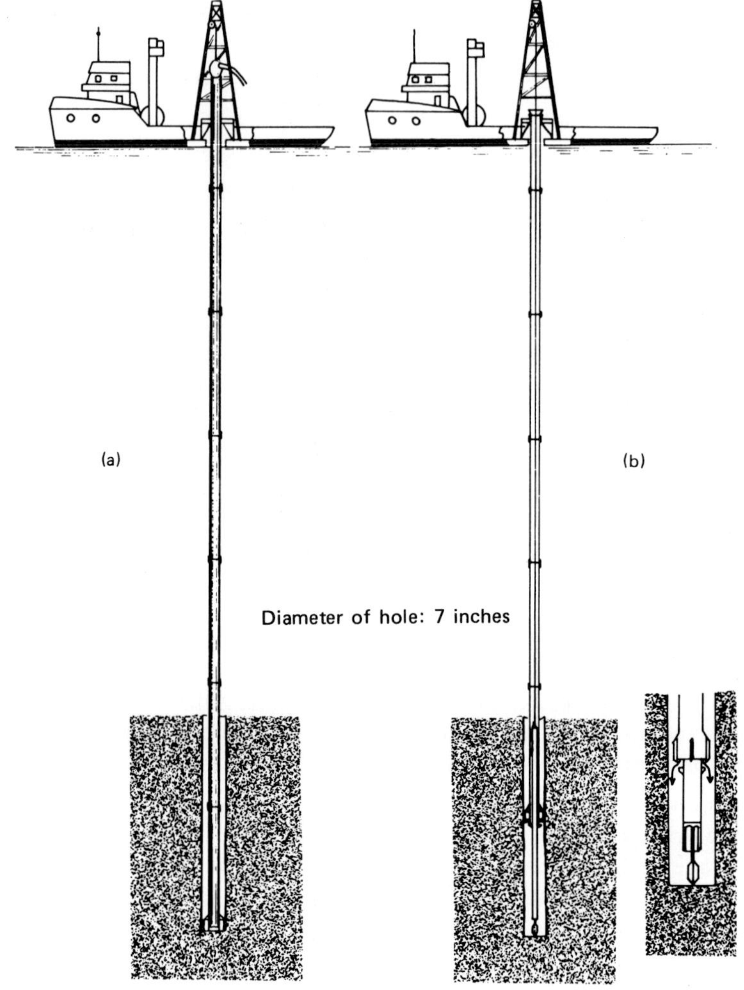

FIG. VI.4.1. – Implementation of the remote wireline vane (scissometer).

(7) The characteristics of the *McClelland* wireline vane have been gradually improved for use in the Gulf of Mexico in water depths of over 300 m with penetrations of up to 120 m. Analysis of obtained results in about 60 soundings has made it possible to establish certain correlations with other methods of measuring shear strenght [14].

(8) A wireline vane of the same type is sold by *Golder Associates* and by *Soil Instruments Ltd.*

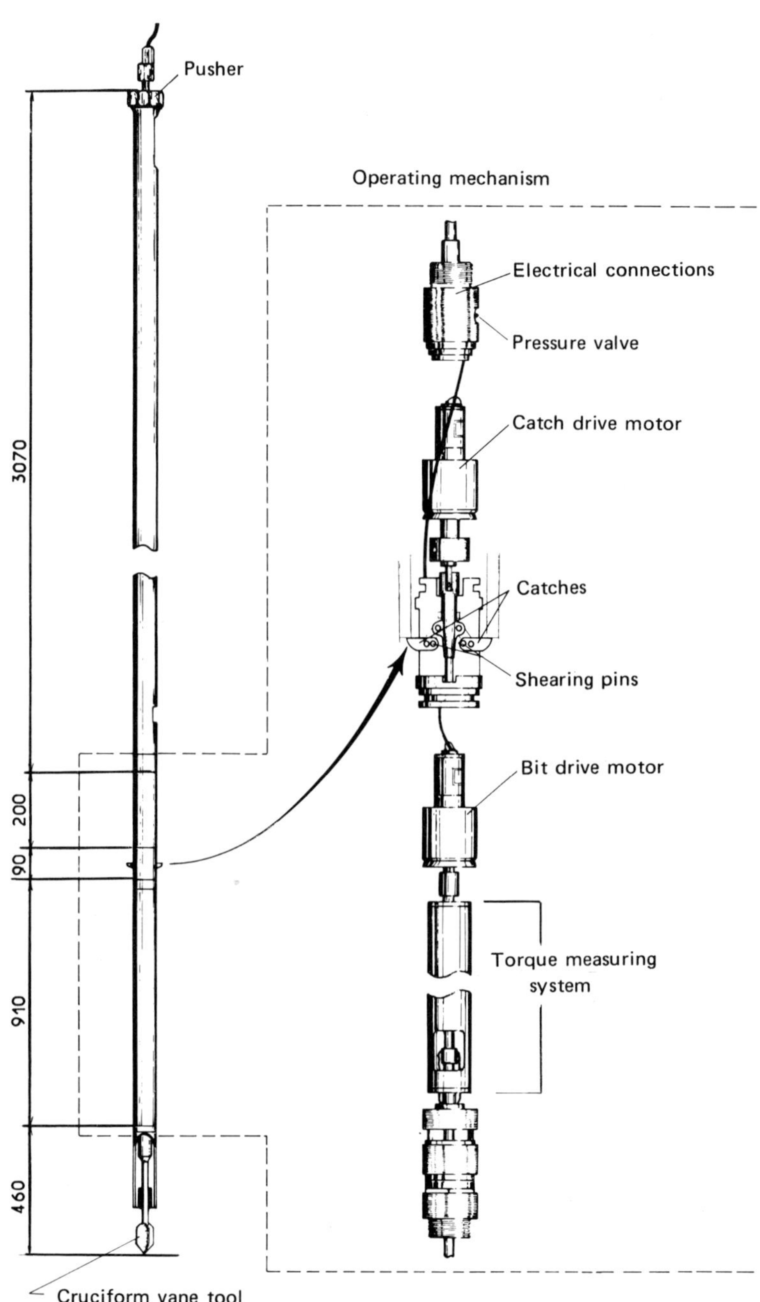

FIG. VI.4.2. – Diagram of McClelland remote vane (scissometer).

The borehole is sunk to about 0.60 m above the elevations selected for measuring the shear strength by means of the vane. After raising the drilling bit by about 2 m and picking up the insert bit, the vane is lowered by wireline through the drilling string (Fig. VI.4.1b). The vane tool, which consists of two rectangular blades (or fins) (Fig. VI.4.2) arranged in cruciform, is then rapidly pushed into the terrain to the desired depth by remote control from the surface. While making the shear test, the vane is detached from the drilling string. The soil is broken up by the rotation of the tool and the shear strength is deduced from the moment of torsion measured.

VI.4.1.2 Description of the wireline vane

The powering and operation of the vane are controlled by a 7-conductor cable. The device comprises two electric motors (Fig, VI.4.2):

– one operates a pawl system for driving the tool into the ground,
– the other rotates the measuring tool (vane).

The motor torque is transmitted to the vane via a steel rod, the torsion torque of which is proportional to the shear strength of the soil.

VI.4.1.3 Surface equipment

The surface equipment needed for implementing the wireline vane comprises:

– the drilling equipment,
– the operating, control and measuring devices of the vane.

The drilling equipment used by *McClelland Eng. Inc.* consists of:

– a Failing 1500 drilling rig,
– the mud circulation pump (open circuit) and mud tanks.

The specific equipment for the vane comprises:

– the "insert-bit" and vane operating winch,
– a remote-control console for the drive motors (penetration and rotation) of the tool,
– a recording device connected to the torque measuring system.

VI.4.2 Characteristics of use of the wireline vane

VI.4.2.1 Drilling characteristics

The drilling characteristics are those already indicated for drilling using the Failing 1500 equipment.

The weight on bit is the result of:

- the weight of the drilling string,
- an additional maximum load of 4 t applied by means of a hydraulic jack.

The **rotating speed** of the drilling string can vary from about 70 to 120 rpm.

The **flow-rate of the circulating pump** (on open circuit) varies from 300 to 500 l/min.

The drilling fluid consists of sea water and natural clay (with or without bentonite) with a density of 1.2 to 1.3 in low consolidation sediments.

VI.4.2.2 Characteristics of the vane

The vane consists of two identical rectangular blades (or fins) forming four straight dihedrons.

Three tools with the following approximate dimensions:

- radius $R \approx 1.5 - 2 - 2.5$ cm,
- height $h \approx 2$ to $3\ R$,

can be mounted on the device.

The maximum **torque** applied to the soil through the fins is 1.15 m × kg. This low value of the driving torque that can be applied to the vane governs the choice of the small size blades.

VI.4.2.3 Characteristics of use of the vane

The vane measures the mechanical characteristics of the soils discontinuously:

- the interval between measurements is normally 1 to 2 m,
- alternating running in of the vane and the corer permits reconnaissance at one and the same time of the nature of the formations and the degree of disturbance caused by coring.

The rapid insertion of the vane into the soil enables disturbance to be reduced but leads to uncontrolled increase in the interstitial pressure, particularly in soft and sensitive clays.

The rotating speed of the tool can vary from 0.14 to 0.89 degree per second. This speed is selected to bring about breakup of the soil within about 2 min.

The vane test is an **undrained shear test** of the soil. Thanks to the speed of the test, the interstitial pressure is not dissipated.

Because $c_u < 10$ t/m² ($c_u < 100$ kPa) (see Section VI.4.3.1), the maximum depth for reconnaissance using the wireline vane can rarely be more than a hundred metres (even in deltaic zones).

The torsion recorded in terms of the angle of rotation θ of the vane at a constant angular speed goes through a maximum, corresponding to the value of the undrained cohesion of the soil c_u (see Fig. VI.4.3).

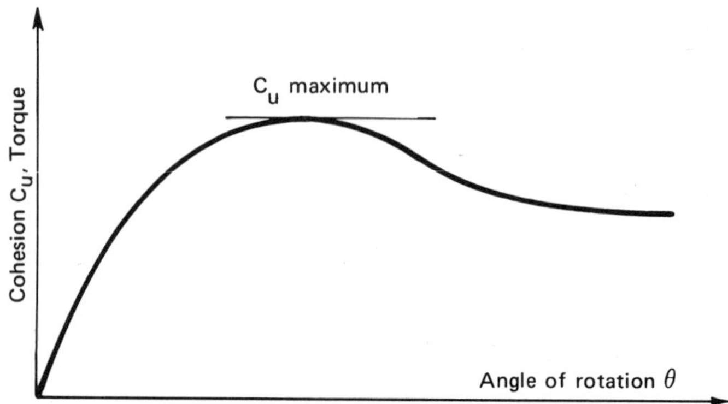

FIG. VI.4.3. – Relationship between the torque measured by the vane (scissometer) and the angle of rotation θ (at constant rotary speed).

VI.4.3 Geotechnical characteristics deduced from vane measurements

VI.4.3.1 Possibilities and limits of use of the wireline vane

The vane measures the undrained shear strength of soft soils where sampling of undisturbed cores is invariably difficult, if not impossible.

Among in situ soil reconnaissance devices, only the vane measures the undrained cohesion of the soils directly without friction (muds, clays). The value of c_u thus obtained does not however represent an intrinsic characteristic of the soil, since it depends on the interstitial pressure which is not measured.

The use of vanes (in general) is limited to the reconnaissance of coherent soils ($\varphi \approx 0$) of relatively low consolidation.

Owing to the low driving torque applied to the tool (1.15 m × kg at the most), the use of the wireline vane is limited to soils the shear strength of which is below 10 t/m^2 (100 kPa).

VI.4.3.2 Undrained cohesion deduced from the vane test

The angle of friction φ of the soil sheared by the vane is assumed to be zero or negligible.

The undrained cohesion c_u of the soil is then calculated on the basis of the maximum torque applied to the area enveloped by the rotation of the blades (lateral surface area of the cylinder + circular cross-section) (Fig. VI.4.3).

Under these conditions, the undrained cohesion c_u ($c_{u\,\mathrm{vert}} = c_{u\,\mathrm{horiz}} = c_u$) results from the relationship:

$$c_u = \frac{M}{2\pi R^2\left(h + \frac{2}{3}R\right)}$$

where:

M : maximum torque measured,
R : radius of blades,
h : height of blades.

In the case of the wireline vane where $h \approx 3R$:

$$c_u = \frac{3M}{22\pi R^3}$$

The maximum shear strength (or cohesions c_u) that can be measured with three tools adaptable to the vane are given in Table VI.4.1.

TABLE VI.4.1

Approximate radius of blades R (cm)	**Maximum cohesion c_u (t/m^2)**	**Classification of clay**
1.5	10	Stiff
2	3	Medium
2.5	1.2	Soft to very soft

The shear strength of soils measured with the vane depends to a certain extent:

– on the rotating speed of the tool, owing to the undrained nature of the test (no dissipation of the interstitial pressure) (see Section VI.4.4.1),

– on the dimensions of the blades, which are selected according to the consolidation of the soil.

The vane measures the shear strength of the soil in the horizontal direction.

Owing to the anisotropy, this strength is normally below the shear strength in the vertical direction as measured on cores or generally involved in the stability of deep foundations.

VI.4.3.3 Disturbance of the soil through implementation of the wireline vane

The soil can be disturbed:

– by circulation of the fluid during the preliminary drilling phase: it is vital to stop drilling at least 0.60 m above the chosen elevation to take the measurements,

– by the introduction of the vane bringing about at one and the same time partial breakup of the terrain and an unmeasured increase in the interstitial pressure.

The shear strengths of soils of low consolidation deduced:

– first, from wireline vane tests,

– second, from laboratory measurements on cores sampled by cable coring during the same drilling,

would appear to indicate that disturbance of the soil resulting from use of the vane is generally less than that caused by coring (see Fig. VI.4.4).

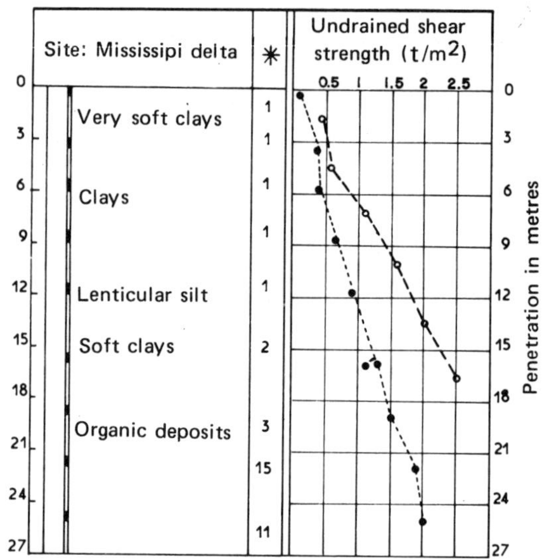

FIG. VI.4.4. – Comparison of shear strength determined by the triaxial test on 2 1/4" core samples and by the McClelland remote vane (scissometer).

VI.4.3.4 Applications of the obtained results by the vane test

Analysis of the results of tests for piles tends to show that the shear strength of clayey soils deduced from wireline vane tests is overestimated by about 50%. These observations confirm the difficulty in appraising the lateral friction force of piles on the basis of the undrained cohesion c_u measured under varying conditions.

According to correlations provided by Matlock, it appears on the other hand that vane tests are the best way of determining load-displacement curves ($p - y$ curves) of marine foundation piles subject to lateral loads.

Studies are now going on to estimate the breakout force of anchors on the basis of the shear strength of surface soils as measured by a low penetration vane set on the bottom.

VI.4.4 Comparison of the values of shear strength measured in situ with a variety of devices

The undrained cohesion of a clayey soil does not represent an intrinsic characteristic of the soil but depends on the type of device and the experimental conditions. All in situ mechanical tests (vane penetrometer or pressuremeter) are undrained tests.

VI.4.4.1 Undrained shear strength measured by the vane test

The undrained shear strength (undrained cohesion c_u) of a clayey soil measured in situ by a vane test is generally considered as a reference magnitude. In fact, the quality of the cores sampled appraised according to the degree of matching between shear strength determined using a vane and those determined in the laboratory. For example, for a soft clay from the Mississippi Delta, the undrained shear strength determined by means of the triaxial test on cores sampled by wireline coring would be about 40 to 50% lower than that obtained by the wireline vane test (Fig. VI.4.4).

The undrained cohesion c_u measured with a vane depends on the rotating speed:

- for vane used ashore the speed is generally set at 0.1 degree per second,
- for wireline vanes, the angular speed varies from 0.14 to 0.89 degree per second.

When operating at a very low angular speed, the vane shear test tends towards a drained test with dissipation of the interstitial pressure. For a very low angular speed of 10^{-4} degree per second, conducted in increasing torque stages, in some clays a reduction of c_u of up to 30% can be observed: the value measured tends towards the effective cohesion c' of the material.

VI.4.4.2 Undrained shear strength deduced from penetrometer measurements

In clayey soils, the undrained shear strength s_u is related to the cone resistance R_p with a penetrometer by the following relationship (see Section VI.1.9.2) [15]:

$$s_u = \frac{R_p - \gamma \cdot z}{N_c}$$

The dimensionless parameter N_c varies from 5 to 20, with an average of around 9. In reality, N_c depends:

- on the undrained shear strength or cohesion c_u,
- on the rate of penetration, permeability and compressibility of the clays,
- on the sensitivity of the clay.

The influence of the undrained compressibility can be explained by the formation of a rigid zone in the clay in the vicinity of the penetrometer cone.

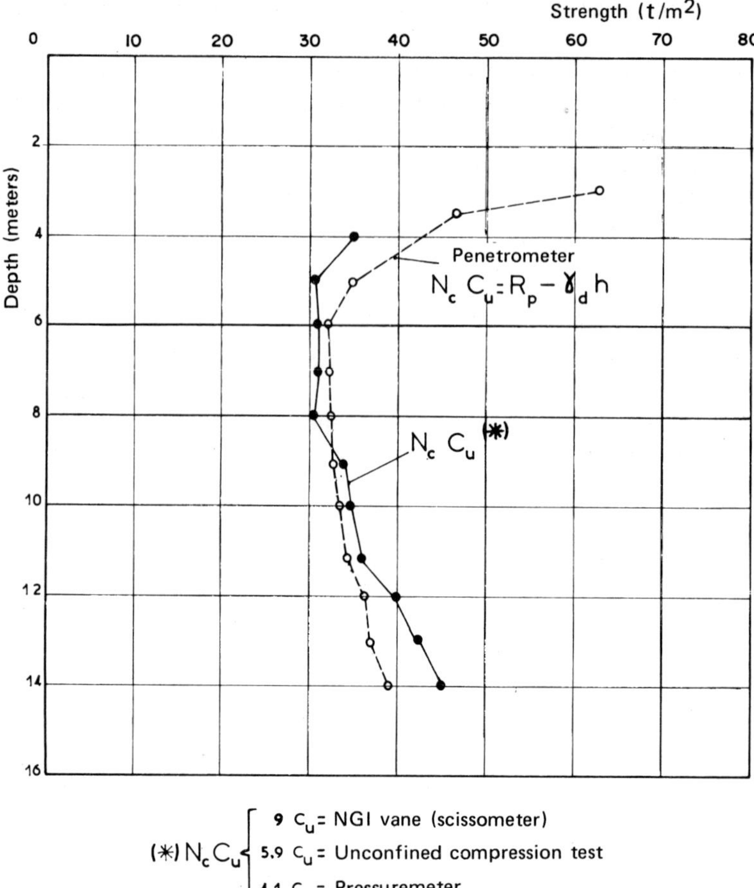

(*) $N_c C_u$ { 9 C_u = NGI vane (scissometer); 5.9 C_u = Unconfined compression test; 4.1 C_u = Pressuremeter }

FIG. VI.4.5. – Comparison of shear strength as measured on various different instruments.

The undrained cohesion c_u of clays of low consolidation, deduced from the penetrometer test, in reality depends on the value of N_c assumed. If one assumes $N_c \doteq 9$ (mean value), the following relationship applies approximately:

$$\frac{c_{u\,\text{penetrometer}}}{c_{u\,\text{vane}}} \approx 1.5 \text{ to } 2$$

For example, Fig. VI.4.5 shows profile $c_u N_c$ in terms of the depth obtained with a clay of medium consolidation. The value of N_c depends in fact on the method of measuring undrained cohesion c_u.

In sands, the formation of a rigid zone near the cone of the penetrometer results in determination of an equivalent angle of friction greater than the real angle of friction of the sand in place.

VI.4.4.3 Undrained shear strength deduced from pressuremeter measurements

Although there is no simple relationship between the limit pressure measured on the pressuremeter and the undrained shear strength of the soil, various methods are at present proposed to deduce the undrained cohesion c_u from the pressuremeter test: *Menard, Ladanyi, Ponts et Chaussées* (see Fig. VI.4.6).

Generally speaking, the undrained shear strength c_u deduced from the pressuremeter test is invariably greater than that measured by the vane test, i.e. the ratio is frequently about 2 to 3.

It will be noticed that the vane test and the pressuremeter probe both measure the shear strength of the soil in the horizontal direction. The considerable difference observed

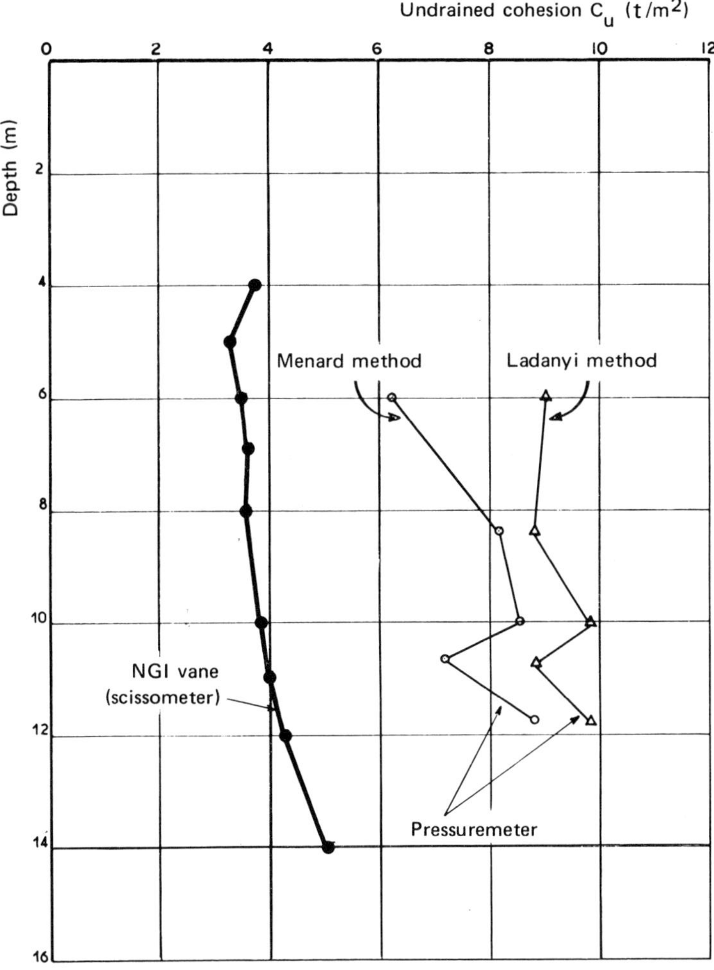

FIG. VI.4.6. – Comparaison of cohesions as measured on a pressuremeter and vane (scissometer).

is therefore explained by the different breakup process with one tool and with the other and not by any anisotropy of the soil.

VI.5 NUCLEAR LOGGING APPLIED TO THE RECONNAISSANCE OF SOILS AT SEA

The firm *Syminex* used nuclear logging probes for reconnaissance of marine soils [16] [3] :

- gamma-ray and neutron,
- gamma-gamma,

implemented through the drilling string.

The drilling logs recorded:

- enable the lithology and the limits of the various strata to be determined and thus the positioning of the cores which are sampled discontinuously,
- provide a continuous profile of the porosity and the specific gravity in place, which can be compared with the specific gravity measured on cores.

Various attempts are also being made to record nuclear logs at the sea bottom continuously in order to determine the nature of the surface sediments.

VI.5.1 Principle of nuclear logging

VI.5.1.1 Natural gamma radiation

The intensity of the natural gamma radiation of formations is determined by their content of radioactive elements (uranium, thorium, potassium). These elements are encountered in particular in clays and fairly rarely in sands and limestones.

Recording this natural radioactivity in terms of depth provides an excellent way of determining the lithology and situating the various strata with great accuracy.

Owing to the high penetrating power of gamma rays, their transmission from the formation to the detector is affected relatively little by the presence of a thin casing or by drill pipes.

VI.5.1.2 Neutrons

A neutron probe comprises a source of high-speed neutrons coupled with a thermal neutron detector situated at a set distance from this source.

The neutrons emitted collide with the elements they encounter and are slowed down and finally thermalized at a certain distance from the source depending on the relative mass of the elements they encounter. The mass of the neutron being the same as that of hydrogen, "neutron-hydrogen atom" collisions result in maximum loss of energy in the neutrons.

In a hydrogen-rich medium, the high speed neutrons emitted are thermalized in the neighbourhood of the source and very little will reach the detector. Conversely, in a hydrogen-lean medium, the neutrons are thermalized farther away from the source, and hence nearer the detector. Accordingly, the number of neutrons counted and recorded on the surface decreases as the hydrogen concentration of the medium crossed through rises.

In marine sediments, hydrogen is present in the pore water. Knowing the relationship between the number of neutrons counted and the corresponding percentage of hydrogen, the porosity of the medium can be deduced.

If N designates the number of neutrons detected per second and n the porosity of the sediments:

$$N = a - b \log n$$

where a and b are two constants dependent on the geometry of the probe.

VI.5.1.3 Gamma-gamma

A source of gamma radiation is placed a set distance from a detector. The gamma rays emitted collide with the constituent elements of the formations and a certain number of them are reflected towards the detector, where they are counted.

Let N be the number of collisions per second furnished by the detector, and
γ the density of the sediments in place,
then it can be shown :

$$\log N = c - d \,.\, \gamma$$

where c and d are constants determined by the activity of the source and the geometry of the sonde.

VI.5.2 Characteristics of Syminex nuclear probes

VI.5.2.1 Description of probes

The characteristics of the "gamma-ray-neutron" probe are:

- diameter = 43 mm,
- length = 2.90 m,
- weight = 17 kg,
- natural gamma-ray detector = 10 cm,
- fast neutron source = 3 Ci of americium-beryllium,
- distance between source and detector = 30 to 40 cm.

The characteristics of the "gamma-gamma" probe are:

- diameter = 43 mm,
- length = 1.20 m,

- weight = 8 kg,
- gamma-ray source = 125 mCi of caesium 137,
- distance between source and detector = 30 to 40 cm.

VI.5.2.2 Characteristics of the use of probes

The probes are calibrated by means of two reference wells:

water: $\gamma = 1$ g/cm^3, $n = 100\%$
sand: $\gamma = 2.11$ g/cm^3, $n = 32.6\%$

The "$N - n$" and "$N - \gamma$" relationships are straight lines when plotted in semi-logarithmic coordinates.

The recordings are made at a probe lifting speed of about 3 m/min.

The surface equipment (weighing 300 kg in all) comprises:

- the winding winch of the power supply/carrier cables,
- the electronic equipment and the recorder.

VI.5.3 Offshore support and implementation of logging probes

VI.5.3.1 Offshore support

Drilling log equipment is used on various soil reconnaissance vessels in the North Sea: the "Surveyor", "Mariner", "Ferder", . . .

The principle of installation aboard ship is shown diagrammatically in Fig. VI.5.1.

The heave compensator is indispensable, just as with all drilling logs, if a valid recording is to be obtained.

The surface equipment (dimensions 1.36 × 1.10 × 0.90 m) must be contained in a sealed cab or room, sheltered from spray and waves.

VI.5.3.2 Implementation of logging probes

At present, recordings are made by lifting the probe up through the inside of the drilling string.

This solution is imposed in practice by the low density of the drilling mud used (on the average 1.05 to 1.10), leading to the risk of collapse and hence jamming of the probe.

Use of the records calls for corrections taking into account the nature of the mud, the thickness of the drill string, the "probe-pipe" annular space and the "formation-pipe" annular space.

A gamma-gamma probe now being developed and equipped with two detectors located

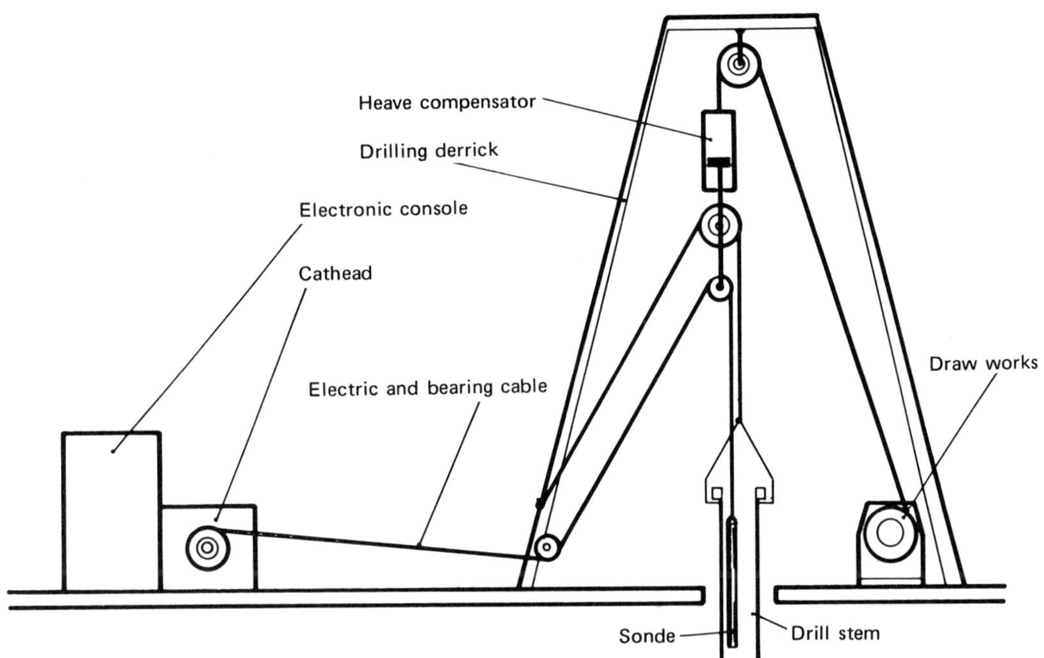

FIG. VI.5.1. – Shipboard installation for recording nuclear logs (*Syminex*).

at different distances from the source, should enable the influence of the annular space variations between the drill string and the formation to be corrected.

Prior raising of the drilling string to a few metres below the sea bottom would enable logs to be recorded in an open borehole and thus reduce the correction needed.

However, this solution could only be envisaged in practice where the density of the mud to be sufficiently high to ensure that the walls of the borehole stand up. The use of high density muds supposes a closed circulating system (with riser) for reasons of cost.

The implementation of logging probes during reconnaissance drilling of soils at sea is a very simple, low-cost operation calling only for:

– an operator aboard the vessel,
– 2 kVA of electric power.

The cost represents no more than about 3% of the total cost of the reconnaissance campaign.

VI.5.4 Interpretation of nuclear logs and significance of measurements made in the borehole

VI.5.4.1 Interpretation of logs

If the formations crossed through consist of sand and clay in alternation, interpretation is generally very simple.

If on the contrary, however, the formation consists of a mixture of sand and clay or silt, interpretation becomes a more difficult matter owing to the lack of exact rules.

Among other things, from the "gamma-ray" log one has to a confine oneself merely to a qualitative appraisal of the clay content, without being able to calculate any percentage.

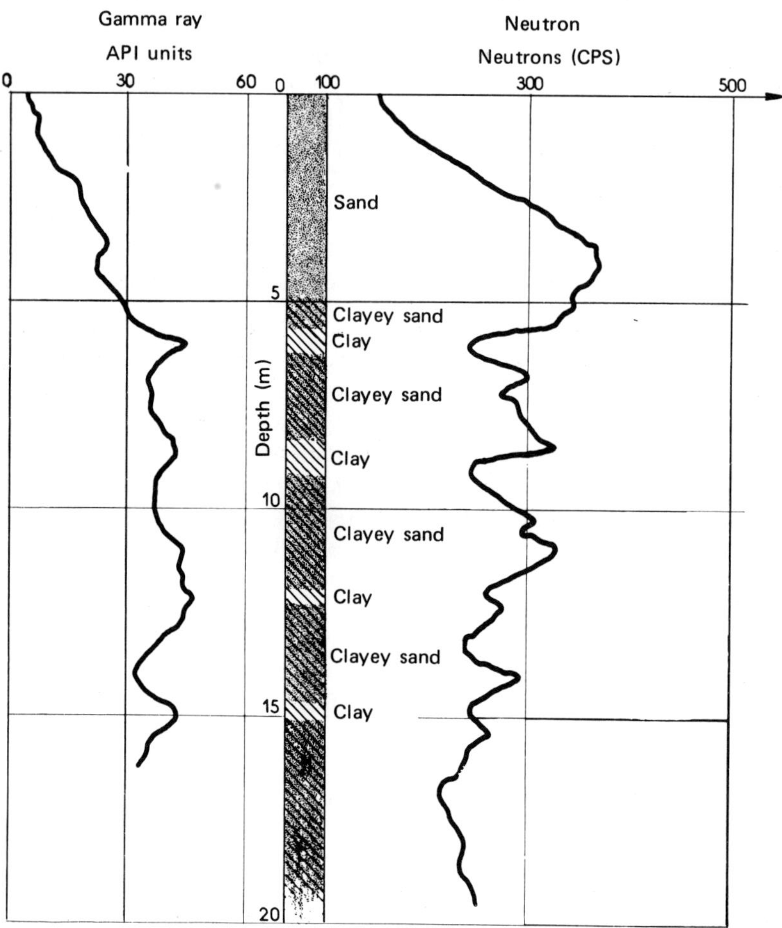

FIG. VI.5.2. – Example of lithological interpretation on the basis of γ-ray and neutron recordings.

Nuclear logs are unaffected by grain size distribution. It is therefore impossible to mention the presence of silts. According to the drilling logs, either clays, sands or mixtures of sands and clay are distinguished. For instance a silty sand and a sandy clay have the same natural radioactivity; the lithological description deduced from the "gamma-ray" log would be "sandy clay" in either case. For instance, Fig. VI.5.2 gives an example of lithological interpretation from a "gamma-ray" and a "neutron" log.

The distance between the source and the detector (spacing) of "neutron ' and "gamma-gamma" probes is about 30 to 40 cm. It follows that thin layers of clay or silt in a sandy formation may be undetected, though they would affect the mean reading.

However, the "gamma-ray" detector, which is 10 cm long, will yield a more detailed curve.

VI.5.4.2 Densities and porosities deduced by logging and by core measurements

The samples used for laboratory measurements are invariably much disturbed by drilling, coring, extraction of the core barrel, etc. Furthermore, the measurements are very much of the spot measurement type.

The densities deduced from "gamma-gamma" logging represent averages over a length of about 30 cm.

It would therefore appear illusory to seek for very close correlations (for instance per foot) between laboratory measurements and deduced measurements from drilling logs. We shall merely confine ourselves to a few general observations.

In view of the disturbance of cores, one can generally expect to obtain higher specific gravities from drilling logs than from laboratory measurements. In reality, the reverse is often observed, and this could be explained by the compacting of the soil under the effects of the penetration of the core barrel, for instance (see Fig. VI.5.3).

Likewise, a higher porosity is observed on the basis of neutron logging than from the measurements made on cores. This phenomenon would appear particularly appreciable in the case of clayey formations (Fig. VI.5.3).

On the basis of measurements made in the North Sea, the relative accuracy of radioactive measurements appears to be:

- about 3% for porosity,
- about 2% for specific gravity.

VI.5.5 Recording of nuclear logs on the sea bottom

Coring or in situ measurement techniques rarely make it possible to obtain precise information as to the nature of the sediments (generally very loose) at the water-soil interface on the sea bottom. The recording of logs on the bottom provides a means of obtaining information which is very often useful.

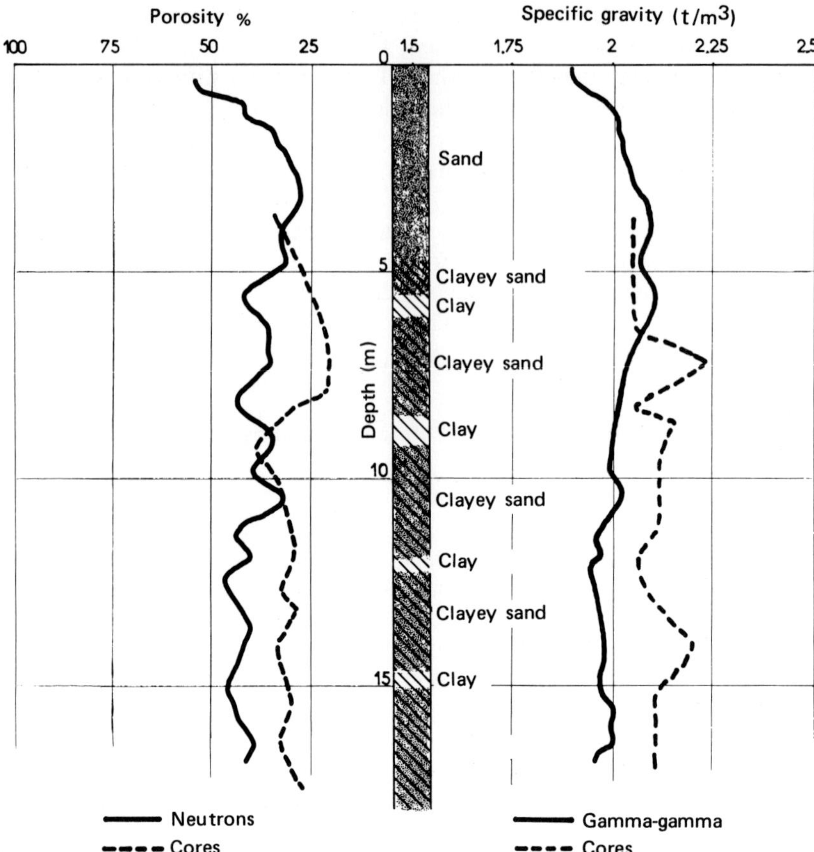

FIG. VI.5.3. – Comparison of porosities and densities measured on core samples and deduced from nuclear logs.

VI.5.5.1 Principle and implementation of the method

A nuclear logging probe (gamma-ray and gamma-gamma) is pulled along the sea bottom by means of a submersible (such as the "TOM 300") or a tubular frame of the "Troïka" type.

The probe provides information on the nature (clays or sands) and consolidation (density) of the sediments to a depth of 10 to 20 cm.

VI.5.5.2 Possibilities and limits of the method

The recording of nuclear logs on the sea bottom offers the advantage of objectivity, which is not always the case for indirect (side-scan sonar) or even visual observations (television cameras).

It should be possible with the method to plot charts, and it could for example be useful for reconnoitering routes for pipelines.

The shallow depths of investigation, however, limit application of the method strictly to surface reconnaissance.

Furthermore, the data obtained may depend on the microtopography of the bottom (contact between probe and soil).

Remark. The recording of electrical logs (spontaneous polarization, resistivity, etc.) also envisaged for surface reconnaissance ($<$ 2 to 3 m) obviously raises difficulties owing to the imposed conditions for instance salinity contrast for *SP*).

BIBLIOGRAPHY

[1] ZUIDBERG (H.). – "Seabed Penetrometer Tests", *Fugro Symposium on Penetrometer Testing.* The Hague, October 1972.

[2] de RUITER (J.). – "The Use of In Situ Testing for North Sea Soil Studies", *Offshore Europe Conference,* Aberdeen 1975, paper No. OE-75-219.

[3] de RUITER (J.). – "Site Investigations for North Sea Forties Field", *OTC,* Houston, 1975, paper OTC 2246.

[4] de RUITER (J.). – "Electric Penetrometer for Site Investigations", *J. of Soil Mech. and Found. Div.,* February 1971.

[5] SANGLERAT (G.). – *The penetrometer and soil exploration. Interpretation of penetration diagrams: theory and practice.* Elsevier Publishing Co., 1972.

[6] BJERRUM (L.). – "Geotechnical Problems Involved in Foundation of Structure in the North Sea". *Geotechnique,* Vol. XXIII. Sept. 1973.

[7] "Submersible Sounding Tool to Test North Sea Floor". *Oil and Gas J.,* Janv. 10, 1972.

[8] FERGUSON (G.H.), McCLELLAND (B.) and BELL (W.D.). – "Seafloor Cone Penetrometer for Deep Penetration Measurements of Ocean Sediment Strength", *OTC,* Houston 1977, paper OTC 2787.

[9] HOEG (K.). – "State of the Art : Foundation Engineering for Fixed Offshore Structures", *Proceedings of BOSS 76,* Trondheim, 2-5 August 1976, Vol. I, p. 39-69.

[10] ALLEN (J.H.), DAYAL (U.) and JONES (J.M.). – "Development of Marine Sediment Impact Penetrometer", *Oceanology Intern. 75.,* 16-21, March, 1975, Brighton.

[11] GIBSON (R.E.) and ANDERSON (W.F.). – "In situ Measurements of Soil Properties with the Pressuremeter", *Civil Eng. and Publics Works Rev.,* London, May 1961.

[12] MENARD (L.). – "Intérêt technique et économique du vibro-marteau hydraulique annulaire pour le prélèvement d'échantillons en mer et la réalisation d'essais géotechniques *in situ", Oceanexpo,* Bordeaux, 1974.

[12 bis] BAGUELIN (F.), JEZEQUEL (J.F.) and SHIELDS (D.H.). – "The Pressuremeter and Foundation Engineering". *Trans. Tech. Publications,* 1978.

[13] DOYLE (E.H.), McCLELLAND (B.) and FERGUSON (G.H.). – "Wireline Vane Probe for Deep Penetration Measurements of Ocean Sediment Strength"., *OTC,* Houston, 1971. Paper No OTC 1327.

[14] KRAFT Jr. (L.M.), AHMAD (N.) and FOCHT Jr. (J.A.). – "Application of Remote Vane Results to Offshore Geotechnical Problems", Offshore Technology Conference, May 3.6, 1976, Houston. Paper No. *OTC,* 2626.

[15] EIDE (O.). – "Marine Soil Mechanics. Applications to North Sea Offshore Structures", *Offshore North Sea Tech. Conf.,* Stavanger, 3-6 Sept. 1974.

[16] *Data Sheet of SYMINEX* on nuclear logs.

3 Installation of Offshore Structures

CHAPTER VII

Fixed Platforms on Piles

CONTENTS

INTRODUCTION

1. For over 25 years, a great many fixed platforms on piles have been installed at sea in ever-increasing water depths, today reaching over 300 m. The metal jacket set on the sea bottom is secured by means of piles which are either driven or drilled down to depths of about 100 m (Fig. VII.1.1). The total length of the piles may therefore exceed 250 m.

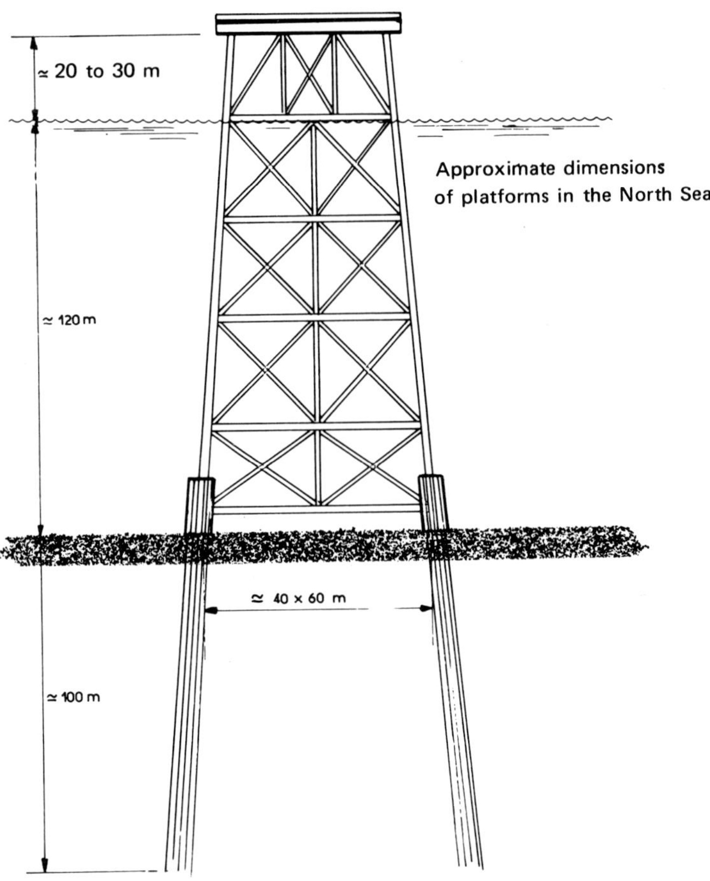

FIG. VII.1.1. – Diagram of a pile-fixed platform.

2. The basic design principles for the foundations of marine-pile structures are similar to those used for the design of deep foundations on land. The problem is nonetheless complicated by three types of consideration [1]:

– the loads supported by each pile of a fixed platform at sea are commonly in the range of 2,000 to 3,000 t, namely distinctly greater than those usually applied to the piles on structures ashore,

– the lengths of the piles are greatly superior to those of driven piles for building structures on land, rendering it not directly possible to transpose practices used for designing the bearing capacity,

– knowledge of the soil at sea is invariably limited owing to the high cost of operations, and the information obtained is often subject to caution owing to the frequently difficult operating conditions of the reconnaissance campaign.

3. The following will be examined in the present chapter:

– the geotechnical problems raised by the siting and stability of platforms on piles,

– the necessary geophysical, geological and geotechnical reconnaissances before siting platforms on piles,

– the principle of the methods of designing the foundations of platforms on piles from the twofold standpoint of vertical static loading and horizontal sollicitations.

VII.1 DEFINITION OF THE GEOTECHNICAL PROBLEMS INVOLVED IN THE SITING AND STABILITY OF FIXED PLATFORMS ON PILES

The geotechnical problems set by fixed platforms on piles concern:

– the siting of the piles by driving and/or drilling,

– the stability of the platform under the effect of the fixed loads and the dynamic sollicitations.

Problems involved in stability will first of all be set forth and then those relating to the siting of the piles.

VII.1.1 Stability of fixed platforms on piles

Study of the stability of a fixed metal platform can be reduced to study of the stability of piles partially inserted into the soil and characterized by the following (see Fig. VII.1.2):

– the compressive and tensile pullout of the piles bearing capacity,

– the behaviour of the piles and the soil under the effect of lateral sollicitations,

– the possibilities of liquefaction of the soil during driving, upon which the initial bearing capacity depends,

– the risk of scour near the piles.

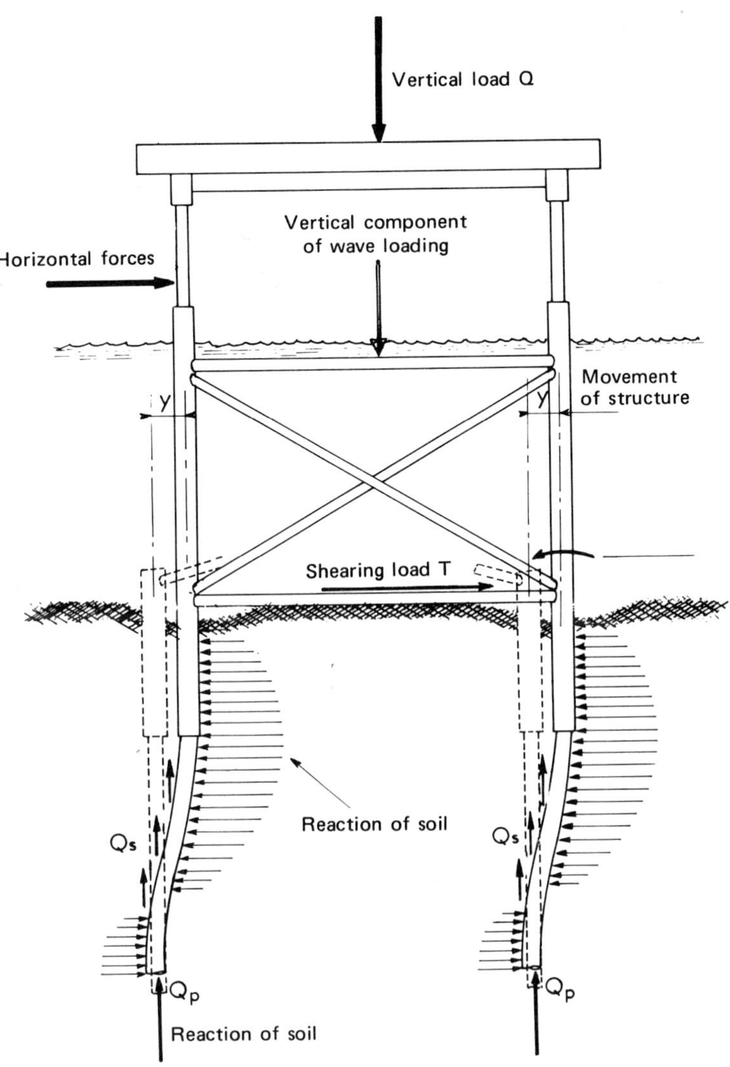

FIG. VII.1.2. – Loads to which a pile-fixed platform is subjected.

VII.1.1.1 Bearing capacity and pullout force of a pile

The bearing capacity of a pile is calculated from the shear strength of the soil. Depending on the techniques used for reconnaissance of the soils, the shear strength is defined by either of the following well known parameters:

– the undrained shear strength s_u measured in the laboratory or in situ,

– the drained shear strength determined in the laboratory after dissipation of the interstitial pressure,

– the tip resistance and lateral friction measured continuously with a penetrometer,
– the limit pressure deduced from the pressuremeter test.

The bearing capacity of a pile depends on its diameter and its depth of insertion into the soil.

The pullout force of a pile (tensile test) is invariably lower than its bearing capacity (under compression). However, it will be seen that while estimation of the bearing capacity is subject to much uncertainty, evaluation of the pullout force is even more imprecise and justifies the highly conservative maximum value applied for the lateral force of friction in particular (see Section VII.3.1).

VII.1.1.2 Stability of a pile under the effect of lateral forces

The reaction of the soil to lateral movement of a pile subject to horizontal loading is characterized by the stress-strain curve of the soil, i.e. for a given stress, its modulus of deformation.

In actual fact, the characterization of the deformation properties of a soil is highly difficult, since the modulus essentially depends on the experimental conditions (triaxial test, pressuremeter, etc.), the disturbance of the soil, rate of application of the loads, etc.

The inserted length affected by the horizontal loads is rarely more than about 20 m (see Section VII.3.2).

The stability of the structure is often increased by putting groups of piles at the corners of the platform. This is the solution generally adopted for the structures located in the North Sea (see Section VII.1.4.2).

The thickness of the piles is a function of the horizontal loads:

– the thickness of the pile over the top and bottom sections varies from approximately 2.5 to 5 cm (1 to 2 in),

– the thickness is considerably greater at the sea bottom where the bending moment is at its maximum.

VII.1.1.3 Liquefaction of the soil under the driving force and recovery

The influence of the impacts when driving the piles may cause loose or sensitive soils to liquefy (soils of low compaction, muds or soft clays) resulting in a very considerable (temporary) reduction in the shear strength and hence the bearing capacity. The difficulties encoutered in siting certain platforms in soft soils can be explained at least partially by liquefaction of the soil when driving the piles.

After driving is stopped, clay soils "recover" at varying speeds, thus fairly quickly regaining a higher shear strength and modulus of elasticity. The duration needed for complete "recovery" of the initial mechanical characteristics most certainly depends on the sensitivity and the thixotropy of the soil.

VII.1.1.4 Scour near the piles

Scour near the piles is a complex function of a number of parameters, the most important of which are:

– the nature of the sediments,
– the speeds of the currents and the amplitude of the waves,
– the diameter of the piles and the distance between them.

The result of scour is:

– to reduce the depth of insertion and hence the bearing capacity of the pile,
– to reduce the stability under lateral loads.

Table VII.1.1 gives a cursory summary of the characterization of the properties of the soil involved in the stability of fixed platforms on piles and how to attain them.

TABLE VII.1.1

CHARACTERIZATION OF THE PROPERTIES OF SOILS INVOLVED IN THE STABILITY OF FIXED PLATFORMS ON PILES

	Characteristics of soil	Means of determining the characteristics of the soil		
		Coring + laboratory tests	**Penetrometry**	**Pressuremetry**
Bearing capacity and pullout force of a pile	Shear strength of soil	Undrained shear strength s_u Drained shear strength (triaxial test)	Cone resistance R_p Lateral friction f	Limit pressure p_l
Stability of a pile under lateral loads	"Stress-strain" relationship of soil, giving *p-y* curve	"Stress-strain" relationship of soil determined by the unconfined compression test (modulus of elasticity E)		Pressuremeter diagram (pressuremeter modulus E_p)
Scour of soil near the piles	Size distribution of soil	Surface soil identification tests		

VII.1.2 Insertion of piles by driving

Steel piles for platforms are:

– either driven into soft or low-consolidation soils,
– or driven and then drilled and cemented into hard soils according to a number of variants (see Section VII.1.3).

VII.1.2.1 Driving design methods. Principle of the wave equation

The numerous **pile-driving formulae** (of the Dutch, of Redtenbacker, Hiley and other builders of pile-drivers) are all empirical in nature [2].

While these various formulae provide information as to the ability of the pile to be driven, they yield knowledge of the state of stress of the pile during driving. Furthermore, application of these formulae:

– would appear increasingly dubious with the increasing dimensions of the piles and driving systems; the results are generally conservative,
– leads to values which often differ very widely (up to a factor of 10).

For very long piles, analysis of the driving phenomena by the **wave equation** is the only satisfactory method of studying the performance of a given "pile-driving system" combination [3].

The impact of the hammer on the top of the pile generates a compressive wave pulse which propagates downwards. In the part of the pile already inserted, the amplitude of the wave gradually dampens through lateral friction: the residual energy causes the end of the pile to sink yet further.

VII.1.2.2 Pile-driving hammers

The hammers used for driving piles are of two types [2] :

– single or double-acting steam-driven hammers. The driving frequency, which is a function of the steam inlet rate, is generally approximately 1 blow per second.
– Diesel hammers (single-acting), the driving frequency of which may attain 2 blows per second.

The main characteristics of a hammer (provided for the design programme) are:

– the striking mass M,
– the energy $E = MH$ (H being the drop height),
– the driving power (for Diesel hammers),
– the efficiency or effectiveness.

The choice of the hammer for driving a pile is governed by the length to be inserted and the maximum acceptable stress in the pile:

– if the hammer is too small, the depth of insertion will not be reached,
– if the hammer is too large, the stresses generated in the pile will be unacceptable.

The ever-increasing dimensions of piles (diameter and length) for setting up platforms, particularly in consolidated soils, has led to the development of increasingly powerful driving hammers, as can be seen in Fig. VII.1.3 [4].

Large dimension piles ($\geqslant$ 1.20 m in diameter is very frequent) for platforms in the North Sea are currently being driven using the Menck 7000 developing an energy of 83.5 t x or the HBM 3000 A developing 110 t x m (1).

(1) Piles of 84″ (2.13 m) in diameter for siting of Cognac platform (Gulf of Mexico) have been driven by HBM 3000 hammer. The HBM 4000 develops 160 t x m energy.

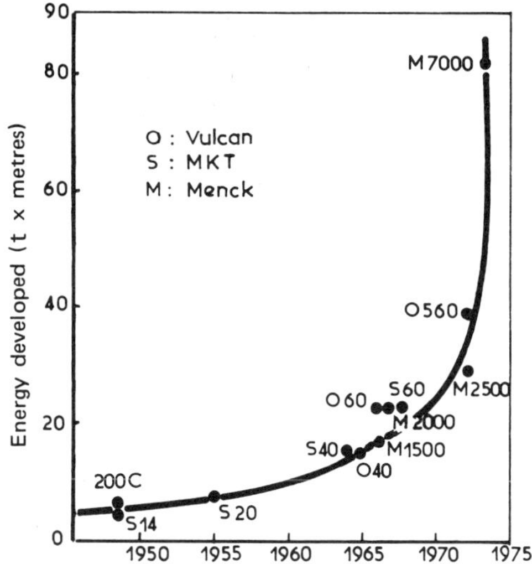

FIG. VII.1.3. – Curve of energy developed by hammers used for driving piles at sea (according to McClelland).

VII.1.2.3 Driving design programme

Numerical design programmes for driving require knowledge of the characteristics of:

– the hammer,
– the driving accessories,
– the pile,
– the soil.

These programmes provide:

– the soil resistance to driving: the graph of R_u in terms of the number of hammer blows per unit length of insertion of the pile (see Fig. VII.1.4) under various driving conditions,
– the stresses in the various sections of the pile. The stress in the pile is at its maximum immediately beneath the hammer.

The results of calculation shows that the maximum strength of the soil attained (bearing strength on refusal) by a given "pile-hammer" combination varies little beyond a certain driving energy characterized by the number of hammer blows per unit length of penetration of the pile. Fig. VII.1.4 shows the results of an analysis made for the "0-20" type hammer, developing an energy of 8.1 t × m [4] [5].

Among the various parameters influencing the driving programme, the thickness of the pile plays a predominant role.

It should be pointed out that the driving programme gives no indications as to the static bearing capacity of the pile, which must necessarily be estimated from static calculation formulae (see Section VII.3.1).

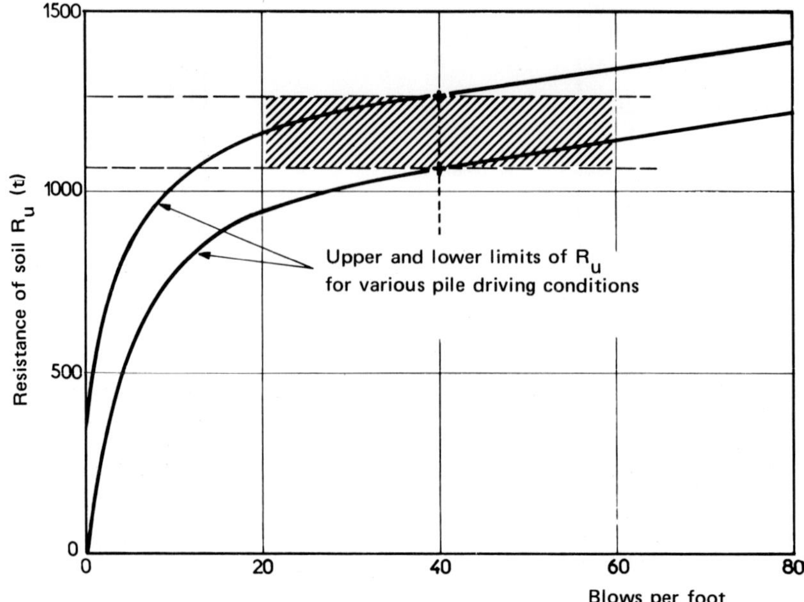

FIG. VII.1.4. – Limit curves of resistance of soil to pile driving.

VII.1.2.4 Liquefaction of the soil under driving and "recovery" of the soil

During driving, certain soft or sensitive soils liquefy around the pile, resulting in a loss (or even the temporary disappearance) of lateral friction. The driving design programme is not capable (at present) of allowing for this phenomenon.

Each time the driving is stopped, to weld a new section of tube for instance, the strength of the soil on resuming driving is considerably increased through the "recovery" of the soil (Fig. VII.1.5). The length of halts for welding a 1.20 m pile (48 in) varies from 4 h (thickness of tube 25 mm) to 7 h (thickness of tube 38 mm) approximately.

The "recovery" of the soil:

– may be beneficial, when it enables a limit bearing force judged sufficient to be obtained [4],

– or on the contrary may make it impossible to attain a designed bearing force, sometimes rendering any further penetration of the pile impossible without changing to a heavier hammer.

The degree of "recovery" depends on the sensitivity of the clayey soil:

– In normally consolidated clays, the driving results in considerable disturbance. After stopping driving, the soil gradually reverts to its initial state. The increase in the bearing capacity (set-up) of a pile driven into normally consolidated clays may commonly reach a factor of 2 to 5.

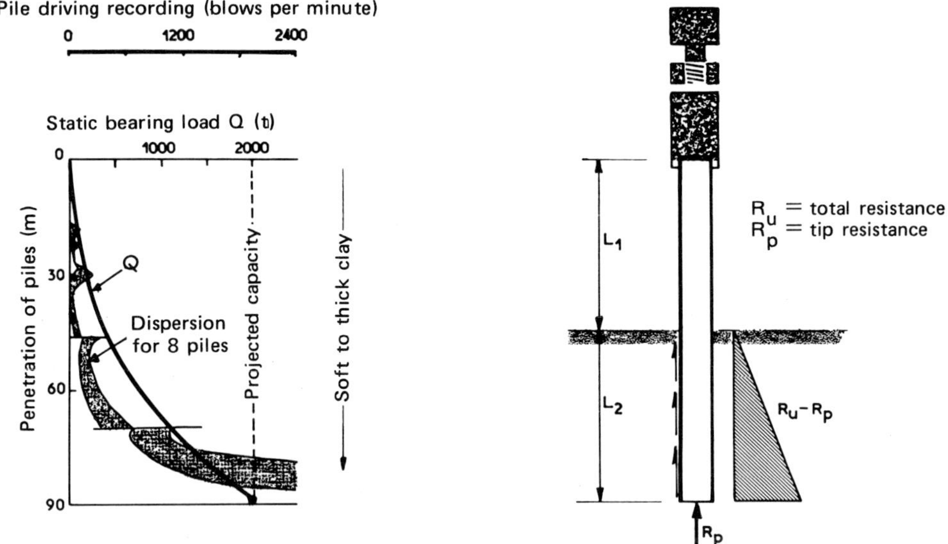

FIG. VII.1.5. – Example of pile driving diagram (according to McClelland).

– In overconsolidated clays, the "recovery" effect is generally less, so the set-up is probably no more than 2.

VII.1.3 Insertion of piles by drilling

In the case of highly consolidated clays, driving alone does not generally make it possible to insert the pile to the necessary length. Various techniques are used, depending on the nature of the soil and particularly on the loads to be transmitted [4] [6].

VII.1.3.1 Techniques with a main pile

The simplest foundation consists merely of one main pile inserted by successive drilling and driving stages in accordance with the following procedure:

– driving to total refusal,
– drilling beneath the foot of the pile,
– resumption of driving down to the depth of insertion intended.

Foundations using drilled and cemented piles (with a 2 to 3 in annular space) are a common technique very widely used for anchoring.

VII.1.3.2 "Insert-pile" techniques

The "insert-pile" foundation technique consists of the following:

– driving of a main pile down to refusal,

– drilling of a pilot hole through the main pile for inserting a smaller diameter tube known as the "insert pile" (see Fig. VII.1.6) sealed into the ground and to the main pile by cementing.

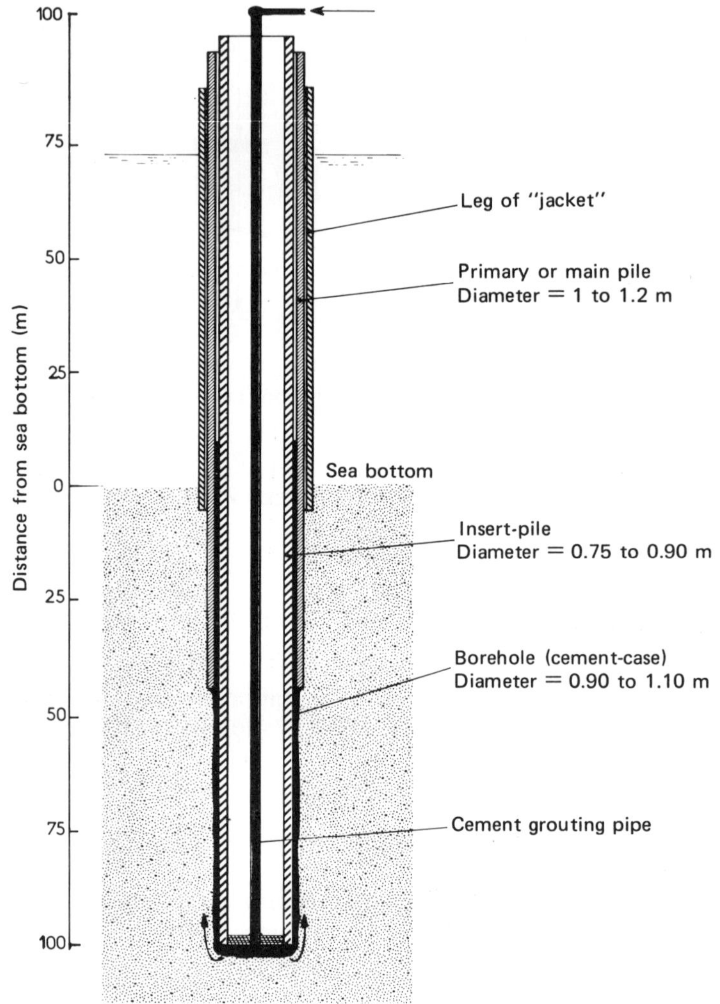

FIG. VII.1.6. – Principles of the "insert-pile".

The "insert-pile technique" with successive pressure groutings of cement into the soil has been developed experimentally by *Soletanche.* The increased lateral frictional force of the pile obtained by the method is approximately the following [7] [8]:

– triple in sands,
– double in clayey soils.

The result is a considerable reduction in the length of the pile that has to be inserted.

VII.1.4 Description of the foundations of platforms on piles

The number of piles, their dimensions (diameter, thickness, penetration) and their layout clearly vary from one structure to another. We shall confine ourselves here to describing a few types of foundations for large structures in terms of the soil and environment conditions.

Two essential problems govern the choice of the type of pile foundation:

– first, the layout of the piles,
– second, the level of the vertical and horizontal loads applied.

VII.1.4.1 Pile-mounted platform foundations of the type built in the Gulf of Mexico

The type of pile-mounted platform foundations built first of all in the Gulf of Mexico and later in other regions of the world are designed for [4]:

– soils of relatively low consolidation,
– average oceanographic conditions (100-year wave amplitude less than 15 m).

In the case of large dimension platforms, these foundations consist:

– first, of 4 to 8 main piles 1 to 1.20 m in diameter penetrating 80 to 100 m into the soil,

– second, a number of secondary "skirt-piles" which generally penetrate to lesser depths than do the main piles. Their length above the sea bottom is no more than 10 to 15 m, and the link with the structure is ensured by cementing the annular space between the pile and the guide tube (see Fig. VII.1.7). The purpose of "skirt-piles" is to ensure better distribution of the horizontal and vertical loads, particularly in the case of soft soils.

The piles are inserted almost exclusively by driving provided the undrained shear strength of the soils is below 5-10 t/m² (50-100 kPa).

The piles consist of sections of tube 30-m long welded together as the driving proceeds.

The acceptable vertical loads for one pile clearly vary with the nature of the soils and are commonly in the range of 2,000 t for a penetration of 90 to 100 m.

Owing to their design, these foundations are characterized by:

– an independent reactive force on each pile; the main piles in fact lie 12 to 20 m apart and the skirt-piles 8 to 10 m from the main piles,

– a considerable concentration of the loads owing to the fewness of the piles; obviously, this results in fairly significant movements of the structure,

– a safety factor of a least 1.5, particularly as regards stability under cyclic loads.

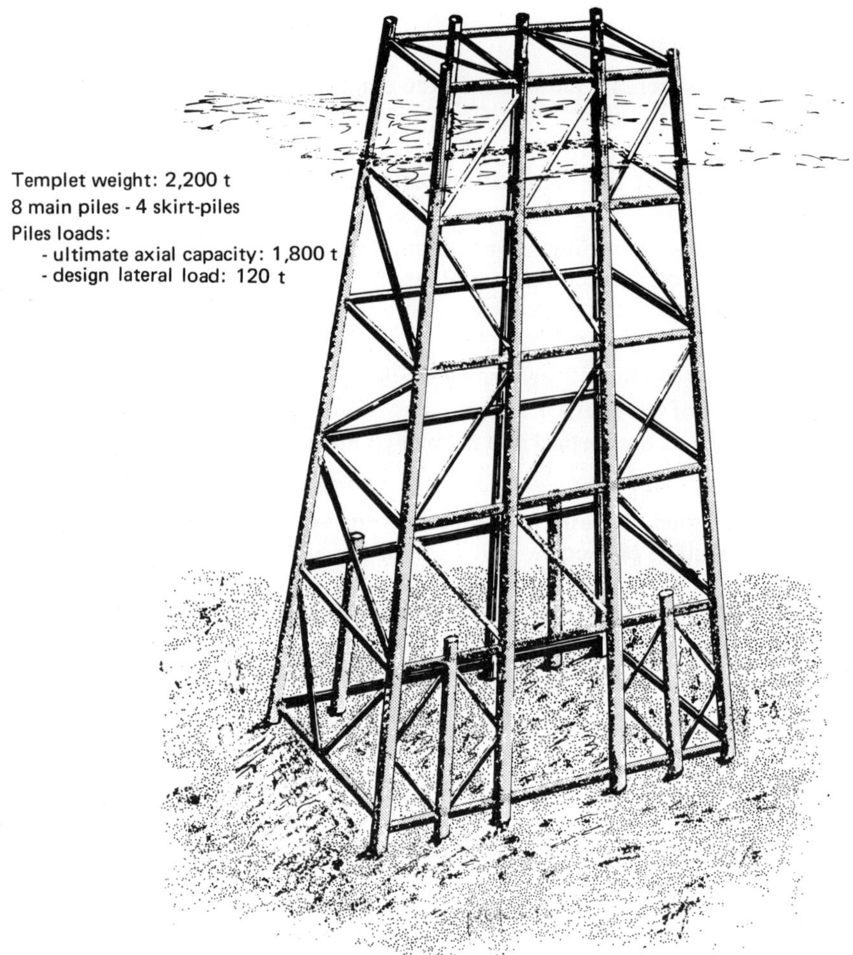

FIG. VII.1.7. – Drilling/production type platform for the Gulf of Mexico.

VII.1.4.2 Pile-mounted platform foundations of the type built in the North Sea

The foundations for platforms located (or planned) in the North Sea are governed by the following two criteria [4] [9] [10]:

– extremely severe oceanographic conditions (100-year wave of about 30 m),
– relatively or highly consolidated soils.

The foundations for structures with 4 main legs almost invariably consist of a group of 4 to 12 piles (see Fig. VII.1.8). These piles have a diameter of 1 to 1.50 m and a thickness of about 5 cm (reinforced to 6 cm at soil level) and are fastened together by cementing over a length of about 20 m above the sea bottom. The overall diameter of the group of 10 to 12 piles is about 10 to 12 m.

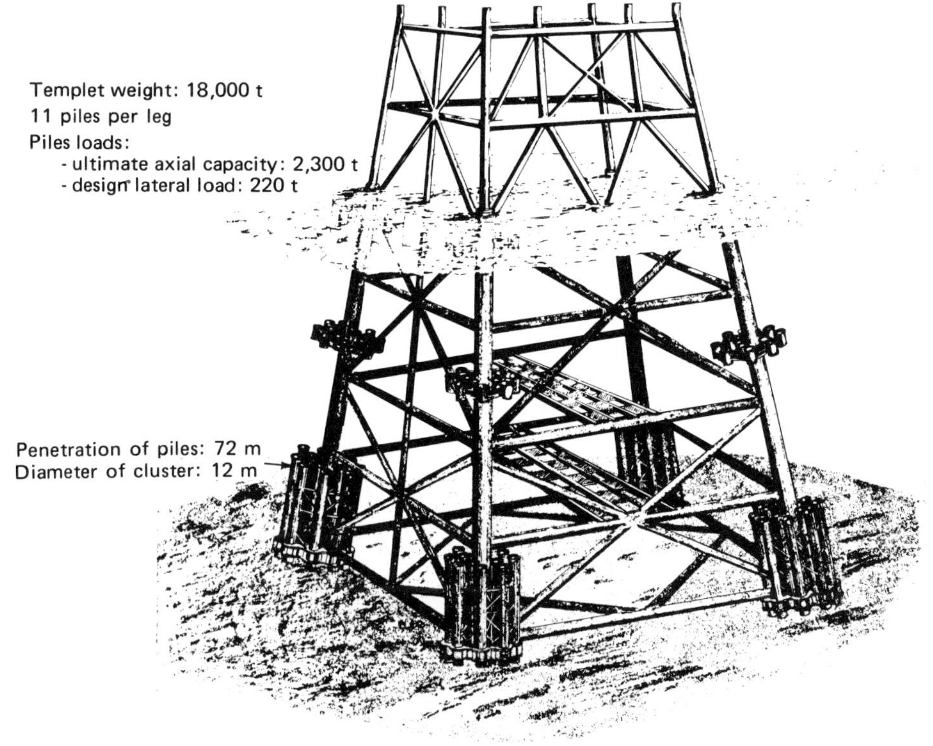

FIG. VII.1.8. – Drilling production type platform for the North Sea.

The intermediate piles (if any) set 15 to 20 m from the main legs enable the loads applied to the structure to be distributed.

Owing to their considerable thickness (and hence welding difficulties) these piles are brought to the site in a single section.

The penetration of the piles varies with the type of foundation built and is chosen in the light of the consolidation of the terrains encountered:

– in the case of soils of medium consolidation, the piles are inserted by driving and then by drilling and cementing (see "insert-pile" technique) (see Section VII.1.2.2, Fig. VII.1.6) [4] [6],

– in the case of highly consolidated soild, the length of the piles can be reduced by using the "bell" foundation technique applied in particular at Ekofisk, consisting:

– first in driving a pile to a depth of 25 to 30 m,

– next, drilling, expanding the hole and cementing steel rods (Fig. VII.1.9) [4] [6].

Total penetration is no more than 30 to 40 m.

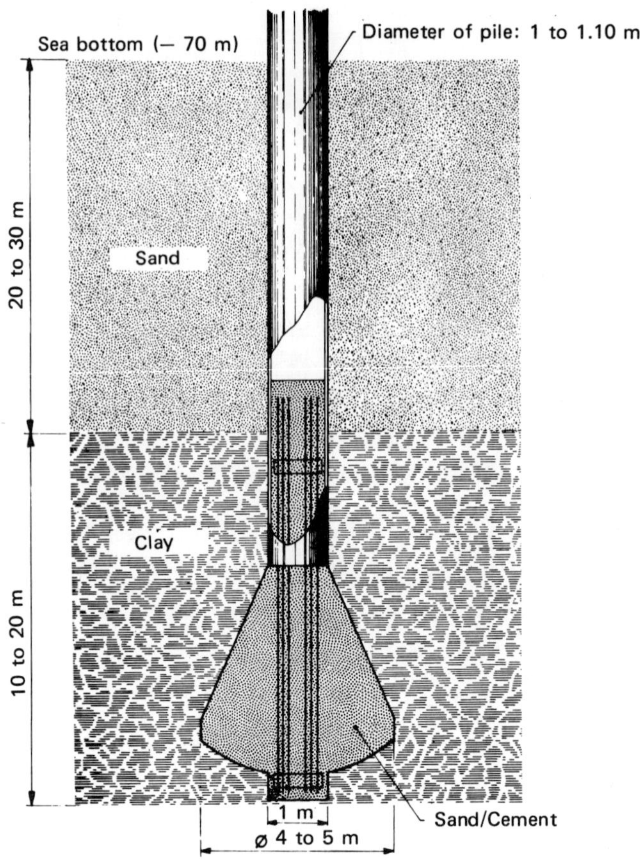

FIG. VII.1.9. – Underreamed footing for platform on Ekofisk ("belled piles").

VII.2 RECONNAISSANCE OF SOILS FOR SITING OF FIXED PLATFORMS ON PILES

Reconnaissance of soils before siting a fixed platform on piles (just as for installing any structure at sea) must proceed by successive approximations [11].

Overall reconnaissance of the site often carried out 2 to 4 years before setting up a structure and covering a considerable area comprises (Fig. VII.2.1):

– an initial geophysical reconnaissance with bathymetry and seismic prospecting,
– a geological reconnaissance.

Detailed reconnaissance of the site of the structure (or site reconnaissance) covering a fairly limited area comprises (Fig. VII.2.1):

– a second geophysical reconnaissance describing the topography of the bottom and the geometry of the strata,

– a geotechnical reconnaissance of the soils.

The indications concerning the reconnaissance to be carried out on the soils are theoretical in character and represent a maximum, the actual reconnaissance programme obviously depending on the soil and the subsoil, conditions of environment, etc.

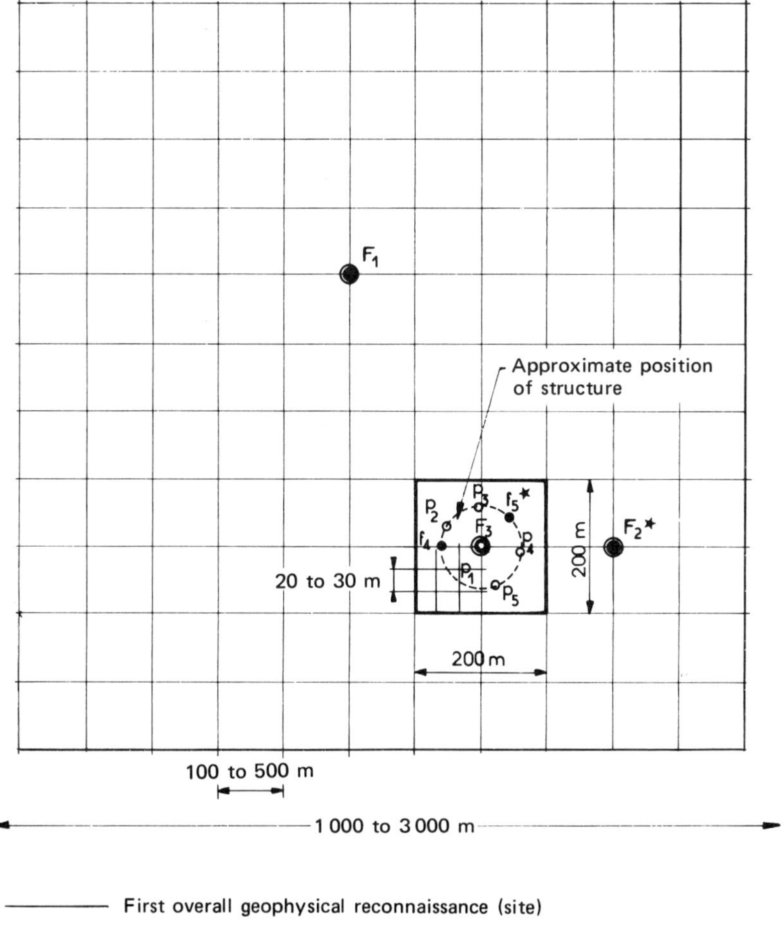

FIG. VII.2.1. – Diagram of soil reconnaissance to be made before siting a pile-fixed platform.

VII.2.1 Initial geophysical reconnaissance (of the site)

The purpose of the first geophysical campaign is to reconnoitre the overall nature of the bottom and soils.

VII.2.1.1 Coverage and unit mesh

Geophysical reconnaissance of the site has to cover:

– preferably the whole extent of the field if it has been sufficiently accurately delineated at the time of this initial campaign,
– in any case, a minimum area with sides measuring 1 to 3 km.

The considerable area to be reconnoitred is justified:

– first, by the uncertainty as to the final position of the structures at the time this initial campaign is conducted,
– second, the need for gross knowledge of the favourable positions for installation which will be studied by the subsequent detailed campaign (s):

The unit mesh to be adopted differs depending on the recordings made:

– for bathymetry, the unit mesh will be from 100 to 500 m and may be different in the two directions, depending on the general topography of the site,
– for seismic prospecting, the unit mesh may vary from 200 to 500 m and also differ in direction.

VII.2.1.2 Bathymetry

The initial bathymetry must provide general indications as to the gradient and irregularities of the relief (mounds and hollows).

This reconnaissance is made by means of an echo sounder, yielding an accuracy of about 1% of the water depth, i.e. an absolute accuracy of the dimensions of ± 0.5 m in a depth of 100 m.

VII.2.1.3 Morphology (or roughness) of the bottom

Recording the roughness of the bottom by side-scan sonar is not indispensable in the initial phase if it does not help to specify the general topography, particularly if the bottom is relatively flat.

It is recalled that sonar recordings make it possible to distinguish:

– clear and uniform images corresponding to stretches of sediments displaying little break up,
– echos indicating obstacles of various kinds: beds of material, rock outcrops, wrecks, etc.

VII.2.1.4 High resolution seismic prospecting

The purpose of initial seismic prospecting is to detect:

– either the continuity of the strata,
– or the possible presence of geological accidents, in the form of major facies modifications.

The possible (or foreseeable) depth of insertion of the piles often attains 100 m or more. As a result, the penetration sought in seismic prospecting would be 100 to 500 m. The resolution obtained is generally not less than 2 m.

This prospecting is carried out by means of:

– a sparker,
– possibly, a boomer (Uniboom).

VII.2.2 Initial geological and geotechnical reconnaissance (of the site)

The purpose of the initial deep coring and drilling campaign, offering at one and the same time geological and geotechnical interest, is:

– to facilitate the interpretation of the seismic prospecting,
– to reconnoitre the nature and consolidation of the terrains.

VII.2.2.1 Emplacement of boreholes

An examination of the geophysical recordings concerning an area of several square kilometres should make it possible to locate the boreholes at the best possible site in the light of:

– the foreseeable conditions of development of the field,
– the topography of the site,
– the uniformity of the seismic profiles (absence of geological accidents) and the minimum gradient of the strata,
– the probable consolidation of the surface soils.

To facilitate the interpretation of the seismic recordings, the boreholes will be located at the intersection of two profiles (within the error due to the positioning of the vessel).

This optimum emplacement of the initial boreholes offers the twofold advantage:

– of acquiring accurate geological (and geotechnical) data concerning the site,
– of reducing the subsequent number of deep boreholes, if possible.

VII.2.2.2 Number and depth of boreholes

The number of boreholes (with core sampling) needed depends on the regularity of the seismic profiles encountered:

– if the seismic profiles are practically uniform over the site, and particularly on the probable zone of location of the structures, one borehole may be enough,

– if on the contrary the seismic recordings display many irregularities, it would seem of advantage to make at least 2 or 3 deep boreholes.

In practice, the principle of at least 1 deep borehole is accepted for the initial geological and geotechnical reconnaissance.

The depths of the initial borehole (s) will depend:

– on the dimensions of the structure envisaged and hence the insertion needed, which is dependent on the vertical and lateral loads applied and the water depths,

– the nature and the consolidation of the formations encountered,

– the type of foundation (to be defined).

In actual fact, it would appear desirable to attain (see Section VII.1.3):

– a minimum penetration of 80 to 100 m if the soils are highly consolidated (dense sands, very stiff clays, etc.),

– a penetration of 120 to 150 m if the soils are of only low consolidation (loose sands, soft clays, etc.).

A priori, in the absence of any information about the site, an average depth of 100 m for the first boreholes will be adopted.

The sequence of corings (wireline corings) through the drilling string varies with the nature, consolidation and heterogeneity of the strata of the formations. As an indication, core samples are often taken:

– every metre (practically continuously) down to about 10 m,
– every 1.50 m from 10 to 20 m,
– every 3 m from 20 to 50 m,
– every 5 m beyond 50 m.

VII.2.3 Second geophysical reconnaissance (of the actual emplacement)

The second geophysical reconnaissance, confined to the approximate emplacement of the structure, will be performed in the light of the previous results obtained (initial geophysical prospecting campaign and deep boreholes) of an area with sides measuring 300 to 500 m.

VII.2.3.1 Bathymetry

The siting of platforms on piles does not call for highly uniform bottoms:

– an accuracy of about 0.50 m on the bathymetry would appear quite adequate,
– a slope of a few percent is no clear hindrance.

The bathymetry will be determined by means of an echo sounder with a unit mesh of about 50 × 50 m.

Correction of the recordings requires:

- calibration of the instrument for depth (temperature and salinity of the sea water),
- recording of the tidal movements on the site,
- recording of the roll, pitch and heave,
- frequent resetting of the positioning system.

VII.2.3.2 Bottom morphology and magnetometry

The side-scan sonar, making it possible to verify topographical accidents in the range of a few metres longitudinally and a few decimetres in height, will be used at intervals of 50 to 100 m (in one direction only), the range used (75 or 150 m) allowing a certain overlap, which is desirable.

A magnetometer makes it possible to identify metal objects on the bottom or buried at slight depths (in principle < 3 m).

The unit meshing of the profiles, of about 100 m, is comparable to that of a side-scan sonar. It should be noted that the magnetometer and the side-scan sonar can be used simultaneously without inducing any mutual interference.

VII.2.3.3 High resolution seismic prospecting

The second seismic reconnaissance campaign carried out at the same time as the bathymetry, will be conducted with a unit mesh of about 50×50 m.

It would be desirable:

- to attain a penetration of at least 100 m (in about the same range as the length of insertion of the pile),
- to obtain a resolution of better than 2 m over the first 20 to 30 m concerned by the lateral loads.

The incompatibility between high penetration and high definition desired at low depths requires the use of two different techniques:

- first, a high energy sparker,
- the second a boomer (Uniboom) or penetrator (if the soils are of very low consolidation) or again a sparker employed at low energy.

VII.2.4 Second geotechnical reconnaissance (of the actual emplacement)

The second soil reconnaissance campaign should make it possible to obtain a detailed description of the subsoil and the approximate position of the final structure:

- first, on the basis of the production requirements,
- second, from the results of previous reconnaissances.

VII.2.4.1 Emplacement of the soundings

After determination of the approximate emplacement of the platform, detailed reconnaissance of the soil must allow at one and the same time for:

– the uncertainty of the positioning system, both as regards the execution of soundings (coring sampling and in situ measurements) and the emplacement of the structure,
– the need to know the properties of the soil in the immediate vicinity of the structure (the dimensions at soil level of some structures in the North Sea are 60 to 90 m).

It can be concluded that the soundings will be situated:

– in a zone with sides measuring about 200 m,
– preferably, at the intersections of two seismic profiles.

VII.2.4.2 Type, number and depth of the soundings

The depth for soil reconnaissance is dictated at one and the same time by:

– the dimensions of the structure to be set out,
– the water depth,
– the nature and qualities of the soils encountered.

The detailed reconnaissance of the emplacement of the structure requires boreholes and coring samples, in situ measurements, and surface coring and sampling.

A deep borehole with core sampling should reach a depth of at least 100 to 120 m. This depth will increase with the depth of the overlying water, i.e. with the foreseeable insertion of the piles.

If the emplacement adopted for the platform coincides fairly well with the deep drilling made during the initial reconnaissance (see Section VII.2.2), it would be superfluous to drill new borehole down to 100 m.

The 3 to 5 (approximate) in situ tests normally made with boreholes of medium depths will consist of:

– preferably penetrometer tests down to refusal in dense sands; the maximum depths reached will vary from a few metres in dense sands to 20 to 25 m in some clays,
– possible, pressuremeter tests in clays of medium stiffness,
– sometimes, in soft soils, wireline vane measurements implemented through the drilling string.

The main purpose of boreholes to medium depths of from 20 to 50 m is to verify the horizontal homogeneity of the strata:

– if the soils are relatively homogeneous, a single borehole plus coring will be enough,
– if the boreholes reveal horizontal variations in the facies, two boreholes will be preferred.

The sequence of coring sampling adaptable to the needs of particular specifications might be the following:

– every metre, in other words practically continuously down to about 10 m,
– every 1.50 m from 10 to 20 m,
– every 3 m from 20 to 50 m,
– for each soft or highly heterogeneous stratum revealed by penetrometry.

The recording of nuclear logs: γ-ray, neutron, γ-γ, by means of sonde lowered through the drilling string will render the following possible, among other things:

– the number of cores to be reduced,
– direct information on the lithology and true density to be obtained,
– the exact level of the layers to be checked.

Perfect knowledge of the upper levels, which are particularly subject to stressing under the lateral loads on the structure, would appear of prime importance in the design of the foundations.

The surface cores and samples:

– serve to identify the exact nature and size distribution of the surface formations in order to counter possibilities of scour near the piles,

– are made down to 3 to 5 m by means of vibrocorers or Kullenberg corers, depending on the consolidation of the formations. It will often prove necessary to sample considerable quantities of sediments by means of grab samplers for testing scour on a model.

Tables VII.2.1, VII.2.2 and VII.2.3 summarize the following respectively:

– the geophysical reconnaissances to be carried out first on the site and then on the actual emplacement of the structure (Table VII.2.1),

– the geological and geotechnical reconnaissances to be carried out first on the site and then on the actual emplacement (Table VII.2.2),

– the common measurements to be made in the laboratory and in situ to identify the soils and determine their shear strength (Table VII.2.3),

before setting up fixed platforms on piles.

TABLE VII.2.1

GEOPHYSICAL RECONNAISSANCE OF SOILS FOR SITING FIXED PLATFORMS ON PILES. APPLICABLE TECHNICAL PERFORMANCES

		Scope	Techniques applicable	Mesh (m)	Transmission frequency	Penetration (m)	Definition (resolving power) (m)	Precision
Initial geophysical reconnaissance (overall) Site reconnaissance	Bathymetry	Whole field (if known) At least 1 to 3 km along sides	Echosounder	from 100 to 500	30 to 50 kH	≈ 0		Absolute value of precision: ≃ 1 m per 100 m of water
	Morphology (roughness index) not indispensible		Side-scan sonar	Spacing ≈150 to 200	≈ 100 kHz	0		Width of scan: 100 to 150 m
	Seismic prospecting		Sparker or boomer (Uniboom)	from 200 to 500	100 to 1,000kHz 500 Hz to 4 kHz	100 to 150 50 to 75	≈ 2 to 3 ≈ 1.5	
Second physical reconnaissance (fine) Reconnaissance of emplacement of structure	Bathymetry	300 to 500 m along sides	Echosounder	50 x 50	30 to 50 kHz	≈ 0		Absolute value of precision: ≈ 1 m per 100 m of water
	Morphology (roughness)		Side-scan sonar	50 x 100	≈100 kHz	0		
	Magnetometry		Magnetometer	≈ 100		< 3		
	Seismic prospecting		Sparker or Boomer (Uniboom) (possibly penetrator)	50 x 50 approx.	100 to 1,000 Hz 500 Hz to 4 kHz A few kilohertz	≈ 100 ≈ 50 to 75 ≈ 10 to 30	≈ 2 to 3 ≈ 1.5 ≈ 1	

TABLE VII.2.2

GEOLOGICAL AND GEOTECHNICAL RECONNAISSANCE OF SOILS BEFORE SITING FIXED PLATFORMS ON PILES. TECHNICAL PERFORMANCES APPLICABLE

		Scope	Type and number of boreholes	Penetration necessary (m)	Techniques applicable	Performances	Implementation
Initial geological reconnaissance (overall): reconnaissance of site	Boreholes + core samples	If possible, to be located in zone of emplacement of structure (s)	1 (or 2) deep boreholes with wireline coring	100 to 150 depending on soils	Techniques implemented from the surface by a drilling string: – drilling and wireline coring by percussion, – nuclear drilling logs	Penetration ⩾ 100 m Rate of penetration: 2 to 3 m/h Continuous recording	From a vessel: wave amplitude < 2 m From a drilling platform: wave amplitude < 5 to 6 m
Second geological and geotechnical reconnaissance (detailed): reconnaissance of actual emplacement of structure	Boreholes + core samples	About 200 x 200 m for each emplacement of structure	A deep borehole with coring (unless already done)	100 to 150	ditto	ditto	From vessel: wave amplitude < 2 m
			1 to 2 medium deep boreholes with core samples	20 to 50	ditto	ditto	ditto
	In situ tests		3 to 5 tests: – Penetrometry	20 to 25	Cable penetrometer (Wison)	Maximum penetration: depends on terrains Discontinuous recording	From a vessel: wave amplitude < 1.2 m
					Penetrometer laid on bottom (Seacalf)	Penetration: 10 to 25 m depending on terrains Continuous recording	From a vessel: wave amplitude < 1.5 m
			– Pressuremetry		Ménard pressuremeter	Techniques used on certain sites Penetration down to 70 to 80 m in certain soils	From a vessel: depending on the various implementation techniques
			– Vane testing		McClelland remote-vane (wireline operated)	Soft formations	Used very rarely
	Sampling and surface coring		A few sampling and coring	3 to 10	Vibrocorers (electric or hydraulic) Kullenberg corers Dredging	Consolidated or soft soils Soft soils only	From a vessel: wave amplitude < 2.5 to 3 m

TABLE VII.2.3

CORE AND/OR IN SITU MEASUREMENTS TO BE CONDUCTED BEFORE DECIDING THE EMPLACEMENT OF FIXED PLATFORMS ON PILES

	Characteristics of soils measured		Remarks
Laboratory analysis of cores	**Identification of soils**	Nature Atterberg limits (W_L, W_P) size distribution grain	For clayey soils
	Water content w Specific gravity γ		Made immediately on the vessel
	Shear strength of soil Undrained shear strength (undrained cohesion c_u)	unconfined compression vane "fall cone" pocket penetrometer triaxial	Preferably, immediately on the vessel In the laboratory
	Deformability of soil Determination of *p-y* curves by repeated loadings	unconfined compression triaxial	In the laboratory
Penetrometer	Cone resistance R_p Lateral friction f		These measurements are indispensable in sands
Pressuremeter	Limit pressure p_l Pressuremeter modulus E_p		Of interest for the *p-y* curves
Nuclear logs	Nature of soil (gamma-ray) Density and water content *in situ*		These measures are highly desirable

VII.3 DESIGN OF THE FOUNDATIONS FOR FIXED PLATFORMS ON PILES

Analysis of the installation and stability problems raised by fixed platforms on piles seems to call for three essential remarks:

– considerable uncertainty remains as to the minimum penetration required for the piles,

– the pile-driving equipment available does not always (depending on the nature of the soils and the water depth) make penetration possible by pile driving alone,

– the boring or jetting used to facilitate penetration increases the factor of uncertainty concerning the bearing capacity of the piles.

The various stresses to which the structure is subjected stem from:

– the loads it carries,

– the action of the wind on the above-water portion,

– the action of current and waves on the submerged portion.

The present section will examine the principle of the methods used to calculate the stability of fixed platforms, namely:

– the bearing capacity of the piles,
– the action of horizontal sollicitations.

VII.3.1 Calculation of the bearing capacity of piles

We shall describe in turn:

– the conventional method of calculating bearing capacity, applying the recommendations of the API RP 2A [12] or *DNV* [13] specifications,
– the calculation method derived from pressuremeter measurements.

VII.3.1.1 Conventional method of calculating the bearing capacity of a pile driven into clay [5]

The static **bearing capacity** Q of a pile driven into the soil to a depth h is the sum of two terms Q_s and Q_p:

$$Q = \underset{\substack{\text{Friction}\\ \text{term}}}{Q_s} + \underset{\substack{\text{Tip}\\ \text{term}}}{Q_p} = fA_s + qA_p$$

where:

$Q_s = fA_s =$ the load mobilized by the lateral friction between pile and soil,
f = the lateral friction stress,
A_s = the lateral area of the inserted section of the pile,
$Q_p = qA_p =$ the load supported by the end of the pile,
q = the stress at the end of the pile,
A_p = the cross-section of the pile tip.

The **pullout force** of the pile is reduced to the friction term:

$$Q_s = f'A_s$$

where f' is invariably less than f and generally assumed $f' = 0.7\,f$ $(0.5 < f' < 1)$.

The application of the calculation method depends on the determination of parameters f and q, the values of which depend on:

– the soils = nature and consolidation,
– the type and dimensions of the piles,
– the method of inserting the piles.

The unit bearing force at the end of the pile is:

$$q = c_u N_c$$

where:

c_u = undrained cohesion of the clay,
N_c = dimensionless bearing-capacity factor.

The cohesion c_u considered is assumed to have been determined with a vane. For values of c_u obtained by other methods, consideration should be given to observations made for methods of measuring cohesion (see Section VI.4).

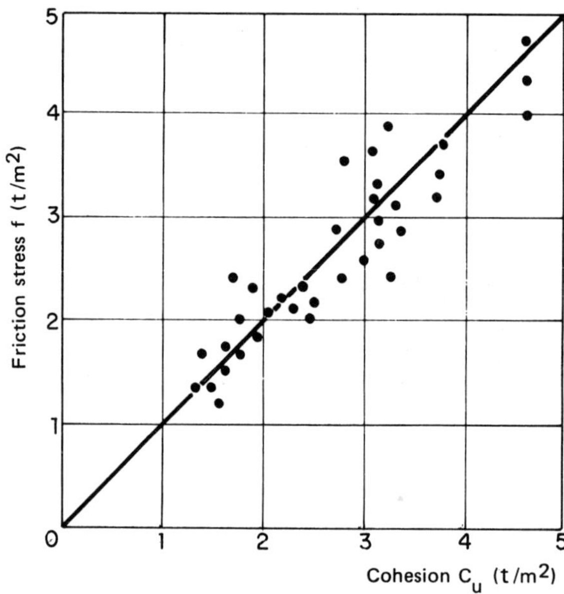

(a) soft to firm clays (according to Mc Clelland)

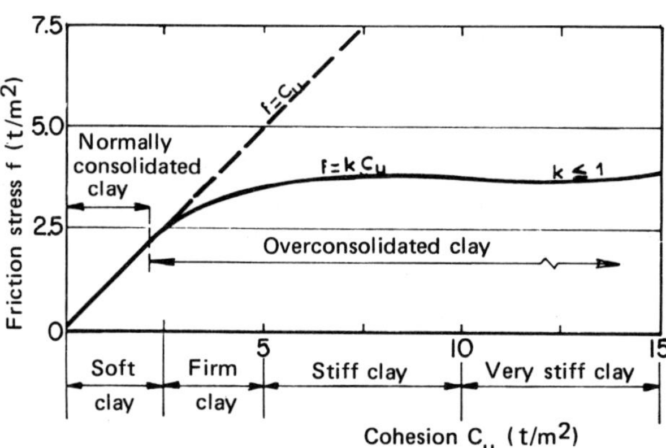

(b) Firm to stiff clays (according to Mc Clelland)

FIG. VII.3.1. – Relationships between the cohesion of the soil and the skin friction stress of the piles.

In soft clayey soils and in the case of a cylindrical pile driven deep into the ground, experience shows that:

$$N_c \approx 9$$

In most cases, the peak term for piles driven into clay will be negligible compared with the lateral friction term.

The lateral friction term, also known as the "soil-pile adherence" is expressed by:

$$f = kc_u$$

where: $k \leqslant 1$.

The main difficulty lies in determining coefficient k:

– in soft to medium clays where $c_u < 5$ t/m² ($c_u < 50$ kPa), the lateral friction stress f should be similar to the cohesion c_u (Fig. VII.3.1),

– in stiffer clays, where $c_u >$ about 5 t/m² ($c_u >$ about 50 kPa), f is distinctly less than c_u, as shown by various experimental equations for short piles (Fig. VII.3.2). The application of these equations leads to a very inaccurate evaluation of the bearing capacity of the piles (Fig. VII.3.3).

In the case of **normally consolidated clays,** an empirical equation suggested by Vijayvergiya and Focht [14]:

$$Q_s = \lambda(\sigma_m + 2c_m) A_s$$

should relate the lateral frictional resistance to a magnitude that can be assimilated to the thrust stress of the soils. In this equation:

$\sigma_m = \gamma_m h$ = effective mean vertical stress resulting from the submerged weight for the depth of soil considered,

c_m = mean undrained shear strength, for the depth of penetration of the pile.

This equation would indeed appear applicable to piles of different dimensions (from 15 to 76 cm in diameter and 2 to 101 m in length), as shown by the graph in Fig. VII.3.4. Tests recently performed in the United States of America on piles driven about 100 m into a clay soil were aimed at verifying the validity of this graph at great depths.

In the case of **overconsolidated soils,** the Tomlinson method may be applied:

$$f = 0.3\, c_u$$

API specifications do not distinguish between soft and stiff clays, though they give upper limits for lateral frictional strength:

– **driven piles:**
  – length of insertion less than 30 m: $f \leqslant 5$ t/m² ($f \leqslant 50$ kPa),
  – length of insertion over 30 m: $f \leqslant 1/3\, p_0$,

where p_0 designates the effective geostatic stress,

– **piles inserted after drilling:**
  – f is lower than the corresponding values for driven piles.

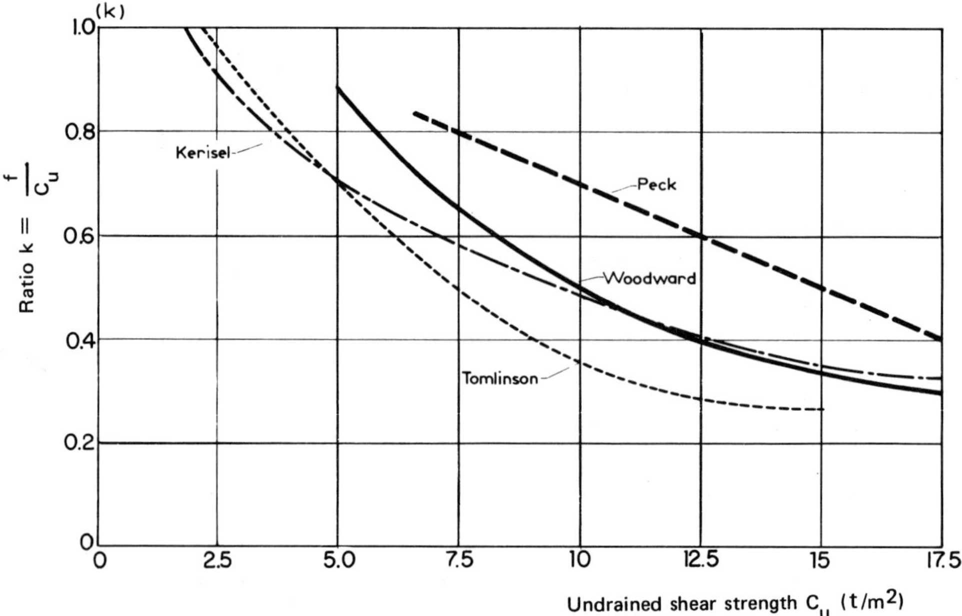

FIG. VII.3.2. – Relationship between the coefficient $k = f/c_u$ and the undrained shear strength of clayed soils.

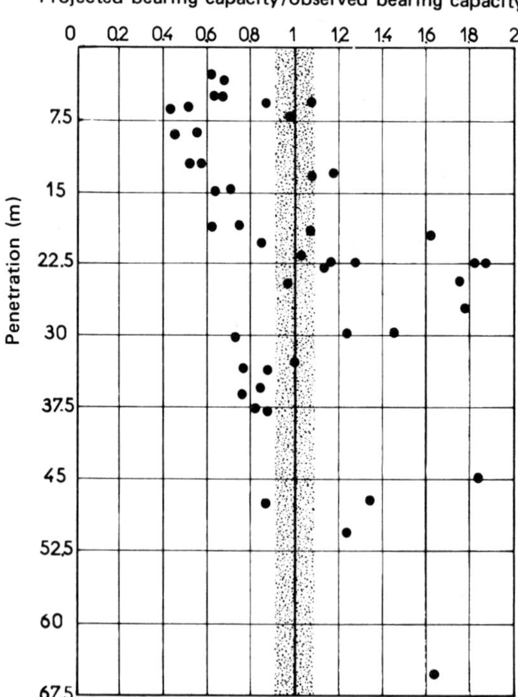

FIG. VII.3.3. – Relationship between the projected and observed bearing capacities of piles driven into clay soils, on the basis of the Tomlinson criterion (according to McClelland).

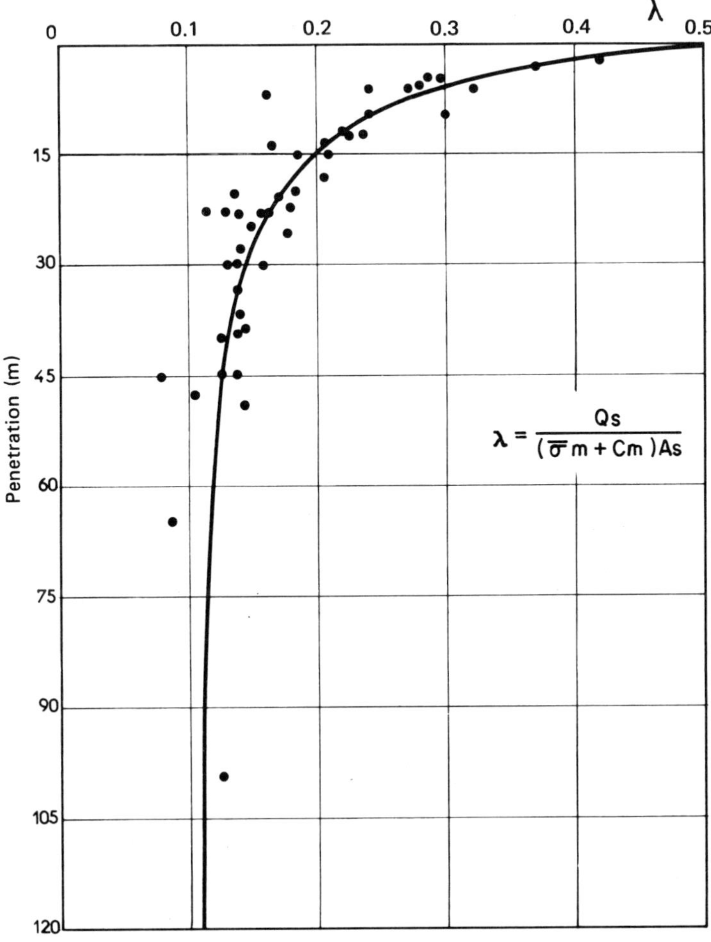

FIG. VII.3.4. – Correlationship between the coefficient λ and the penetration of the pile (according to Vijayve Γ giya and Focht).

The value of the cohesion deduced from the penetrometer tests is generally taken as:

$$c_u = \frac{R_p}{20}$$

where R_p is the cone resistance measured on the penetrometer.

The following formula is often accepted for the bearing force of the end of the pile:

$$q = 4c_u$$

(with q no more than 400 t/m²) (4 MPa),

and for the lateral friction:

$$f = 0.4\, c_u$$

(with f no more than 40 t/m^2) (400 kPa).

VII.3.1.2 Conventional method of calculating the bearing capacity of a pile driven into sand [5]

The unit bearing force of the end of the pile can theoretically be calculated by the following formula:

$$q = p_0 N_q = \gamma' h N_q$$

where:

γ' = apparent (i.e. submerged) density of the sand ($\gamma' = \gamma - 1$) (2),
h = depth of insertion of the pile,
p_0 = effective geostatic stress,
N_q = dimensionless factor of the bearing capacity of circular piles driven deep into the soil.

According to this formula, the term q would be directly proportional to p_0, i.e. the depth of penetration. In actual fact, all experimental studies show that q tends towards a maximum value reached at a critical depth h_c (see Fig. VII.3.5), dependent in particular on the density of the sand, the vertical stress due to the upper layers of clayey soils, etc.

The coefficient N_q increases very rapidly with the internal angle of friction φ of the sand but differs very widely according to different authors (Fig. VII.3.6). For instance, $\varphi = 35°$, N_q varies from 41 according to Terzaghi to 300 according to Meyerhof. In fact, it would appear that q is more related to the relative density D_r of the sand than to its internal angle of friction.

The lateral friction term is equal to:

$$f = K p_0 \operatorname{tg} \delta$$

where:

K = lateral pressure coefficient of the soil,
δ = angle of friction between pile and sand (*API* recommends the adoption of $\delta = \varphi - 5°$).

In this relationship, coefficient K is the main factor of indetermination since it in fact depends on:

– the initial conditions of the sandy formation, i.e. the density of the sand and the

---

(2) The density of a 40 % porosity sand is:

$$\gamma = (2.65 \times 0.60) + (1 \times 0.40) = 2 \text{ t/m}^3.$$

The apparent or submerged density is:

$$\gamma' = \gamma - 1 = 2 - 1 = 1 \text{ t/m}^3.$$

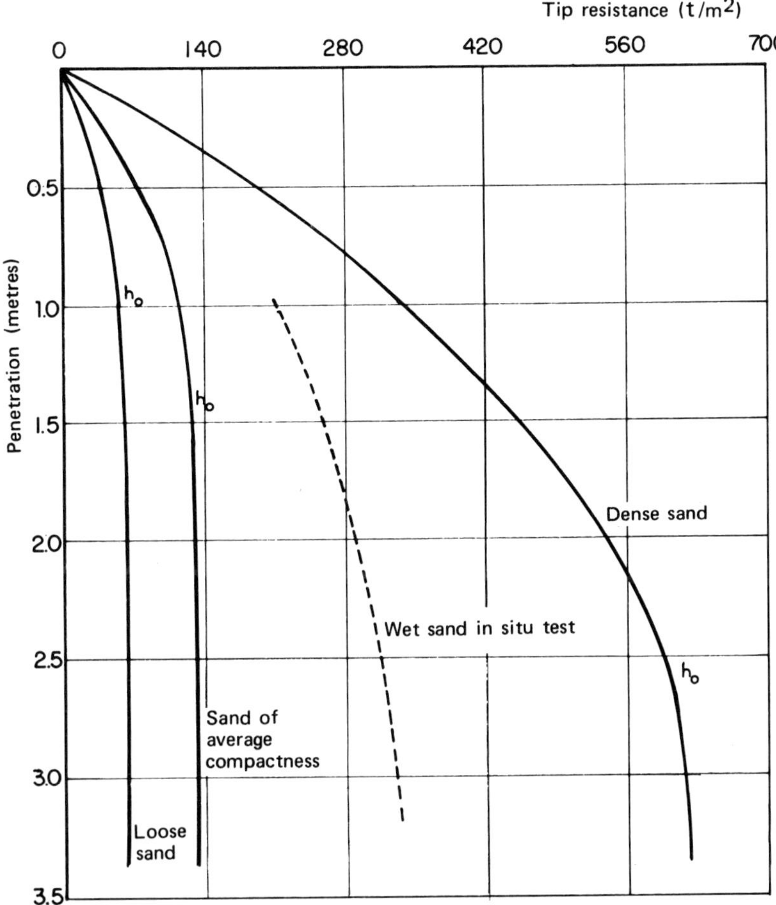

FIG. VII.3.5. – Variation of the tip resistance in terms of the penetration into sands ; 4" (100 mm) pile model (according to Vesic).

thrust coefficient of the soil at rest:

$$K_0 = \frac{\text{effective horizontal stress } p_h}{\text{effective geostatic stress } p_o}$$

(generally around 0.5),

– the changes in volumes that occur when inserting the pile, these being dependent on the shape of the pile and the method of insertion,

– the direction of loading of the pile: compression (bearing capacity) or tension (pull-out force).

For example, values of K lie within the following limits:

driven pile	$K = 0.7$ to 3,
driven pile with pilot hole	$K = 0.4$ to 0.7,
drilled pile	$K = 0.1$ to 0.4.

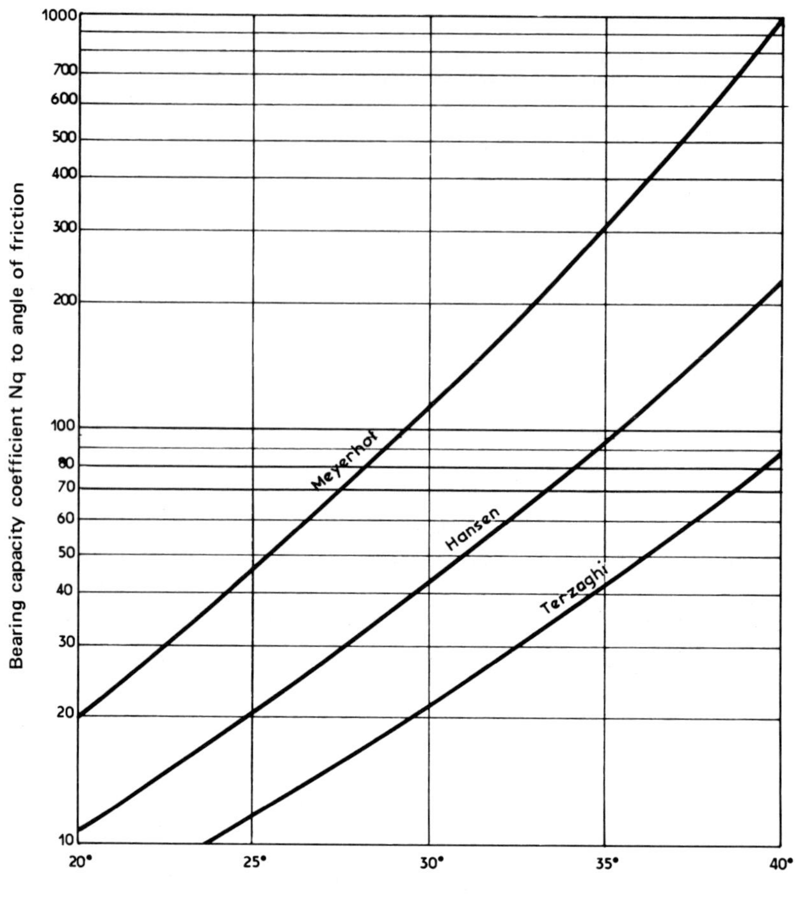

FIG. VII.3.6 – N_q coefficients for deep foundations.

Owing to the considerable degree of uncertainty in the calculation of the bearing capacity of piles in sand, *API* [12] recommends maximum values both for:

– the lateral friction, and
– the bearing capacity of the extremity.

These limit values, given in Table VII.3.1, are reached for a maximum penetration of about 25 m.

For clean sands ($c_u = 0$), the following are often adopted:

– the bearing force of tip

$$q = \frac{R_p}{4}$$

– the lateral friction:

$$f = \frac{R_p}{400}$$

where R_p designates the value of the resistance measured on the tip of the penetrometer.

TABLE VII.3.1

Type of soil	δ (degrees)	Maximum lateral friction, f_{max} (t/m²)	N_q	Maximum unit bearing capacity (t/m²)
Clean sand	30	10	40	1000
Silty sand	25	8,5	20	500
Sandy silt	20	7	12	300
Silt	15	5	8	200

The maximum values advocated for siliceous sands should absolutely not be applied to the case of **limestone sands**, for which the information would appear yet more imprecise.

Until further results are obtained, McClelland [4] recommends adopting the following values for calculating the bearing capacity of piles driven into limestone sands:

– maximum bearing force of extremity: $q_{max} = 500$ t/m² (5 MPa),
– maximum lateral friction: $f_{max} = 2$ t/m² (20 kPa).

VII.3.1.3 Examples of calculation of the penetration of driven piles

The uncertainty in the calculation of the bearing capacity of piles stems at one and the same time from:

– the scatter of the values of the mechanical characteristics measured in the laboratory and in situ,
– the uncertainty of the above-mentioned parameters: K, N_q and k.

As an example [5] the necessary penetration has been determined for a 36 in (90 cm) pile driven:

– either into a normally consolidated clay,
– or into a dense sand,

reaching a bearing capacity of 1,800 t.

In **normally consolidated clay** (Fig. VII.3.7a), calculation of the bearing capacity is carried out under 2 assumptions:

– $f = c_u$ $(k = 1)$,
– $f = 5$ t/m² for $c_u > 5$ t/m², $(k < 1)$ ($f = 50$ kPa, for $c_u > 50$ kPa).

The 1,800 t bearing capacity is reached for a penetration of:

– 83 m under the first assumption,
– 143 m for the second.

In dense sand (Fig. VII.3.7b), the calculation of the bearing capacity is based on the assumption:

$$\varphi = 35° \quad \text{and} \quad \varphi = 30°$$

Under these 2 assumptions:

– $K = 1$, $N_q = 94$ (according to Hansen), f and q directly proportional to p_0 (i.e. the depth),

– $K = 0.7$, $N_q = 41$ (according to Terzaghi), $f_{max} = 10$ t/m² (100 kPa), $q_{max} = 1{,}000$ t/m² (10 MPa).

The bearing capacity of 1,800 t is reached for a penetration of:

– 23 m under the first assumption,
– 53 m under the second.

Both in clay and sand, it can be seen that the penetrations calculated differ considerably with the assumptions adopted (proportionality of f and q with depth or limit values) (3).

– for sand, the limit values (for a certain critical depth) have been approximated thanks to various experimental studies,

– for clay, the limit values accepted still appear quite arbitrary, thus leaving the safety factor completely undetermined.

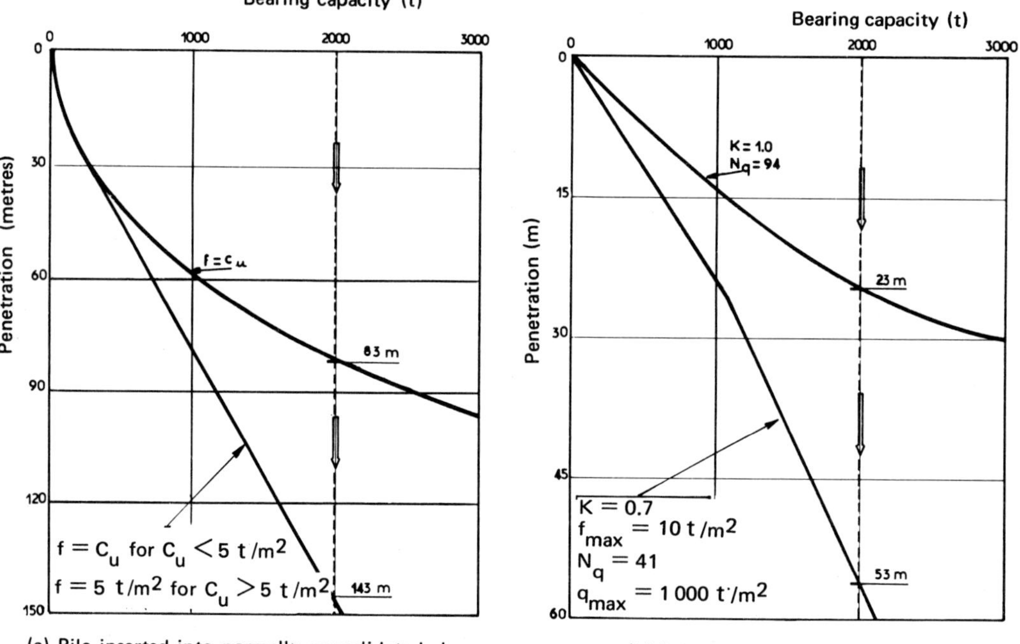

FIG. VII.3.7. – Variations in the penetration of a pile for a bearing capacity of 2,000 t, depending on the design assumptions (according to McClelland).

(3) Tendency is now to calculate bearing capacity of piles from effective stresses.

VII.3.1.4 Calculation of the bearing capacity of an insert-pile

While considerable uncertainty prevails as to the true bearing capacity of driven piles, the estimation becomes even more difficult in the case of drilled and grouted piles [15] as insert-piles.

Experience with concrete piles poured in place is limited to piles less than 20 m in length, as is shown by the results published by the *NGI* (see Fig. VII.3.8).

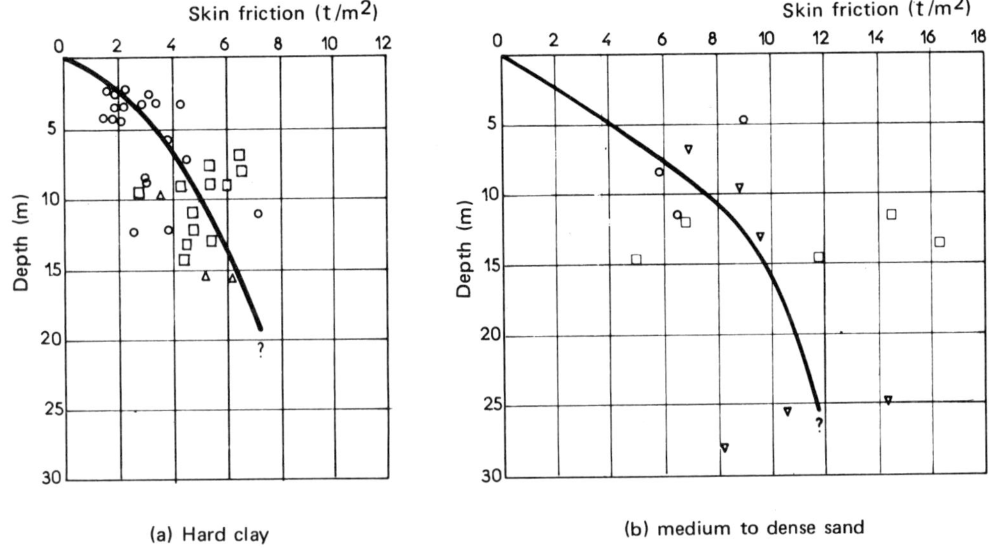

FIG. VII.3.8. – Skin friction stresses derived from cast pile tests (according to Ove Eide).

On the basis of experience gained, *NGI* [6] recommends use of the criteria summarized in Table VII.3.2 for the bearing capacity of insert-piles in sands and clayey soils.

TABLE VII.3.2

	Bearing capacity of extremity q	Lateral friction f
Sandy soils	$q = p_0 N_q$ $N_q(\varphi) = 40$ to 500 $q_{max} \approx 500$ to 600 t/m² in dense sands	$f = K p_0$ tg δ $K = 0.7$ (compression) $K = 0.5$ (tensile) $f_{max} \approx 10$ t/m² in dense sands
Clayey soils	$q = s_u N_c$ $N_c = 9$	$f = k \cdot s_u$ $k = 0.3$ in hard clays $f_{max} \approx 10$ to 12 t/m²

In reality, the bearing capacity of an insert-pile depends on a great many factors, among which the following may be mentioned:

- the nature and exact characteristics of the soil,
- the drilling technique used,
- parameters linked to the drilling mud: nature, pressure, displacement by the cement,
- parameters linked to the cement: pressure, swelling or contraction, etc.

Selective and repeated cementing under pressure in zones on the terrain increases the bearing capacity of piles compared with simple sealing by gravity (primary cementing).

This technique has been the subject of recent experiments showing that the lateral frictional strength is at least doubled compared with the conventional insert-pile technique (see Section VII.1.3). A considerable reduction in the necessary insertion can therefore be expected [15].

VII.3.1.5 Method of calculating the bearing capacities of a pile from pressuremeter measurements

The *TLM* method of calculating the bearing capacity [16] is based on pressuremeter measurements and knowledge of the lithology.

The unit bearing force of the tip, proportional to the limit pressure measured on the pressuremeter, results from the following relationship:

$$q_l - q_0 = k\,(p_l - p_0)$$

where:

q_l = unit bearing force,

p_l = limit pressure of the soil. In the case of heterogeneous soils, the equivalent limit pressure p_{el} is defined as the geometric mean of the values measured near the level of the tip,

p_0 = horizontal stress of the undisturbed soil,

q_0 = vertical stress of the undisturbed soil,

k = bearing capacity factor which is a function of the nature of the soil and the type of foundation (driven or drilled piles).

The mean values of k are given in Table VII.3.3.

TABLE VII.3.3

Type of soil	Limit pressure p_l (t/m²)	Bearing capacity factor k	
		Driven piles	Drilled piles
Soft soils	< 10 – 20	< 1	
Clayey soils	10 to 100	2	1.8
Limestone sands and chalk		< 2	
Loose to dense sands	70 to 100	3.6	3.2
Dense to highly dense sands	100 to 200	5.8	5.2

The term lateral friction is deduced from the limit pressure by means of the diagram shown in Fig. VII.3.9.

The upper limit of the lateral friction, which is equal to the shear strength s_u of the soil, is expressed (in principle) by the following approximate formulae:

$$f = s_u = \frac{p_l - p_0}{5.5}, \text{ in clayey soils}$$

$$f = s_u = \frac{p_l - p_0}{8}, \text{ in sands}$$

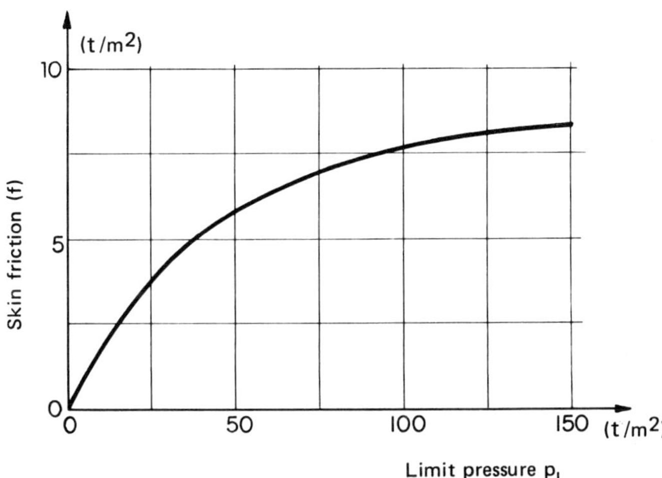

FIG. VII.3.9. – Skin friction in terms of limit pressure measured on a pressure-meter (according to Menard documentation).

VII.3.1.6 Change in the bearing capacity of piles with time

The bearing capacity of a pile is calculated from the shear strength of the soil measured during a rapid (undrained) test subject to a varying degree of disturbance (depending on the techniques used), and invariably unknown. Driving the pile also leads to disturbance, all the greater the more sensitive the soil, i.e. the greater the thixotropy or tendency to liquefy.

Furthermore, the choice of the limit values of the bearing strength terms q and f is, to say the least, arbitrary.

It may then be reasonably asked what the real bearing capacity of the pile is and how this value changes with time, depending on the nature of the terrains.

In the case of clayey soils, the shear strength, which is at a minimum after disturbance, increases more or less slowly with time, reaching the true shear strength of the soil [5]. This increase in the strength of the soil owing to recovery occurs (see Fig. VII.3.10):

– quickly, after interrupting pile driving,
– subsequently, more gradually with time after driving is completed.

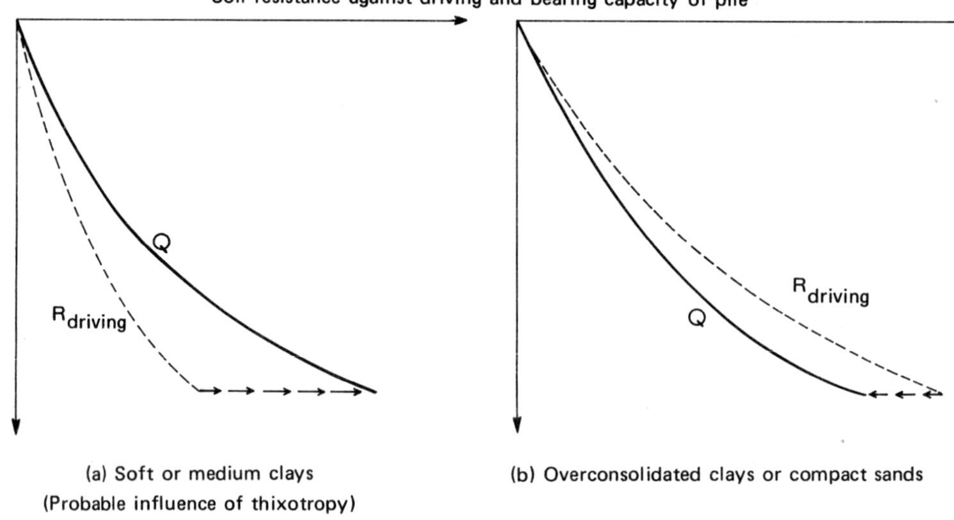

FIG. VII.3.10. – Curve of the bearing capacity of a pile in terms of time after driving the pile (according to McClelland).

The bearing capacity (static) of the pile calculated from the above-indicated formulae naturally presumes that the "recovery" of the soil is complete.

The effect of the horizontal loadings results in a reduction in the friction of the soil on the upper section of the pile, leading to a loss in its bearing capacity:

– in soft or medium consolidation clays, the loss will probably be temporary owing to subsequent gradual "recovery",

– in overconsolidated clays or compact sands on the contrary, the reduction in friction and consequently of the bearing capacity may be permanent.

The effect of scour around the piles is to reduce the depth of insertion and hence the bearing capacity (see Section VII.1.1.4).

Two main types of solutions have been used to counter scour around piles:

– first, enrockments arranged immediately around them,

– second, sheets of plastic 5 to 10 m in diameter (Fig. VII.3.11) secured to the pile at the centre and held down around the edges by enrockments [17].

VII.3.2 Design of piles subject to horizontal loads

While the static load Q to be applied to each pile determines the penetrations h, the action of the horizontal loads enables the thicknesses of the pile necessary at different depths to be calculated.

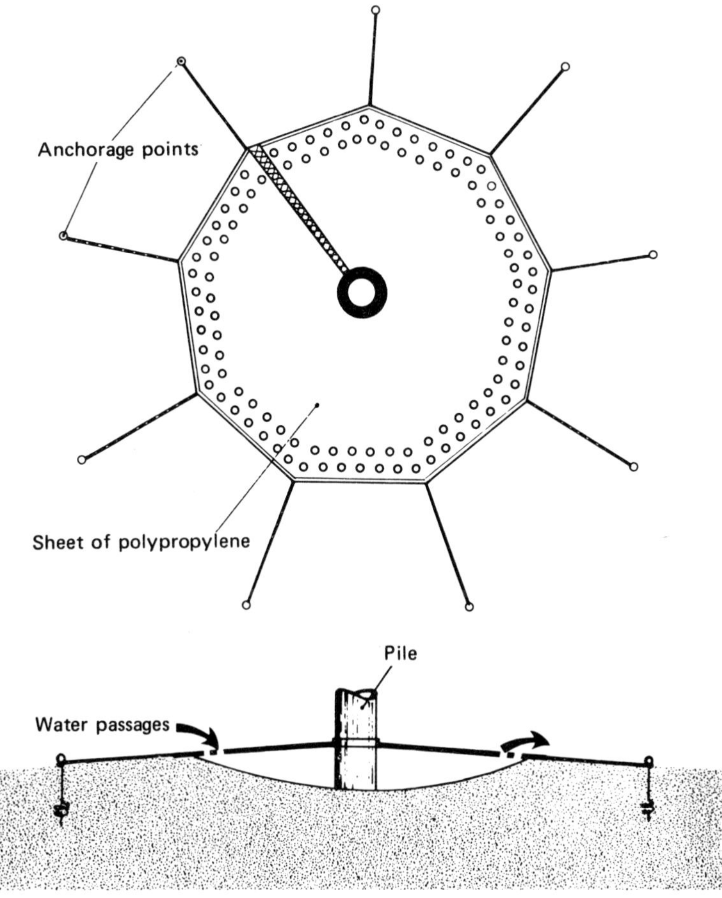

FIG. VII.3.11. – Anti-scour device by means of a sheet of polypropylene (*Seditech*).

All the various calculating methods proposed the solution of the following differential equation [18]:

$$E_a I \frac{d^4 y}{dz^4} = - p(y, z)$$

where:

y = lateral displacement of pile in the soil,
z = the variable of the depth in the soil (beneath sea level),
p = the reaction of the soil to lateral movement,
$E_a I$ = the rigidity of the pile (E_a being the modulus of elasticity of steel and I the moment of inertia).

Solution of this differential equation presupposes knowledge of function $p(y,z)$, i.e. the behaviour of the soil, since $p(y,z)$ represents the reaction of the soil to a lateral shift y of the pile at a depth of z.

The various methods of calculation advocated differ in the assumptions as to function $p(y,z)$ and the limit conditions. The following will be examined in turn:

– the two methods of Matlock and Menard which make the simplifying assumption of a flexible soil, though offering the advantage of permitting a quick approximate approximation,
– the more elaborated method of curves (p,y), requiring the use of a computer.

VII.3.2.1 The Matlock method

The reactive force exerted by a soil against lateral loads is defined in terms of the deflection of the pile in accordance with the following relationship [1]:

$$p = E_s y$$

where:

p = reaction of soil per unit length,
y = lateral deflection,
E_s = modulus of deformation of the soil.

The method proposed by Matlock and accepted by McClelland assumes that the modulus of deformation of the soil increases in proportion to the depth:

$$E_s = kz$$

where:

k (a constant) is homogeneous with a density.

The values of k given by McClelland are given in Table VII.3.4.

TABLE VII.3.4

Type of soil	k (t/m³)
Normally consolidated clay:	
very soft	15 to 90
consistent	60 to 300
Slightly overconsolidated clay ($c_u > 2$ t/m²)	150 to 600
Loose sand	150 to 450
Dense sand	450 to 1,500

The characteristic length resulting from the solution of the differential equation is expressed as follows:

$$l_0 = \sqrt[5]{\frac{E_a I}{k}}$$

The bending moment in the pile inserted in the soil is given by the relationship:

$$M = M_0 \left[A \frac{T_0 l_0}{M_0} + B \right]$$

where:

M_0 = bending moment of the pile at soil level (i.e. the sea bottom),
T_0 = shear load in this section,
A and B = coefficients defined as a function of the reduced depth: $Z = z/l_0$ (see Fig. VII.3.12a).

In this method of calculation, the depth of insertion subject to horizontal loads is equivalent to about $4l_0$.

Numerical application. As an example, let us consider:

– a pile 1 m in diameter and 3.8 cm (1 1/2") thick,
– a soil characterized by a constant k of 100 t/m^3.

A characteristic length l_0 of about 5 m is deduced.

The depth of insertion subject to horizontal loads is therefore about 20 m.

The bending moment M_0 and the shear load T_0 at the level of the sea bottom are calculated from the horizontal and vertical loads applied to the structure. The following are assumed:

$$M_0 = 300 \text{ t} \times \text{m}$$
$$T_0 = 10 \text{ t}$$

The bending moment in the pile M then reaches its maximum M_{max} of about 315 t × m at a depth of $Z = 0.6$, i.e. $z = 3$ m.

The corresponding maximum bending stress is 22.5 kg/mm^2.

VII.3.2.2 The Menard method

The method used by Menard [19] makes the following assumption:

$$p = k_M y$$

where:

k_M = a constant known as the horizontal modulus of reaction of the soil and expressed in t/m^2 per centimetre where p is in t/m^2 and y in centimetres.

The modulus of horizontal reaction k_M is related to the pressuremeter modulus E_p by a complex relationship involving a rheological parameter α of the soil and the radius of the pile.

Depending on the nature and consolidation of the soil, a pressuremeter modulus varies from a few tens of t/m^2 to several hundred t/m^2. The modulus of reaction k_M then varies from about 1 to 100 t/m^2/cm (10 to 1,000 kPa/cm).

The characteristic length l_0 related to the diameter d and the rigidity E_aI of the pile results from the relationship:

$$l_0 = \sqrt[4]{\frac{4\, E_a I}{k_M\, d}}$$

The bending moment in the pile inserted in the soil is then given by the expression:

$$M = M_0 \left[A_M \frac{T_0 l_0}{M_0} + B_M \right]$$

where:

M_0 and T_0 = respectively, the bending moment and the shear load of the pile at soil level,

A_M and B_M = coefficients defined in terms of the reduced depth $Z = \frac{z}{l_0}$ (Fig. VII.3.12b).

In the Menard method, the depth of insertion subject to horizontal loads is equivalent to $3\, l_0$.

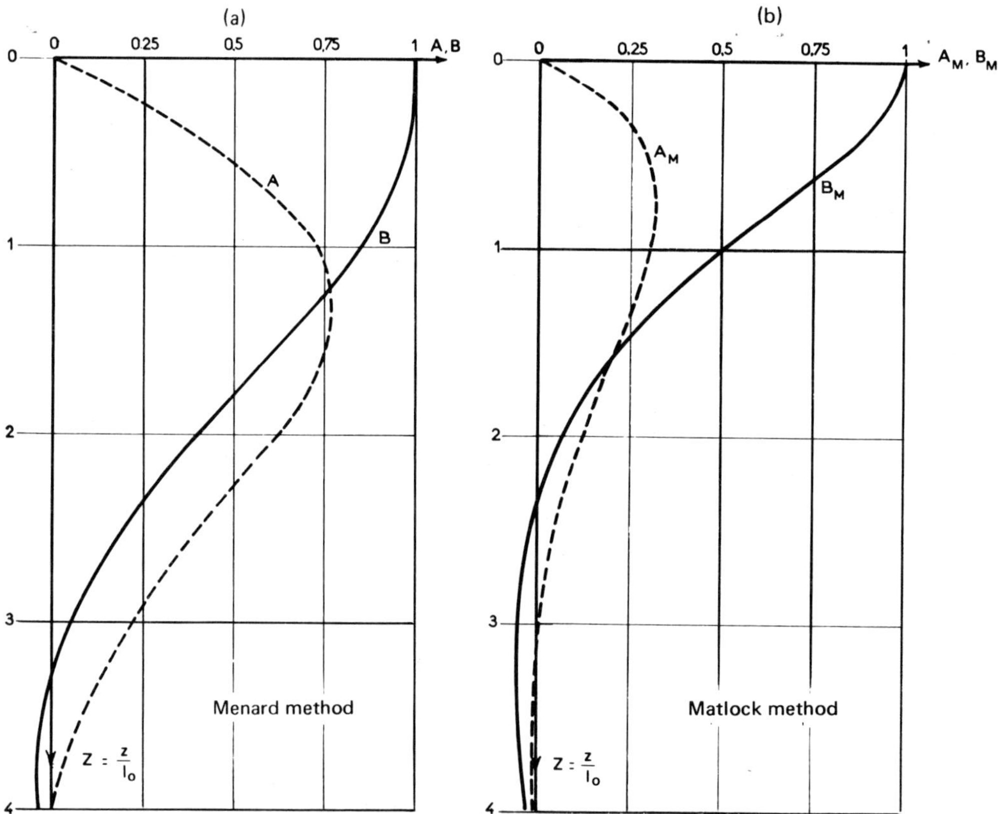

FIG. VII.3.12. – Curves of the bending movement of piles (Matlock, Menard).

Numerical application. As an example, let us consider:

– a pile with a diameter of 1 m and a thickness of 3.8 cm (1 1/2"),

– a soil displaying a pressuremeter modulus E_p of 300 t/m^2, (3 MPa) i.e. a modulus of reaction k_M of 10 t/m^2/cm (100 kPa/cm).

A characteristic length l_0 of about 6 m is deduced.

The depths of insertion subject to horizontal loads is therefore about 18 m, or in other words very much the same as that obtained earlier (see Section VII.3.2.1).

Assuming as before (see Section VII.3.2.1) that:

$$M_0 = 300 \text{ t} \times \text{m}$$

$$T_0 = 10 \text{ t}$$

the bending moment in the pile reaches a maximum in the immediate neighbourhood of sea level, i.e. M_{max} is about 300 t.

The result is a maximum bending stress in the pile of 21.5 kg/mm^2.

It will therefore be observed that for the same depth of insertion subject to horizontal loads, the bending stresses in the pile as calculated by the two methods are quite identical.

VII.3.2.3 Method of calculation from the $p - y$ curves

As indicated earlier, the two preceding methods offer the advantage of simplicity, making it possible by manual calculation to make an estimate at the preliminary project stage.

The more sophisticated method proposed by *API* [12] [20] considers the $p - y$ curves (Fig. VII.3.13) corresponding to various stresses deduced from the "stress-strain" curves of the soils (at various levels z) determined by conventional tests (according to *ASTM*) namely:

– the in situ vane test,

– unconfined compression,

– the laboratory vane test.

The soil is broken down into n superimposed layers which are not intercoupled. In addition, each layer is assumed to display an elastic/plastic behaviour or possibly an even more complex one.

Determination of the $p - y$ curves requires evaluation of the limit pressure p_u and the modulus of reaction E_s in terms of the effective geostatic stress p_0 and the characteristics of cohesion c_u or angle of friction φ of the soil.

This method of calculation obviously requires a computer.

In the case of **sands** [21], *API* standard RP 2A recommends adoption of the following relationships for the evaluation of:

– **the limit pressure** p_u:

$$p_u = \frac{p^*}{d} = 3p_0 \frac{1 + \sin\varphi}{1 - \sin\varphi}$$

where:

p^* = pressure above which the soil reaction remains constant,
p_0 = effective geostatic stress,
d = diameter of pile,
φ = internal angle of friction.

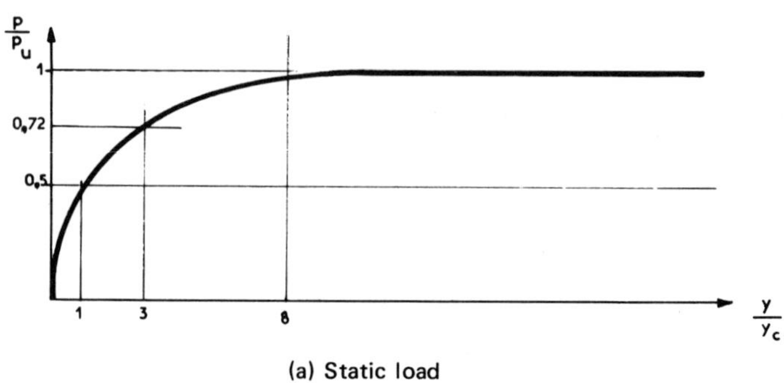

(a) Static load

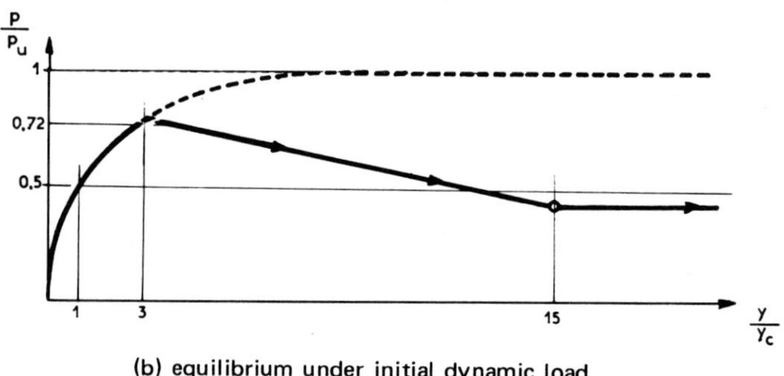

(b) equilibrium under initial dynamic load

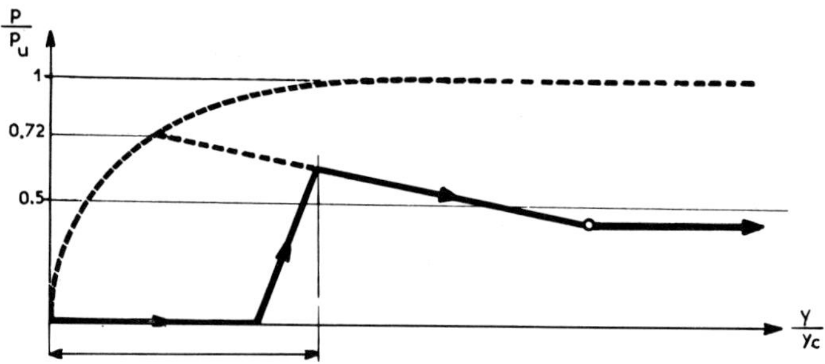

(c) Reloading after preliminary cyclic loads

FIG. VII.3.13. – Determination of the $p - y$ curves for clays.

– **the modulus of reaction** E_s:

$$E_s = \frac{p^*}{y} = \frac{A\, p_0}{1.35\, d}$$

where:

A = coefficient of density varying from 200 for loose sand to 2,000 for dense sand.

The $p - y$ curve is of the elastic-plastic type.

In the case of **soft clays** ($c_u < 10$ t/m²) ($c_u < 100$ kPa), [20] , the *API* standard recommends the following method for determining the $p - y$ curves:

– **limit pressure** p_u :

– p_u increases linearly from 0 to 9 c_u as the depth increases from 0 to h_r,
– $p_u = 9\, c_u$ for a depth $h > h_r$,
– h_r is determined by the equation:

$$h_r = \frac{6\, d}{\dfrac{\gamma' d}{c_u} + 0.5}$$

where:

d = diameter of pile,
γ' = submerged density of the soil,
c_u = apparent cohesion of the soil.

– **plotting of the** $p - y$ **curve resulting from a static test:**

– the $p - y$ curves can be plotted from the values in Table VII.3.5 (see Fig. VII.3.13a).

TABLE VII.3.5

$\frac{p/d}{p_u}$	$\frac{y}{y_c}$
0	0
0.5	1
0.72	3
1	8
1	∞

where:

$y_c = 2.5\, \epsilon_c \,.\, d$

ϵ_c represents the relative displacement which occurs on a sample subjected to an undrained triaxial test for a compression equal to half that causing rupture in the sample.

– **plotting of the *p-y* curves resulting from the action of cyclic loads:**

When the soil is subjected to cyclic loads, the *p-y* curves can be obtained from the values given in Table VII.3.6 (see Fig. VII.3.13b).

TABLE VII.3.6

$h > h_r$		$h < h_r$	
$\frac{p/d}{p_u}$	$\frac{y}{y_c}$	$\frac{p/d}{p_u}$	$\frac{y}{y_c}$
0	0	0	0
0.5	1	0.5	1
0.72	3	0.72	3
0.72	∞	$0.72\ \frac{h}{h_r}$	15
		$0.72\ \frac{h}{h_r}$	∞

If after cyclic application of loads the soil does not completely recover, the *p-y* curves corresponding to reloading are of the type shown in Figure VII.3.13c.

In the case of **stiff clays** ($c_u > 10$ t/m²) ($c_u > 100$ kPa) the *API* standard recommends the application of coefficients of less than 1 to the p/p_u ordinates (calculated above for soft clays) in order to allow for their greater fragility when subjected to cyclic loads.

BIBLIOGRAPHY

[1] McCLELLAND (B.). – "Fixed Structures, Foundations", *Handbook of Ocean and Underwater Engineering.*, Myers (J.,) Holm (C.) and MacAllister (R.), Section 8. McGraw Hill, 1969.

[2] GIRARD (J.M.). – "Battage des piles", *Forages*, Bull. No. 68, July September 1975.

[3] JANSZ (J.W.), VAN HAMME (V.), GERRITSE (A.) and BOMER (H.). – "Controlled Piledriving Above and Under Water with a Hydraulic Hammer", *OTC*, Houston 1976, paper OTC 2477.

[4] McCLELLAND (B.) – "Design of Deep Penetration Piles for Ocean Structures", *J. of the Geotech. Eng. Division*, July, 1974.

[5] McCLELLAND (B.), FOCHT (J.A.) and EMRICH (W.J.). – "Problems in Design and Installation of Offshore Piles". *J. of the Soil Mech. and Found. Division*, Nov. 1969.

[6] EIDE (O.). – "Marine Soil Mechanics. Applications to North Sea Offshores Structures", *Offshore North Sea Tech. Conf.*, Stavanger 3-6 sept. 1974. (Preprint NGI. Publication No. 103).

[7] GABAIX (J.C.), BUSTAMANTE (M.) and GOUVENOT (D.). – "Essais de pieux scellés par injection sous pression", *Annales de l'ITBTB*, No. 331, Sept. 1975.

[8] GOUVENOT (D.) and GABAIX (J.C.). – "A New Foundation Technique Using Piles Sealed by Cement Grout under High Pressure"., *OTC*, Houston, May 5-8, 1975. Paper No OTC 2310.

[9] "First Platform Set in 400,000 bpd Forties Field", *World Oil,* 1 August 1974.

[10] TAYLOR (D.M.). – "BP Sets the First Platform Jacket in Forties Field", *Ocean Industry.* August 1974.

[11] KRAUSE (E.R.) and MOSCH (J.M.). – "Sound Basis for North Sea Platforms. Programming of Essential Geotechnical Surveys". *SPE European Spring Mtg,* London, 14-15 April. 1975, Paper SPE 5265.

[12] *API* – "Recommended Practice for Planning, Designing and Constructing Fixed Offshore Platforms", *API,* RP2A, Fifth Edition, January 1974.

[13] *Det Norske Veritas* – "Tentative Rules for the Design Construction and Inspection of Fixed Offshore Structures", 1974.

[14] VIJAYVERGIYA (V.N.) and FOCHT (J.A.). – "A New Way to Predict the Capacity of Piles in Clay"). *4th Ann. OTC,* Houston, vol. II. 1972.

[15] KRAFT (L.M.Jr.) and LYONS (C.G.). – "State of the Art : Ultimate Axial Capacity of Grouted Piles". *OTC,* Houston, 1974. Paper OTC 2081.

[16] *TLM* – "Règles d'utilisation des techniques pressiométriques et d'exploitation des résultats obtenus pour le calcul des fondations". *General description.* D 60., Ed. Jan. 1975.

[17] WATSON (T.). – "Scour in the North Sea" , Paper SPE 4324, *Second Annual European Mtg of SPE.* London, April 2-3, 1973.

[18] PARRY (R.H.G.). – "Pile Design". *Offshore Soil Mechanics.* Published by Cambridge University, Engineering Department on Lloyd's Register of Shipping, September 1976, pp. 177-223.

[19] MENARD (L.), BOURDON (G.) et GAMBIN (M.). – "Méthode générale de calcul d'un rideau ou d'un pieu sollicité horizontalement, en fonction des résultats pressiométriques" , *Sols-Soils,* Nos. 22 and 23, 1968.

[20] MATLOCK (H.). – "Correlations for Design of Laterally Loaded Piles in Soft Clays" , *2nd Ann. OTC,* Houston, 1970.

[21] REESE (L.C.), COX (W.R.) and KOOP (F.D.). – "Analysis of Laterally Loaded Piles in Sand". *OTC,* Houston, 1974. Paper OTC 2080.

CHAPTER VIII

Gravity Structures: Platforms and Storage Tanks

CONTENTS

INTRODUCTION

1. Gravity structures, generally built of reinforced or prestressed concrete, used as production platforms and/or tanks for storage at sea, are built on a deep water coastal site and then towed to the immersion point.

These structures are essentially characterized by the fact that they are rendered stable by means of gravity thanks to the dimensions and weight of the base (generally of concrete), thus not requiring any anchoring to the bottom of the sea. Accordingly, the time needed for installation on site is reduced to a minimum (1 or 2 days) which, under the conditions prevalent particularly in the North Sea, is an obvious advantage compared with fixed platforms on piles.

From the standpoint of the foundations, gravity structures:

– are similar to surface foundation structures consisting of a concrete floor,

– though nonetheless differing by the absence of any excavation pit which is generally executed before such a floor is laid.

2. Several types of gravity structures, differing considerably in their geometries and the dimensions of their bases, have at present been set up, or are now being built or in the project stage. From the standpoint of stability, these various structures fall into three main categories (1) [1] [2].

All-concrete structures resting on the soil on a large dimensioned slab comprising particularly (see Fig. VIII.1.1):

– the CG Doris structures protected by alveolar screen (Jarlan wall), like the Ekofisk tank (in place since July 1973), the CDP1 platform on the Frigg field, the platform on the Ninian field,

– platforms comprising a Condeep storage tank with a hexagonal base consisting of cylindrical cells forming a "flexible" structure, now in place, on the Brent field, Beryl field, Frigg field, etc.,

– the Sea Tank Co/McAlpine square-base platforms in place on the Frigg field, Brent field, etc.,

– the square-base Andoc platforms, the hexagonal-base Taylor Woodrow, or Subtank, etc. types, complete with storage tanks.

---

(1) Gravity bases, oscillating columns gravity laid on the bottom, etc., can be categorized from the standpoint of stability with the gravity structures considered in the present chapter.

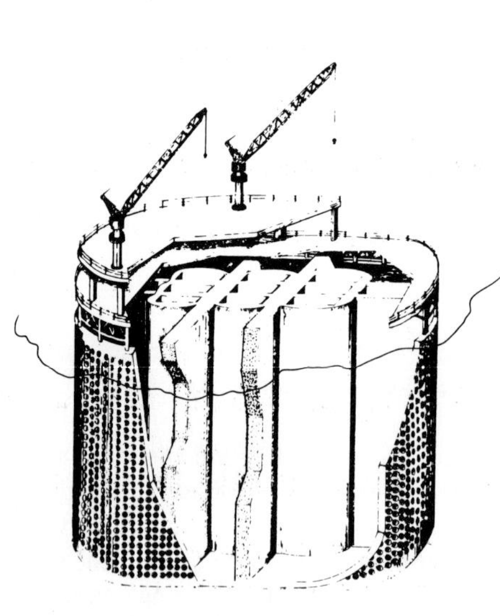

C.G. Doris (Ekofisk)

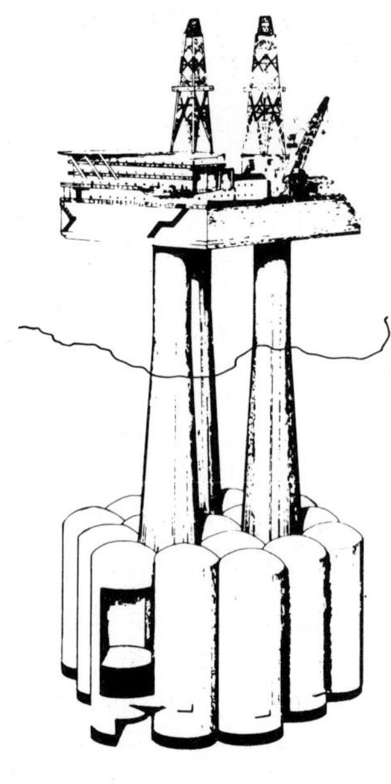

Condeep

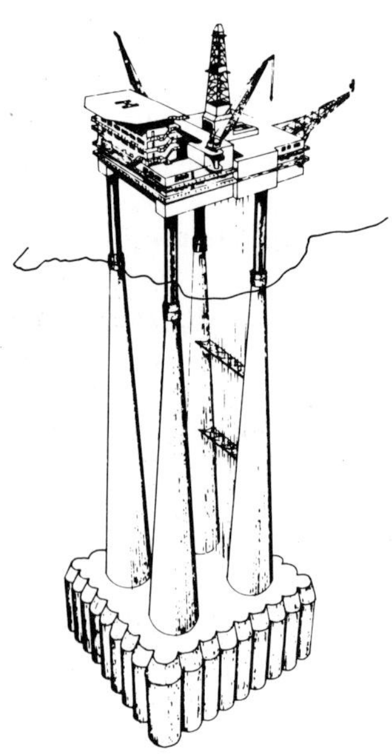

Andoc

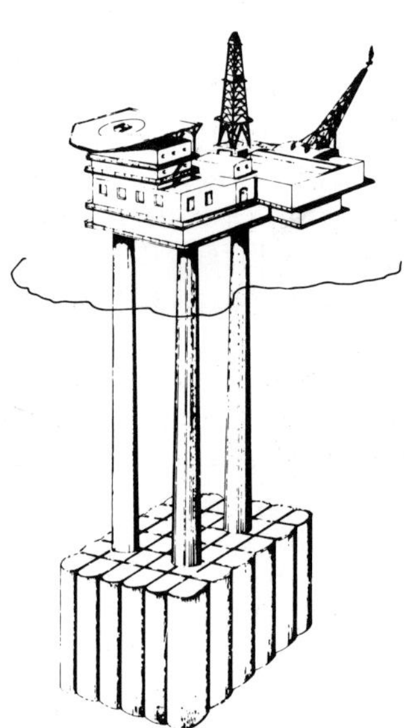

Sea Tank Co.

FIG. VIII.1.1. – Various gravity structures.

Hybrid structures, with a concrete base carrying a steel lattice structure: the North Sea RDL, Christiani and Nelsen, Brown and Root, etc., are still in the project stage.

Tripod structures, also in the project stage, resting on the bottom on three concrete or steel bases, comprise:

– the all-concrete Selmer-Tripod platform with a storage tank,
– the steel and concrete platforms of Chicago Bridge, Technomare, Tristan Engineering Ltd,
– the "Mobil Monopod" ETA jack-up platform with a single column and three concrete bases.

The square-base GSB (*CFEM*) structure is to be built entirely in steel.

3. Reconnaissance of the sea bottoms and soils before installing a gravity base structure calls for particular care:

– bathymetry must give a highly precise description of the topography of the bottoms (flatness and horizontality),
– seismic prospecting calls for rigorous interpretation to determine any geological anomalies that may exist near particularly the surface (in the first 20 or 30 m),
– geotechnical reconnaissance must comprise coring and in situ measurements (in particularly penetrometry) down to a depth of a few tens of metres, together with deep coring to about 100 m.

Although gravity structures can theoretically be laid on more or less consolidated bottoms, their installation would appear to have been considered so far only on the highly consolidated soils (sands or clays) of the North Sea.

4. In this chapter, the following will be examined:

– the geotechnical problems raised by the installation, stability and movement of gravity structures, together with the depths of reconnaissance to be reached,
– the geophysical and geotechnical reconnaissances needed before siting gravity structures,
– the significance of the geotechnical characteristics, i.e. bearing capacity and distortion of highly consolidated soils, determined in the laboratory and in situ, in terms of the quality of the cores and the application of the measuring instruments,
– the use of the results of reconnaissance of soils for calculating the stability of the foundations of gravity structures.

VIII.1 DEFINITION OF THE GEOTECHNICAL PROBLEMS INVOLVED IN THE INSTALLATION AND STABILITY OF GRAVITY STRUCTURES

The geotechnical problems involved in the implantation of gravity structures with different geometries, concern:

– the installation of the structures,
– the stability of the structure under the effect of its weight and repeated loads,
– the vertical and horizontal movement of the structure,
– the settlement of the soil beneath the structures,
– the scour around the structure.

VIII.1.1 Installation of a gravity structure

The slab of a gravity structure may or may not be fitted with skirts arranged around its periphery or beneath each cylindrical compartment. In Paragraph VIII.3 the possible advantages of skirts, depending on their height and the nature of the surface soils, will be examined. Generally speaking, skirts will be capable of reducing:

– the risk of the structure sliding on the soil,
– the effects of erosion (scour) near the structure.

The installation of a gravity structure, governed by the topography of the sea bottom and the mechanical characteristics of the surface soil, raises two major problems related to [3] [4]:

– the penetration of the skirts into the soil,
– the contact stresses between the soil and the slab, depending on the irregularities of the bottom.

VIII.1.1.1 Penetration of the skirts into the soil

The geometry and the dimensions of the skirts or similar devices differ with the type of structure and the nature of the soils:

– skirts a few metres long mounted around each cylinder of the Condeep structures (Fig. VIII.1.2) or beneath other structures,

– skirts 0.30 to 0.40 m deep placed beneath the floor (with different radii and around the periphery), as for the Ekofisk tank,

– various rugosities mounted beneath the floor.

It is important for the stability of the structure that the entire height of the skirts must penetrate into the soil. Calculation of the penetration force must therefore be made, considering the maximum shear strength of the soil (determined either in the laboratory or in situ).

The penetration of the skirts is obtained:

– under the effect of the weight of the structure, ballasted with water (and if necessary with sand),

– thanks to the reduction of the water pressure inside the skirts (particularly in the case of very low permeability clays).

In the case of structures set on irregular or sloping bottoms (Fig. VIII.1.3a), it is indispensable to fill in the free voids within the skirts, for instance by means of cement grouting. Means must also be provided for evacuating the water.

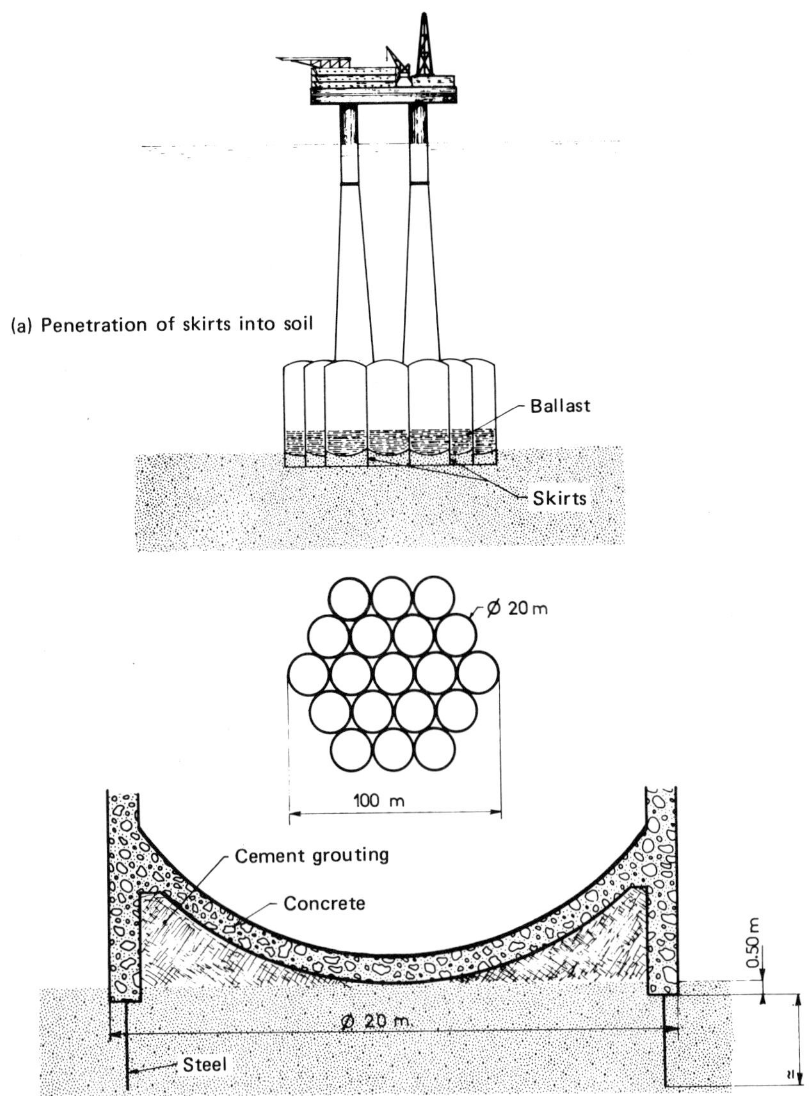

(b) Arrangement of skirts beneath each compartment

FIG. VIII.1.2. – Base of a Condeep platform.

The flattening of the surface of the soil by addition of material before installing the structure may also be envisaged. It would however appear that this preparation can be avoided in the North Sea in view of the sites for structures at present recognized (irregularities of only a few decimetres at the most and slopes often less than 1 %, etc.). However, the prior removal of boulders or erratic blocks is sometimes indispensable on certain sites [5].

VIII.1.1.2 Contact stresses between the soil and the slab

The contact stresses between the soil and the slab depend at one and the same time on the topography of the site and the characteristics of the soils [6].

In the case of **highly dense sands or very stiff clays**, the installation of a slab structure (floor) requires statistical knowledge of the topography of the soil, and this knowledge should be as perfect as possible. The estimation of the contact stresses between the soil and the structure will be accomplished thanks to this statistical knowledge, together with formulae for the bearing capacity and elastic deformation. In actual fact, it is never possible to obtain a highly accurate "map" of the irregularities on the surface, and the structure is invariably designed for more or less broken up surface conditions on the bottom (Fig. VIII.1.3a). It would appear important for the installation of the structure:

- that the height of the domes is no more than a few decimetres,
- and that the slope of the bottom is less than 1%.

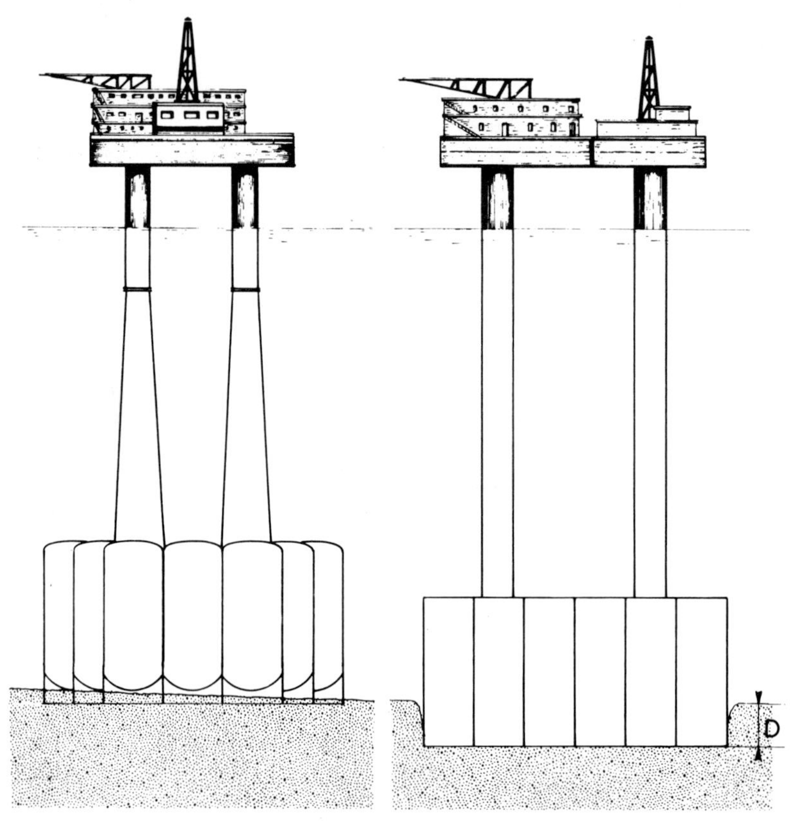

FIG. VIII.1.3. – Gravity structures : "soil-mat" contact.

The contact stresses may be more uniformly distributed once the structure has been installed by cement grouting beneath the slab. The presence of the skirts offers an obvious advantage by limiting the amount of concrete that has to be grouted to the absolute minimum.

In the case of **soft or low consolidation soils** on the surface, the irregularities of the bottom are of relatively little importance: the installation of the structure will result in consolidation of the soil (lateral creep) until a balance is reached between the weight of the structure and the bearing capacity of the soil (Fig. VIII.1.3b).

The installation of gravity structures on soft soils so far has remained only at the project stage, and the choice of the type of platform (mounted on piles or gravity) will be made from the initial results of soil reconnaissance on the site.

VIII.1.2 Stability of a gravity structure under static loads and cyclic sollicitations

Any surface foundation must be designed from the twofold standpoint of:

– stability, governed by the bearing capacity of the soil under the various static or dynamic loading conditions,

– the acceptable deformation, controlled by the total or differential stettlement.

After analysis of the stability, it would appear that rupture of the foundations of gravity structures could occur (Fig. VIII.1.4) [7] [8] [9]:

– by slip,

– by plastic rupture of the soil,

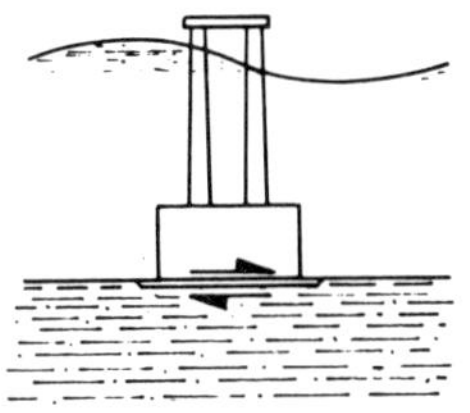

(a) Sliding of the structure along the soil

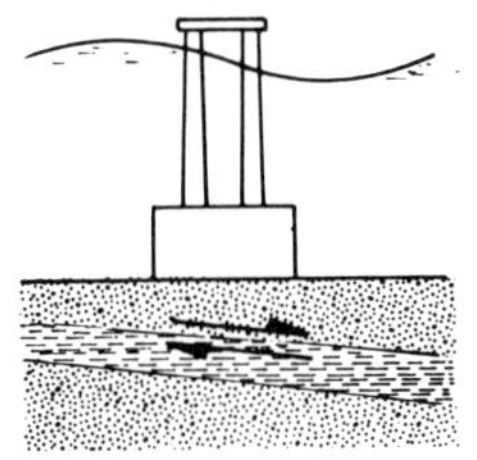

(b) Sliding of the foundation soil at the sand-clay interface

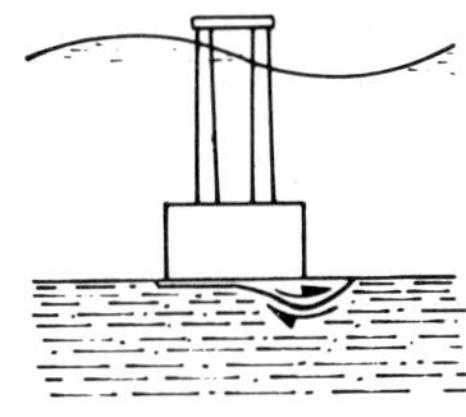

(c) Rupture of the foundation soil

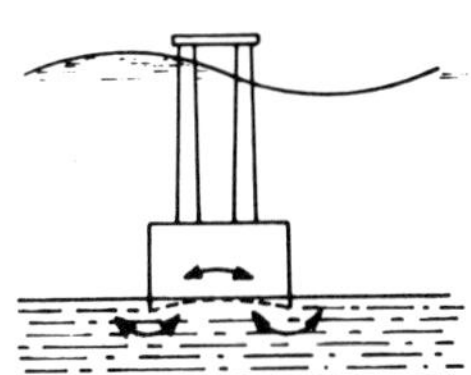

(d) Rupture of the foundation through "rocking"

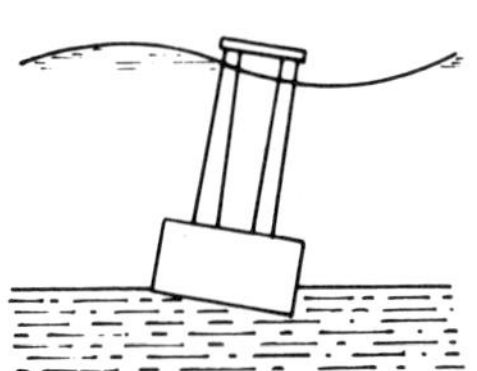

(e) Rupture of the foundation by liquefaction of the soil

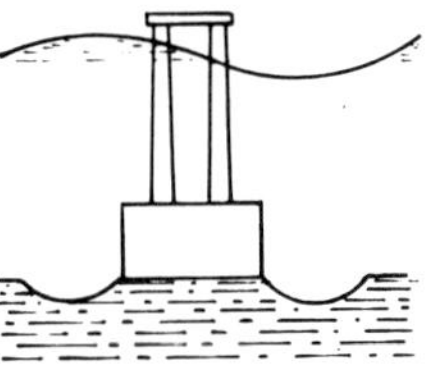

(f) Instability of the foundation caused by scour

FIG. VIII.1.4. – Stability of a gravity structure. Various possibilities of rupture of the foundation.

– by local stress concentration beneath the structure,
– by liquefaction of the soil (in theory),
– as a result of scour.

VIII.1.2.1 Slip of the structure or of the underlying layers of soils

Slip between the base of the structure and the surface of the soil can occur as a result of horizontal stresses, particularly if the shear strength at the interface is less than that of the soil itself (Fig. VIII.1.4a).

The risk of slip of the structure would appear to be reduced by fitting skirts (penetrating several metres into the soil). It is however considered that slip of a layer of clay can occur even at the level of penetration of the skirts.

Slip is also liable to occur between a layer of sand and a layer of clay (Fig. VIII.1.4b), particularly if the strata gradient is considerable ($> 10°$ for instance) at relatively low depths (a few metres) where the stresses induced by the structure are still high (see Section VIII.1.5). The risk is increased by the development of negative pressures in layers of clay of low permeability. This problem is then similar to that of the stability of slopes, which has been studied considerably in civil engineering. By carrying out shear tests on samples taken in the levels formed by alternating clay and sand, the possible risk of slip between layers can be checked.

VIII.1.2.2 Rupture of the soil beneath the gravity base

Rupture of the soil beneath the structure occurs if the shear stress exceeds the shear strength of the soil (Fig. VIII.1.4c). Movement of the structure under wave action (see Section VIII.1.4.2) obviously increases the shear stress in the soil by reducing the effective bearing area of the structure.

In **clays**, this method of rupture is typical beneath surface foundations. The decrease in the shear strength as a result of repeated loads (fatigue effect?) accentuates the risk of rupture.

In **sands**, possibilities of rupture under repeated loads depend particularly on the undrained shear strength. However, the risk of rupture of the foundation assuming drained and undrained shear strength will nonetheless be examined.

VIII.1.2.3 Local stress concentrations beneath the structure

Two very different cases must be considered.

Under the effect of repeated loads, high shear stresses can lead to considerable distortions along the edges of the foundation (Fig. VIII.1.4d). The result is a softening of the soil around the periphery of the foundations, leading to apparent hardening (rocking) in the central section, and accordingly to a local increase in the stresses and instability of the structure.

No rigorous analytical solution of this problem would yet appear to have been developed. An acceptable approximate solution consists in determining the bearing strength of the soils and the conditions of distortion and stresses at the edges of the foundation.

The existence of local humps related to the topography of the bottoms and the distortion properties of the top layers of the soils (see Section VIII.1.1.2) results in stress concentration beneath the structure. The level of these stresses depends on the statistical distribution of the humps and their differences in level.

VIII.1.2.4 Liquefaction of the soil

According to certain hypotheses, rupture of the foundation soils could occur under repeated loadings on the structure during a storm (Fig. VIII.1.4e). However, actual reality would appear to differ very widely, depending on the nature of the soils.

In **sands**, the rise in interstitial pressure in the absence of drainage results in gradual reduction of the effective or intergranular stress. In the limit, the sand behaves as a liquid and loses all its shear strength.

The uniform size distributions of the sands on different sites in the North Sea would lie within the size distribution limits of sands liable to liquefy, as can be verified under certain experimental conditions (Fig. VIII.1.5).

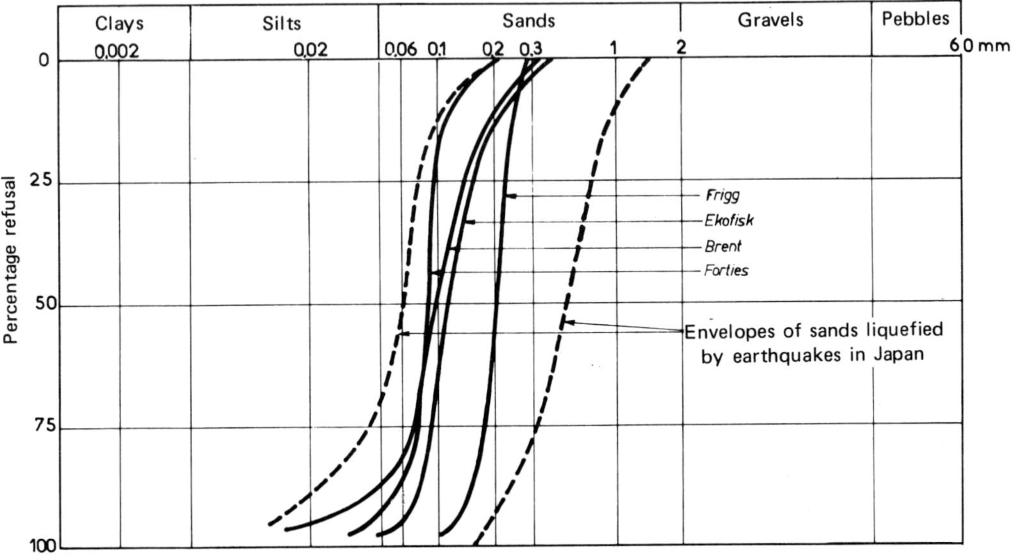

FIG. VIII.1.5. – Size distributions of surface sands in the North Sea.

In actual fact, while the assumption of local liquefaction of the sands beneath the gravity base structure appears probable, it seems difficult to reach any experimental verification of the possible extent of such liquefaction.

Elsewhere (see Section VIII.3.4.1), it will be seen that if the interstitial pressure effectively rises, the foundation may rupture as a result of slip well before the sand liquefies.

In **clays**, the assumption of liquefaction differs, depending on the level of consolidation:

– in the case of soft clays, rupture of the foundation soil by liquefaction appears more probable than in sands, in the absence of drainage,
– in the case of stiff or highly stiff clays, experimental studies are still in their infancy.

VIII.1.2.5 Scour of the soil near the structure

Scour is governed by two essential factors:

– the speed of the bottom currents,
– the grain size distribution of the sediments.

Scour at the base of the structure (Fig. VIII.1.4f) may underlie the rupture of foundations either by slip or by loss of the bearing capacity of the soil.

Experimental study of scour is generally conducted in the testing tank on a scale model. One may however doubt the representivity of such experiments owing to the difficulty of achieving correct similitude of the various parameters of the problems (and in particular the grain size distribution of the sediments).

VIII.1.3 Settlement of the soil beneath a gravity structure

In the case of most of the sites considered so far in the North Sea with a view to installing gravity structures, the soils are almost invariably highly consolidated. It would therefore appear that the foundations of these structures are governed particularly by problems of stability and fairly little by the deformation of the soil. It should however be recalled that the stability (bearing capacity of the soil) and deformation criteria are not independent, particularly in the case of soils of low consolidation.

VIII.1.3.1 Settlement of soil under static loads

Settlement of the soil beneath the structure is the result of volumetric compression and shear distortion under the effect of the submerged weight of the structure.

In the case of **dense sands or consolidated clays**, settlement:

– comprises the instantaneous deformation and primary consolidation, of very short duration [10],
– and can be estimated on the basis of the theory of elasticity.

The predictable settlement beneath a gravity structure is probably no more than a few tens of centimetres (for example the Ekofisk tank, see Section VIII.3.5.1).

In the case of **soft soils** (on the surface or at low depths), the preconsolidation pressure of which is less than the stress induced in the soil by the structure, settlement:

– comprises both the instantaneous deformation, the primary consolidation and the secondary (very long duration – several years or decades) consolidation,
– and can be estimated only very approximately from empirical formulae.

VIII.1.3.2 Settlement of the soil under the effect of cyclic loading

The action of waves on the structure results in repeated variations of the vertical loads combined with horizontal stresses. The result is a discontinuous settlement of the soil beneath the structure, the amount of which is difficult to state with exactness in the light of experience currently gained.

In the case **of dense sands (or highly consolidated clays)**, on the basis of laboratory experiments, it can be expected that the settlement of the soil resulting from wave action will be of the same order of magnitude as the elastic settlement calculated under the effect of the weight of the structure. It can also be assumed that most of the settlement will occur with the first major storms following the installation of the structure.

In the case of **soft soils or loose sands**, very little experimental data exists. It can however be considered that settlement of the soil in the central section of the foundations will be accompanied by a softening, followed by a possible liquefaction of the periphery of the structure (see Section VIII.1.2.3).

VIII.1.4 Movement of a gravity structure

A gravity structure can be displaced under the effect of several stresses:

– vertical displacement resulting from settlement of the soil,
– vertical and horizontal displacement under wave action,
– response to vibrations.

VIII.1.4.1 Vertical displacement resulting from settlement of the soil

The consequences of total and differential settlement on the behaviour of the structure should be examined.

Vertical displacement is assumed to be uniform and progressive:

– a low settlement (a few decimetres) will probably not result in any damage to the structure, conductors or risers (which are designed for this purpose),

– too great a settlement under the effect particularly of repeated loads, would on the contrary ruin the foundations and hence damage the structure and the links.

Vertical displacement of the structure under the effect of differential settlement of the soil can lead:

– either to exaggerated leaning, incompatible with operations,
– or to damage to the links (conductors and risers).

VIII.1.4.2 Vertical and horizontal displacement under wave action

The vertical and horizontal movement of the structure under wave action can be estimated by the method of finite elements, making certain simplifying assumptions as to the rheological properties of the soil.

For example, the foreseeable repeated movements of a Condeep type structure (weight 200,000 t) under the action of a 30-metre 100-year wave would be (see Fig. VIII.1.6):

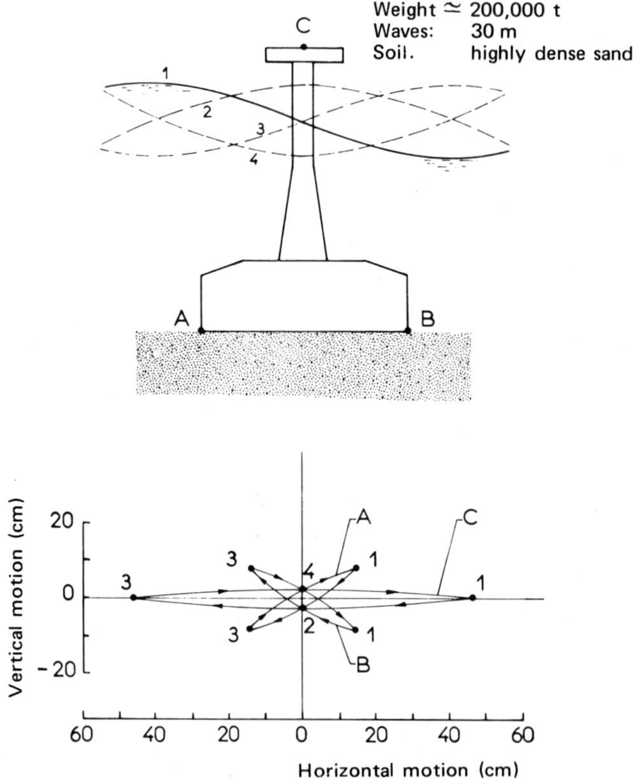

FIG. VIII.1.6. – Horizontal and vertical motion of a gravity structure.

– 10 to 15 cm vertically,
– 15 to 20 cm horizontally, which would certainly detract from the behaviour of the lines.

VIII.1.4.3 Response of the structure to vibrations

The action of waves and the operating conditions (drilling, etc.,) induce in the structure vibrations of widely differing frequencies and amplitudes which may have unforeseeable consequences on its behaviour.

The fact is:

– the experience gained on structures ashore is not applicable,
– the vibrations induced in the structures by earthquakes, which have very different frequency, duration and amplitude, are difficult to compare.

The problem therefore calls for research at very great depths in order to determine the dynamic response of a structure to wave action through a wide range of frequencies.

Table VIII.1.1 summarizes the problems involved in the installation, stability and movement of a gravity structure, together with the characteristics of the soil which have to be determined in each case.

VIII.1.5 Stresses distribution beneath a gravity structure

After listing the geotechnical problems to be solved before siting a gravity structure, the soil reconnaissance depth to be reached should be specified in terms of the distribution of the stresses in the soil, i.e. the dimensions of the structure.

VIII.1.5.1 Distribution of the stresses beneath a base slab

As we know, Boussinesq's solution gives the following stresses beneath a rigid circular foundation (Fig. VIII.1.7a) [11]:

– a contact stress in the centre equal to half the mean stress,
– a theoretical infinite stress at the edge. In actual fact, owing to rupture of the soil by local shear in this high stress zone, a new plastic state of equilibrium results in the adaptation of the contact pressure to the boundary conditions.

In the case of a saturated homogeneous isotropic and elastic clay, where the foundation is partly buried, the distribution is parabolic in nature, with minimum pressure at the centre and of the following form (see Fig. VIII.1.7b):

$$\alpha p + \frac{\beta}{2} p = p$$

On the basis of measurements made, one can generally assume:

$$\alpha = 0.70 \text{ and } \beta = 0.60$$

TABLE VIII.1.1

CHARACTERIZATION OF THE PROPERTIES OF SOIL INVOLVED IN THE INSTALLATION, STABILITY AND MOVEMENT OF GRAVITY STRUCTURES

	Problems to be solved	**Characterization of the bottoms of soils**
Installation of the structure	Penetration of the skirts into the soil	Maximum shear strength
	Contact stresses between the soil and the slab (floor)	Precise bathymetry and morphology of the bottoms
Stability of the structure	Slip of the structure on the bottom Slip between layers of different kinds	Effectiveness of the skirts Strata slopes Shear strength between strata
	Rupture of the soil beneath the structure (static load and stresses induced by the waves)	Minimum shear strength (drained and undrained) of the soil
	Liquefication of the soil under the action of the stresses induced by the waves	Loss of the shear strength of the soil under cyclic loads
	Scour of the soil near the structure	Size distribution of the sediments
Settlement of the soil beneath the structure	Settlement of the soil under the effect of static loads	Modulus of deformation of the soil Consolidation and creep characteristics of the soil
	Settlement of the soil under the effect of cyclic sollicitations	Characteristics of the soil to be specified
Movement of the structure	Vertical shift resulting from settlement of the soil	See above
	Vertical and horizontal shift under the action of the waves	Deformation of the soil Influence of the skirts Friction between layers
	Response of the structure to vibration	Dynamic deformation characteristics of the soil

In the case of homogeneous and elastic sand, the modulus of elasticity of which increases with depth, the pressure would on the contrary be a maximum at the centre and a minimum on the edge of the foundation (see Fig. VIII.1.7c). However, when the dimensions of the foundation become considerable, the distribution of the contact pressure approaches a uniform distribution. It is this assumption which is generally accepted ([2]) when designing a project for a foundation subjected to a static load.

The instantaneous elastic deformation of a sand with a modulus of elasticity of 20,000 t/m^2 (200 MPa) beneath a gravity structure 100 m in diameter weighing 200,000 t is about 10 cm ([3]).

(2) In the absence of more exact information.

(3) It will be noticed that this value closely approaches the settlement observed when installing the Ekofisk tank (see Section VIII.3.5.1).

(a) Infinitely rigid soil and slab

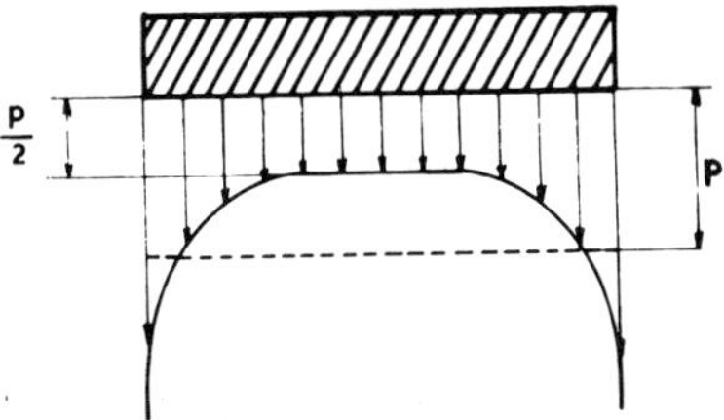

(b) Rigid mat on saturated clay

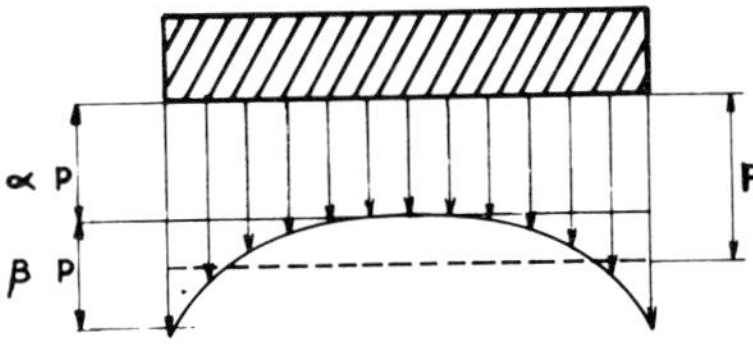

(c) Rigid mat on sand

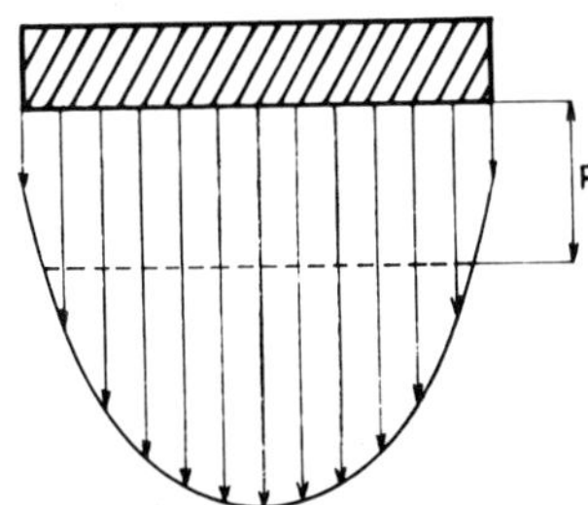

FIG. VIII.1.7. – Stress distribution beneath a rigid foundation.

VIII.1.5.2 Stresses distribution beneath a surface foundation as a function of depth

In the case of a **rigid circular foundation** resting on a **homogeneous and isotropic saturated clay**, the above-indicated parabolic law will be assumed (with $\alpha = 0.70$ and $\beta = 0.60$).

Given $\frac{\Delta\sigma_1}{p}$ and $\frac{\Delta\sigma_3}{p}$ namely the relative vertical and horizontal distributions induced in the soil at the centre of the foundation:

$\Delta\sigma_1$ is independent of the mechanical characteristics of the soil,
$\Delta\sigma_3$ depends on the Poisson coefficient of the soil ν.

Figure VIII.1.8 represents the theoretical distributions of $\frac{\Delta\sigma_1}{p}$ and $\frac{\Delta\sigma_3}{p}$ in terms of the ratio $\frac{R}{z}$ for $\nu = 0.25$ and 0.50,

where:
R = radius of foundation,
z = depth considered.

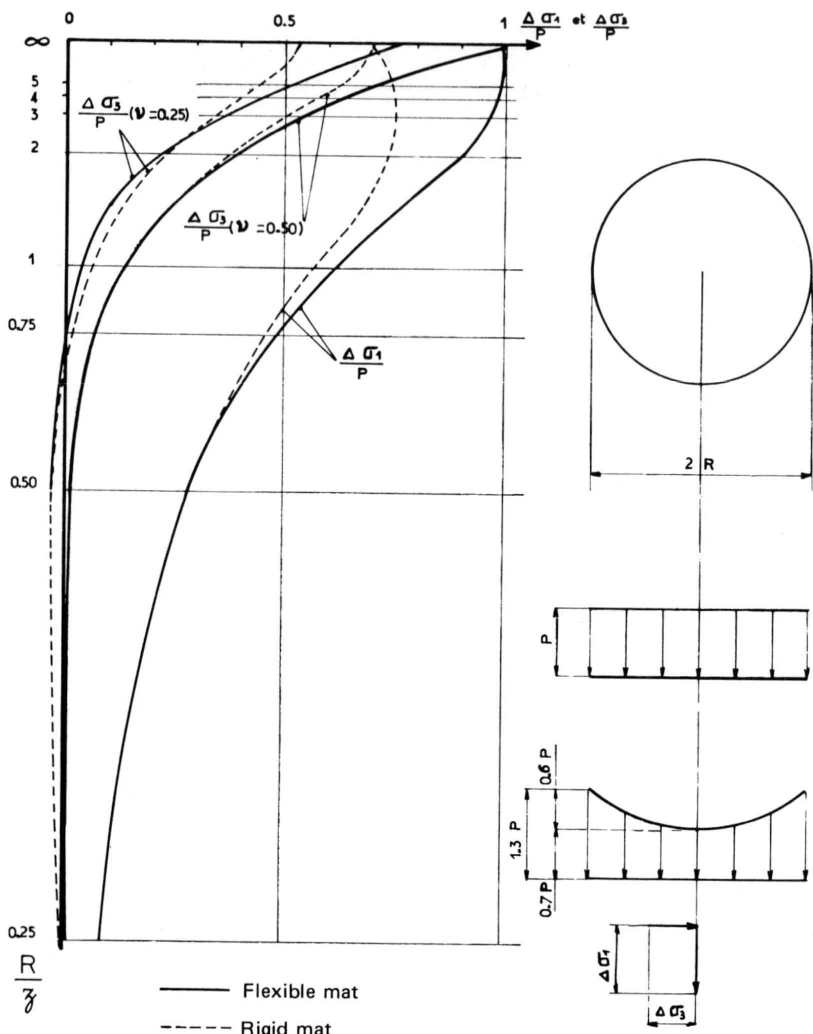

FIG. VIII.1.8. – Stress distribution beneath the axis of a circular mat.

According to this theoretical example, it would appear that:

– the vertical stress $\Delta\sigma_1$ decreases with the depth, first slowly and then more quickly; for $z = 2R$, $\dfrac{\Delta\sigma_1}{p}$ would still be about 0.25. Generally speaking, the stress $\Delta\sigma_1$ is inversely proportional to the square of the depth.,

– the horizontal stress $\Delta\sigma_3$ decreases very quickly as the depth increases but can become a tensile stress when $z > R$, depending on the values of the Poisson coefficient ν.

In **actual formations** consisting of layers of widely different kinds and mechanical characteristics (sands, clays of varying consolidations, etc.), the distribution of the stresses with depth does not conform to this theoretical pattern. It essentially depends:

- on the relative position of the soft and rigid layers,
- on the ratio of the moduli of elasticity on the successive layers.

As a result, the vertical stress distribution with depth in particular is very difficult to predict.

As an example, Fig. VIII.1.9 shows the vertical stress "bulbs" induced beneath a square foundation with a side B for (Table VIII.1.2) [12]:

- a homogeneous, elastic and isotropic terrain (according to Boussinesq),
- a homogeneous elastic and stratified terrain (according to Westergaard).

TABLE VIII.1.2

Depth $\frac{z}{B}$	**Vertical stress distribution** $\frac{\Delta\sigma_1}{p}$ **beneath the centreline of the foundation**	
	Isotropic soil (according to Boussinesq)	Stratified soil (according to Westergaard)
0.5	0.7	0.5
1	0.4	0.2
1.5	0.2	0.13
2	0.12	0.08

To summarize:

– in isotropic formations, the pressure bulb is fairly well defined by a truncated cone forming an angle of 45° with the surface of the soil,

– in stratified formations, the angle of the bulb depends on the ratio of the moduli of the successive layers. However, in the general case, where the modulus increases with depth, the bulb angle is less than that indicated for an isotropic medium.

VIII.1.5.3 Applications to the gravity structures of the North Sea

The dimensions of the base of most of the structures built for the North Sea are in the region of 80 to 140 m, corresponding to foundations covering 6,000 to 14,000 m^2.

The forces acting on these structures under the following conditions:

water depth = 150 m,
100-year wave = 30 m with periodicity of 15 s,

are estimated as approximately:

200,000 t for the maximum vertical static load $F_{v\,max}$,
200,000 t for the minimum vertical static load $F_{v\,min}$,

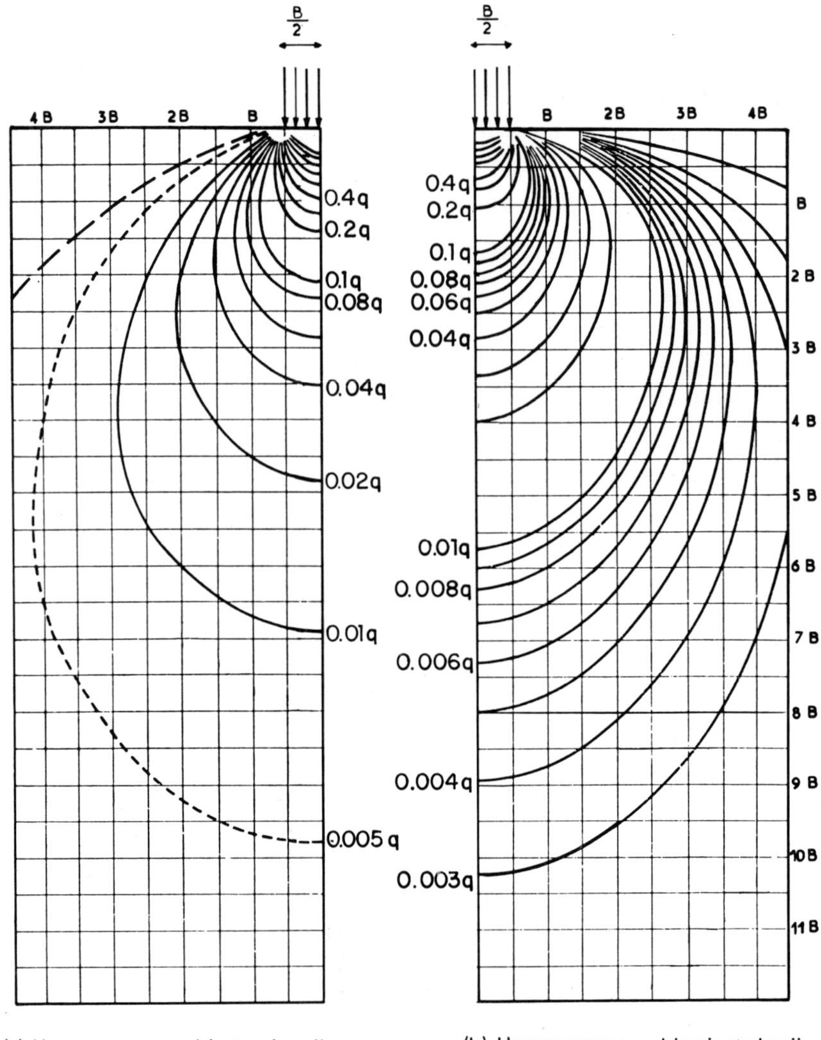

(a) Homogeneous and isotropic soil (according to Boussinesq)

(b) Homogeneous and laminated soil (according to Westergaard)

FIG. VIII.1.9. – Stress distribution beneath a square foundation.

– ± 10,000 t for the vertical dynamic load ΔF_v,
– 50,000 to 70,000 t for the horizontal load F_h,
– approximately 2.10^6 to 3.10^6 t/m for the overturning moment M,
– 4 to 5 t/m² (40 to 50 kPa) for the variations in water pressure at soil level near the structure under the effect of wave action Δp.

According to these figures, an approximate mean estimate of the stresses in the soil beneath the structure can be given as follows:

- at rest:
  - 25 to 35 t/m^2 (250 to 350 kPa) at the level of the sea bottom,
  - 5 to 10 t/m^2 (50 to 100 kPa) at a depth of 50 to 100 m,
- under the action of the 100-year wave (see Section VIII.1.3):
  - at least 35 to 40 t/m^2 (350 to 400 kPa) at the level of the sea bottom,
  - yet more difficult to estimate at 50 to 100 m.

VIII.2 SOIL RECONNAISSANCE FOR THE INSTALLATION OF GRAVITY STRUCTURES

A priori, one cannot set any hard-and-fast framework for the various soil reconnaissance phases, which are dependent on many factors related to the development of a field. The purpose of the pointers given in the present paragraph is merely to set down the main lines of the reconnaissance programme in order to optimize the operations and limit the number of tasks involved [13] [14].

VIII.2.1 General soil reconnaissance programme before siting gravity structures

Just as for siting fixed platforms on piles, a clear distinction should be drawn:

- first, between the overall reconnaissance of the site,
- second, the reconnaissance of the **emplacement**(s) for the approximate location of the structure(s).

VIII.2.1.1 General remarks concerning geophysical and geotechnical reconnaissance

Geophysical reconnaissance, which is indispensable before any geotechnical reconnaissance of the soils, is of particular importance before laying gravity structures, owing to the requirements stipulated both as to the relative position of the bathymetry and the high definition of the seismic profiles of the layers down to depths of several tens of metres.

The indications given in this paragraph, which are applicable particularly to the North Sea, are of theoretical and maximum character, with the true programme for each particular case obviously depending on the conditions of the soil and subsoil, the location and the weather conditions when the recordings were made. It should be mentioned that the characteristics of the subsoil generally vary less than those of the topography of the bottoms. The result is that the meshing for the seismic profiles may be less tight than that for the bathymetric profiles.

In addition to the geophysical techniques proper, visual methods of observation (television camera), pressure sensor bathymetry, etc. are also considered.

Thorough geotechnical reconnaissance of the emplacement of the structure must precede the starting of any construction work in order to ensure the optimum type and dimensions for the foundations (skirts, shape and thickness of the floor slab, etc.) in the light of the nature and mechanical characteristics of the surface layers of the soils.

As with geophysical reconnaissance, the recommendations made in the present paragraph are merely pointers (theoretical and maximum) which may guide the elaboration of a programme of coring sampling and in situ measurements necessarily depending on the local conditions and even modifications made in the project of the structure.

VIII.2.1.2 Overall reconnaissance of the site

An initial overall geophysical reconnaissance on the scale of the site as a whole should generally extend well beyond the overall emplacements on which the structures are to be set out.

An initial geological and geotechnical reconnaissance comprising at least one or two deep boreholes with core samples taken from the presumptive zone of the emplacements for the structure will facilitate interpretation of the seismic prospecting and yield initial information concerning the consolidation of the soils, this being an essential condition for choosing gravity structures.

VIII.2.1.3 Detailed reconnaissance of the emplacement(s)

A second geophysical and detailed visual (fine study) survey limited to each emplacement planned for installing a gravity structure will provide a detailed description of the topography and the subsoil.

A second detailed geotechnical reconnaissance covering the approximate emplacement of each structure will comprise:

– a deep borehole (if not already made on this emplacement),
– in situ measurements and boreholes down to a few tens of metres.

Table VIII.2.1 summarizes the various geophysical and geotechnical reconnaissance phases of the soils which are enlarged upon in Paragraphs VIII.2.2. and VIII.2.3.

VIII.2.2 Initial geophysical reconnaissance (overall survey of the site)

The purpose of the first geophysical survey is to define the approximate emplacement(s) which are possible for the structure and then to set the position of the first reconnaissance boreholes.

VIII.2.2.1 Coverage and meshing

The initial overall geophysical reconnaissance survey must cover:

TABLE VIII.2.1

THE VARIOUS STAGES OF THE RECONNAISSANCE OF BOTTOMS AND SOILS BEFORE INSTALLING GRAVITY STRUCTURES

	Bottom and soil reconnaissance stage	**Coverage**	**Reconnaissance to be carried out**	**Depth of reconnaissance to be reached (m)**
Overall reconnaissance of the site	Initial geophysical reconnaissance (overall survey)	Area of the field (or square with sides measuring 1 to 3 km)	Bathymetry Morphology (if necessary) High resolution seismics	 Approx. 100
	First geological and geotechnical reconnaissance	Probable zone of emplacement of structure	One or two deep boreholes with core samples	50 to 150 (average 100)
Detailed reconnaissance of the emplacement of each gravity structure	Second geophysical reconnaissance (detailed study)	Square with sides measuring 300 to 400 m	Detailed bathymetry Precise morphology of the bottoms Magnetometry Very high resolution seismics	 Approx. 50
	Second geotechnical reconnaissance	200 x 200 m^2	One deep borehole Several medium depth boreholes In situ measurements Surface core sampling	Approx. 100 Approx. 20 to 50 Approx. 10 to 30 A few metres

– preferably, the whole area of the field (if this is well delineated during this reconnaissance),

– at all events, an area of at least a few square kilometres, i.e. generally a square with sides from 1 to 3 km long.

This considerable area to be reconnoitred is justified by the following:

– the uncertainty as to the final emplacements for the structures on the site during this reconnaissance, which often occurs 3 or 4 years before construction,

– the need to reconnoitre all favourable emplacements which are to be studied during the second geophysical (detail) reconnaissance.

The recommended meshing varies (Fig. VIII.2.1):

– for bathymetry from 100 to 500 m, differing with the direction depending on the topography,

– for seismic (sparker) from 200 to 500 m, also differing with the direction.

The length of the profile for an average survey is about 50 to 100 km.

VIII.2.2.2 Bathymetry

Overall reconnaissance of the topography of the site (bathymetry and morphology) is the main purpose of the initial geophysical survey.

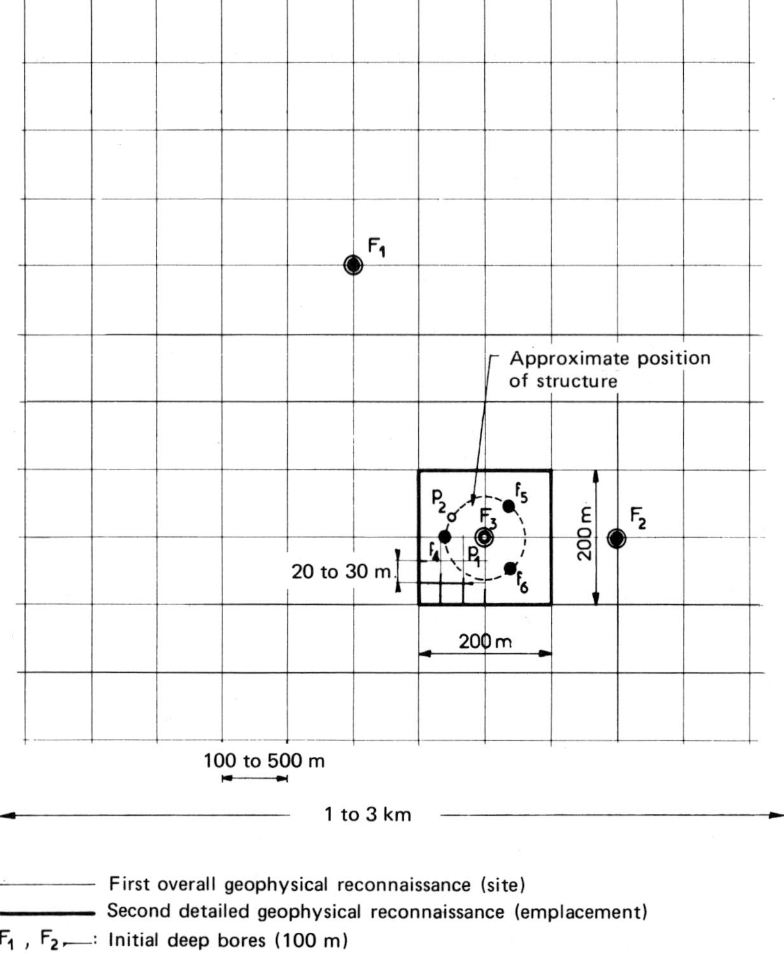

First overall geophysical reconnaissance (site)
Second detailed geophysical reconnaissance (emplacement)
F_1 , F_2 : Initial deep bores (100 m)
F_3 : Deep bore (100 m)
f_4 , f_5 , f_6 : 20 to 50 m bores (3 to 4)
P_1 , P_2, ... P_n: Penetrometer operations (5 to 10)
} at position of structure

FIG. VIII.2.1. – Diagram of soil reconnaissance required before siting a gravity structure.

The initial bathymetry must provide general indications concerning the slope and the irregularity of the relief (humps and depressions).

An echosounder is used for this reconnaissance, providing an accuracy of about 1% of the water depth, i.e. an **absolute** accuracy of the measurements of ±0.50 m per 100 m of water.

VIII.2.2.3 Morphology

If the bottoms are relatively flat, there may be no need to use the side-scan sonar in this initial reconnaissance stage.

On the other hand, if certain doubts subsist, the topography of the site as obtained by bathymetry must be completed by an image of the roughness of the bottom which will be provided by the side-scan sonar.

The recordings of the sonar make it possible to distinguish:

– clear and uniform images which correspond to stretches of sediment (sands, clays, etc.) of relatively flat profiles,

– echos indicating obstacles of various kinds: beds of materials (pebbles, gravel, coral, boulders, etc.), outcrops (rock, indurated clays, etc.), wrecks, etc.

VIII.2.2.4 High resolution seismic prospecting

The purpose of the initial seismic reconnaissance is to detect:

– either the continuity of the strata,

– or the possible presence of geological accidents such as the existence of paleovalleys (frequent in the North Sea) filled by clays of low consolidation and covered with sands.

This reconnaissance is done using a sparker or possibly a boomer (Uniboom).

The penetration sought is at least equivalent to the dimension of the side (or diameter) of the structure, i.e. at least 100 m.

The resolution obtained is generally not less than 2 m.

VIII.2.3 Second geophysical reconnaissance (detailed study of the emplacement)

The second geophysical reconnaissance, much more limited and detailed, will be made in the light of the previous findings (from the initial geophysical and deep drilling campaign):

– over an area with sides measuring about 300 to 400 m,

– with a close mesh, to be defined for each recording, the tightest however being of the order of accuracy of the positioning system (Fig. VIII.2.1).

VIII.2.3.1 Bathymetry

Laying of large dimension rigid gravity structures (about 100 × 100 m) calls for very thorough knowledge of the topography of the sea bottom. The second bathymetry campaign, which must be very precise in relative value, must hence make it possible to check all the irregularities in the soil beneath the future emplacement of the structure. The following are commonly recommended:

– a **relative** accuracy of measurements to within ±10 cm, this being the maximum theoretically achievable,

– execution of the bathymetry with as tight a mesh as the equipment and location will permit, i.e. generally 20 to 30 m. The ideal would be to allow for the profiles determined during the initial reconnaissance.

A satisfactory solution can be obtained under certain conditions (weather, sea state, implementation, etc.) using the precision echo sounder with an expanded scale and operating:

– in calm seas (wave amplitude < 1.5 m),
– with the vessel advancing at < 3 knots,

and by proceeding with:

– twice daily calibration of the instruments in order to allow for temperature variations,
– frequent resetting of the positioning system,
– recording of the profiles in one direction of advance only (in order to reduce noise effects),
– checking the intersections of the profiles aboard the vessel,
– in situ recordings of the tide,
– recordings of the roll, pitch and heave of the vessel ([4]).

A recent solution consists in determining the precise bathymetry from a submarine lying on the bottom:

– measurements (relative) compared to a reference point are made by means of a pressure sensor (absolute or differential) (with the necessary corrections for the tide, temperature, etc.),

– the position is obtained by means of acoustic beacons pingers laid on the bottom (Fig. VIII.2.2),

– the meshing described must make it possible to check the reliability of the measurements (intersections).

VIII.2.3.2 Morphology

The morphology (roughness) of the emplacement for the structure can be determined accurately using the side-scan sonar and the television camera in turn.

The side-scan sonar makes it possible to "see" topographical accidents of the order of a few metres in extent and a few centimetres in height or depth.

The normal conditions of use of the sonar are:

– range = 250 ft (75 m) or 500 ft (150 m),
– interval between profiles = 50 or 100 m in order to ensure a certain coverage of the recordings made,
– recordings in one direction only.

(4) The remarks concerning the tide and various corrections mentioned obviously apply to any bathymetry determination (see Section III).

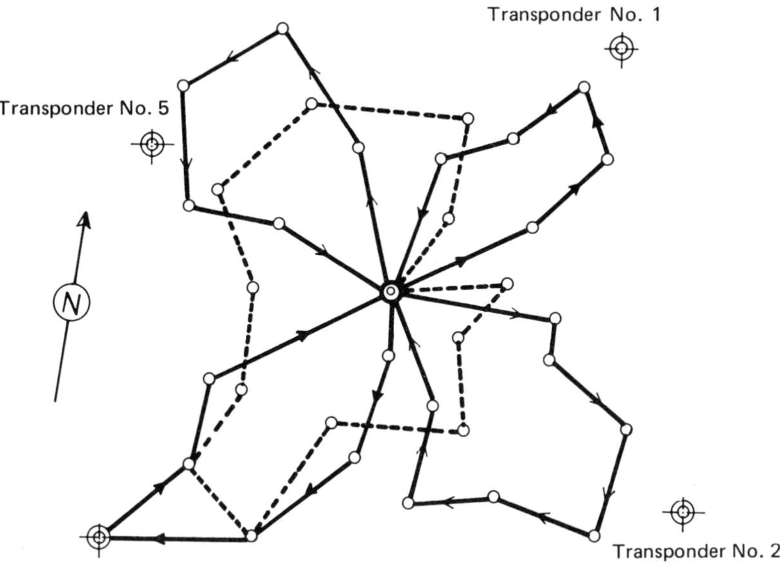

FIG. VIII.2.2. – Diagram of path following for precise bathymetry using a differential pressure gauge, from a submarine (according to *Intersub*).

Visual examination of the bottom, for instance using a television camera, would always appear necessary in order to interpret or specify certain details of the side-scan sonar recordings.

The television camera will be implemented:

– either by means of a tubular frame pulled along on a bottom sled,
– or more generally, from a submersible.

The image obtained will make it possible to discern details of a few centimetres and generally to check the nature of the sediments.

VIII.2.3.3 High resolution seismic prospecting

Only rarely is the soil homogeneous throughout any thickness, and it generally consists of an alternation of thin clayey and sandy layers. The second seismic reconnaissance:

– will be conducted with a meshing of 20 to 40 m, i.e. appreciably the same as for the bathymetry; if the soil is homogeneous, only one seismic recording will be made for every two bathymetric recordings,

– will aim at obtaining a very good resolution over the first 20 to 30 m, with a penetration in the range of 50 m.

The incompatibility between great penetration and high definition in the first few metres will call for the simultaneous application of two different techniques:

– first, high energy sparker in order to attain a penetration of ⩾ 50 m, with a resolving power of 2 to 3 m,

– second, boomer (Uniboom) with resolution power of about 1 to 2 m, for the surface levels (5).

Deep boreholes (with core sampling) down to about 100 m located preferably at the intersection of two seismic profiles, will facilitate the interpretation of the recordings. However, in the case of alternating layers of sands, silts and clays presenting rapid lateral variations, remote correlations invariably remain subject to caution.

VIII.2.3.4 Magnetometry

The purpose of magnetometry is to identify all metal objects lying on the bottom or buried slightly beneath it depth of about $\leqslant$ 3 m).

A magnetometer will be implemented:

– along profiles set about 100 m apart, equivalent to those of the side-scan sonars,
– possibly with one or two intersecting profiles.

It should be noted that there is no interference between a side-scan sonar and a magnetometer. On the other hand, it is not possible to use a sparker and a magnetometer simultaneously.

Table VIII.2.2 summarizes the stages of the geophysical reconnaissance of the soil to be made before installing gravity structures, together with the characteristics and performance of the main bathymetry and seismic techniques applicable.

VIII.2.4 Initial geological and geotechnical reconnaissance (of the site)

The first campagin of deep core sampling and boreholes is of both geological and geotechnical interest, the purpose being:

– to facilitate the interpretation of the seismics,
– to identify the nature, heterogeneity and consolidation of the terrains.

VIII.2.4.1 Location of the boreholes

An examination of the geophysical recordings covering an area of a few square kilometres should make it possible to find the optimum location for the boreholes, in the light in particular:

– of the planned emplacement zone for the structures and the topography of the bottoms,
– of the minimum slope angle of the strata and the uniformity of the seismic profiles (absence of geological accidents),

(5) Sediment sounders, which have a definition of $\leqslant$ 1 m, would not appear appropriate owing to the considerable consolidation of the soils; the penetration would not be more than a few metres.

TABLE VIII.2.2

GEOPHYSICAL AND VISUAL RECONNAISSANCE OF BOTTOMS AND SOILS BEFORE INSTALLING GRAVITY STRUCTURES
CHARACTERISTICS AND PERFORMANCES OF THE APPLICABLE TECHNIQUES

Reconnaissance stages	Recordings necessary	Coverage	Techniques applicable	Meshing or spacing	Transmission frequency	Penetration	Definition (resolution power)	Precision
Initial geophysical reconnaissance (overall study)	Bathymetry	Preferably: the entire area of the field Minimum: a square with sides measuring 1 to 3 km	Echosounder	Meshing from 100 to 500 m	30 to 50 kHz	≈ 0		Absolute precision about 1 m per 100 m depth
	Morphology (not indispensable provided the bottoms are fairly flat)		Side-scan sonar	Spacing = 100 to 500 m	≈ 100 kHz	0	Relief < 1 m	
	High resolution seismic prospecting		Sparker	Meshing from 200 to 500 m	100 to 1,000 Hz	≈ 100 m	≈ 2 to 3 m	
Second geophysical reconnaissance (detailed study)	Bathymetry	Square with sides measuring 300 to 400 m (per gravity structure)	Echosounder	Meshing = 20 to 30 m	30 to 50 kHz	≈ 0		Absolute precision about 1 m Relative precision ±20 to 25 cm
			Relative measurement made from a submarine	Meshing = 20 to 30 m				Relative precision claimed = 10 cm
	Morphology		Side-scan sonar	Spacing = 50 to 100 m	100 kHz	0	Relief < 1 m	
			Television camera	Profiles or stations			Differences in level about 5 to 10 cm	
	Magnetometry		Magnetometer	Spacing about 100 m		< 3 m		
	High resolution seismic prospecting		Low energy sparker	Meshing = 20 to 40 m	100 to 1,000 Hz	≈ 50 m	⩽ 2 m	
			Boomer (Uniboom)		500 Hz to 4 kHz	≈ 50 m	≈ 1.5 m	

– of the probable consolidation of the surface soils.

In order to facilitate the interpretation of the recordings, the boreholes will be located at the intersection of two seismic profiles (see Fig. VIII.2.1).

This optimum location of the boreholes offers the twofold advantage:

– of gathering information on the area concerned by the emplacement of the structures,
– of reducing, if possible, the number of subsequent boreholes (see Section VIII.2.5).

VIII.2.4.2 Number and depth of boreholes

The number of boreholes (with core sampling) needed depends on the seismic profiles recorded:

– if the seismic profiles are practically uniform over a relatively extensive area, one or two boreholes will be enough,
– if on the other hand the seismic recordings display a number of irregularities, it would appear of advantage to drill more boreholes.

On the average, the principle of 1 or 2 boreholes (F_1, F_2) will be adopted for the initial geological and geotechnical reconnaissance (Fig. VIII.2.1).

The depth of the first boreholes will depend:

– first, on the dimensions of the structures envisaged and consequently the stresses induced in the soils,
– second, the nature and consolidation of the formations encountered.

In actual fact, it would appear desirable to reach:

– a minimum penetration of 50 to 70 m, if the soils are highly consolidated, i.e. dense sands, very stiff clays, etc.,
– a maximum penetration of 120 to 150 m if the soils are only slightly consolidated (loose sands, medium clays, etc.) or highly heterogeneous (alternating layers of sands and clays).

A priori, in the absence of any information on the site, one will therefore adopt a mean depth of 100 m for the boreholes, approximately equivalent to the dimensions of the base of the structure.

The sequence of core sampling through the guide tube formed by the drilling string varies with nature, consolidation and heterogeneity of seabed formations. For instance, it will be admitted one coring:

– every metre (practically continuous) down to about 10 m,
– every 1.50 m from 10 to 20 m,
– every 3 m from 20 to 50 m,
– every 5 m deeper 50 m.

VIII.2.5 Second geotechnical reconnaissance (of the emplacement(s))

The second survey must make it possible to obtain a detailed description of the subsoil at the actual emplacement of each structure.

VIII.2.5.1 Location of the soundings

The emplacement of a gravity structure is determined at one and the same time by:

– the production requirements,
– the conditions of the bottom (bathymetry and planimetry) and soil.

However, the location of the reconnaissance soundings must also allow for:

– the uncertainty of the positioning system, both for the soundings and the installation of the structure,
– the need for knowledge of the properties of the soil in the immediate neighbourhood of the structure (dimensions about 100 × 100 m).

To conclude, the soundings will be located:

– over an area with a diameter of approximately 200 m, straddling the planned emplacement for the structure (Fig. VIII.2.1),
– at the intersections of the seismic profiles.

VIII.2.5.2 Types, number and depth of soundings

The detailed reconnaissance of the emplacement of each gravity structure requires:

– a deep borehole (F_3) down to about 100 m, with coring sampling. If one of the deep boreholes drilled during the initial campaign coincides with the emplacement retained, obviously, a new deep borehole would be superfluous,
– 2 to 5 boreholes with coring sampling (f_4, f_5, f_6 ...), of medium depth (20 to 50 m) [6],
– 5 to 10 in situ tests (P_1, P_2. . .); core or surface sampling.

The in situ measurements normally made before medium depth boreholes consist of the following:

– preferably, penetrometer tests made down to refusal in dense sands; the maximum depths reached vary from a few metres in dense sands to 20 to 25 m in certain clays [7],

– if applicable, pressuremeter tests in clays of medium stiffness (undrained cohesion c_u approximately 10 t/m^2 (100 kPa)).

---

(6) One or two medium depth boreholes are enough before setting up a pile-mounted platform.

(7) The simultaneous recording of nuclear logs (γ-ray and γ-γ) and penetrometer diagrams, currently in the course of development, would be highly interesting for the reconnaissance of soils before setting up gravity structures.

The depth of the boreholes will vary from 20 to 50 m depending on the nature, heterogeneity and consolidation of the formation layers encountered.

The sequence of core sampling, which would be adapted to suit each case, could for instance consist of:

– one core sample every metre (practically continuous) down to about 10 m,
– one every 1.50 m from 10 to 20 m,
– one every 3 m from 20 to 50 m,
– one for each soft or highly heterogeneous layer as revealed by the penetrometer tests.

The recording of nuclear logs (γ-ray, neutron, γ-γ) by means of a sonde lowered through the drilling string makes it possible:

– to reduce the number of corings,
– to obtain direct and continuous information concerning the true water contents and densities, thus permitting interpolations between the various levels cored,
– to check the exact depths of the layers,
– to identify errors on the cores resulting from caving in of the soil or falling back of excavated material.

Surface coring using vibrocorers (or Kullenberg type corers) ([8]) is conducted down to:

– 1 to 3 m in consolidated soils,
– 3 to 5 m in soils of relatively low consolidation.

Tables VIII.2.3 and VIII.2.4 summarize the following respectively:

– the stages of geotechnical reconnaissance of soils to be conducted before setting up gravity structures, together with the main characteristics and operational performances of the devices or techniques for coring and in situ measurements that can be used in consolidated terrains,
– the main measurements to be made in the laboratory and in situ, together with the parameters to be determined.

(8) It would not appear possible to use the Kullenberg corer in the highly consolidated terrains (dense sands or stiff clays) of the North Sea. It generally penetrates only to a very shallow depth.

TABLE VIII.2.3

GEOTECHNICAL RECONNAISSANCE OF SOILS BEFORE SETTING UP GRAVITY STRUCTURES
PERFORMANCE OF APPLICABLE TECHNIQUES

Reconnaissance stage	Operation carried out	Coverage	Type and Number of soundings	Penetration necessary	Techniques applicable	Performance	Implementation
Initial geological and geotechnical reconnaissance	Deep boreholes with core sampling	In the probable emplacement of the structures	1 or 2 deep boreholes with core sampling	Approximately 100 m (from 50 to 150 m depending on soils encountered)	Drilling and percussion wireline coring Nuclear logs	Average rate of penetration = 2 to 3 m/h Continuous recording	From a vessel: wave amplitude < 2 m From a platform: wave amplitude < 5 to 6 m
Second geotechnical reconnaissance (detailed survey emplacement of each gravity structure)	Boreholes with core sampling	About 200 x 200 m	1 deep borehole with core sampling (unless already done)	≈ 100 m	*ditto*	*ditto*	From a vessel: wave amplitude < 2 m
			2 to 5 boreholes with core sampling at medium depth	20 to 50 m	*ditto*	*ditto*	From a vessel: wave amplitude < 2 m
	In situ tests		5 to 10 tests: penetrometry and/or pressuremetry	Down to 20-25 m	Seacalf penetrometer laid on bottom Cable penetrometer (Wison) Ménard pressuremeter (if necessary)	Maximum penetration depends on the consolidation of the soils This technique is as yet used only little in the North Sea	From a vessel: wave amplitude < 1 to 1.5 m From a vessel: wave amplitude < 1.2 m From a vessel using the vibroannular pressuremeter
	Surface coring		A few surface core samples	From 1 to 5 m depending on soils	Electric or hydraulic vibrocorers Kullenberg type corer (if necessary)	Penetration < 5 m Cannot be used in highly consolidated terrains	From a vessel provided wave amplitude < 2 to 3 m

TABLE VIII.2.4

CORE AND/OR IN SITU MEASUREMENTS TO BE MADE BEFORE INSTALLING GRAVITY STRUCTURES

	Characteristics of soils measured		Remarks
Laboratory analysis of cores	**Identification of soils**	Nature Atterberg limits (W_L, W_P) Size distribution	For clayey soils
	Water content w Density γ		Made immediately aboard the vessel
	Shear strength of soils		
	Undrained shear strength (undrained cohesion c_u)	unconfined compression vane fall cone pocket penetrometer	Preferably, immediately aboard the vessel
		triaxial	In the laboratory
	Drained shear strength	triaxial	Long duration tests in the laboratory
	Liquefaction or fatigue	triaxial and/or shear box	Cyclic loading tests
	Settlement of soils		
	Compressibility	oedemeter triaxial (drained tests)	In the laboratory
Penetrometer	Cone resistance R_p Lateral friction f		These measurements are indispensable in sands
Pressuremeter	Limit pressure p_l Pressuremeter modulus E_p		Of interest for estimating settlement
Nuclear logs	Nature of soils (γ-ray) Density and water content in situ		These measurements are highly desirable

VIII.3 CALCULATION OF THE STABILITY OF GRAVITY STRUCTURES INSTRUMENTATION

From the standpoint of the foundations, gravity structures (generally concrete) can be assimilated to structures built on surface foundations the stability of which must be verified with respect to two criteria:

– the bearing capacity of the soil,
– the settlement of the soil.

The problem nonetheless differs from that of common surface foundations in several respects, namely:

– the dimensions (and weight) of a gravity structure far exceed those of structures built on land on this type of foundation. Any reference to existing structures is hence practically out of the question,

– the absence of any prior excavation before installing the structure presupposes a uniform and horizontal soil surface,

– the high level of the horizontal forces (caused by wave and wind action) exerted on structures at sea bears no relation to those normally encountered ashore. Because of this, the stability of the structure under the "100-year wave" has to be verified ([9]),

– the repeated stressing of the structure and foundation soil owing to the wave action results in a loss of shear strength and an increase in the deformation of the soil (settlement).

VIII.3.1 Installing the structure

From the geotechnical standpoint, the installation of the structure concerns:

– penetration of the skirts into the soil, where applicable,

– local stresses caused by the irregularities of the soil.

VIII.3.1.1 Penetration of the skirts into the soil

The reason for using skirts seems in fact to differ with the nature of the surface soils:

– in sands, skirts offer first and foremost the advantage of reducing the effects of scour,

– in clays, skirts can limit the danger of slipping.

Furthermore, the advantages of skirts are certainly a function of their penetration into the soil:

– were skirts to be too short (1 m), they would probably not be capable of preventing slipping,

– were skirts to be too long (several metres), on the other hand, difficulties would be encountered in setting up the structure.

As an example of the determination of the penetration of skirts, let us consider a Condeep platform with the following approximate characteristics:

– 100 m (overall) hexagonal base consisting of 19 vertical cylindrical cells, each 20 m in diameter and equipped with a steel skirt with a length h,

– maximum weight in water about 200,000 t. Each skirt penetrates into the soil under a vertical load of about 10,000 t.

The force opposing penetration of the skirt is expressed by the following formula:

$$F = \pi de R_{p\,\max} + 2\pi dh f_{\max}$$

(9) The "100-year wave" is that which will probably occur once every century.

where:

- $R_{p\,max}$ and f_{max} = maximum cone resistance and lateral frictional strength measured (for example) using a penetrometer in the surface layers,
- d = diameter of skirt (d = 20 m).
- e = thickness of skirt (approx. 2 cm for the steel sections),
- h = penetration of skirt into soil (to be determined).

Assuming:

- $R_{p\,max}$ = 2,000 t/m^2 (20 MPa) and f = 20 t/m^2 (200 kPa) in dense surface sands,
- F = 10,000 t (per cell),

we obtain h = 3 m.

In actual fact, determination of the length h of the skirts would appear to involve a certain amount of uncertainty, and in practice there is often a tendency to reduce the calculated length h.

VIII.3.1.2 Contact stresses between the structure and the soil

Should the sea bottom be perfectly flat, the stress distribution depends on the flexibility of the sea floor and the mechanical characteristics of the soil (see Section VIII.1.5.1).

Assuming a perfectly rigid floor, the structure should be designed on the assumption of:

– a stress at the edge (of foundation raft) twice that at the centre if the cone resistance R_p (measured with the penetrometer) is less than 200 t/m^2 (2 MPa) i.e. if the soil undergoes considerable deformation,

– a uniform stress distribution beneath the raft if R_p > 600 t/m^2 (6 MPa), in which the soil is only slightly deformable.

The presence of "hard points" or local bumps or humps on the surface of the soil introduces considerable stress concentrations beneath the structure in contact with the soil when it is set into position.

In fact, it is particularly important to determine the statistical distribution: density, height and lateral dimensions of the bumps or humps, thus justifying the advantages of as precise a bathymetry as possible (in principle, to the nearest 10 cm in relative value).

The most critical situation would clearly be that where an infinitely rigid structure would rest on a single central hump with a height h and a diameter D. According to Bjerrum [6], in the case of the Ekofisk tank, if h = 0.50 m and D = 8 m, the local stress would reach 200 t/m^2 (2 MPa) i.e. a value approaching the maximum bearing capacity of the sand as calculated in Paragraph VIII.3.3.

A certain number of remarks would nonetheless appear necessary as to the true contact stresses between the soil and the structure:

– stress concentrations result in settlement of the soil over a certain zone and consequently a gradual reduction of the average contact stress,

– grouting of cement as soon as the structure is installed enables the surface of the soil to be rendered uniform, thus reducing stress concentrations,

– distribution of the stresses beneath the structure depends both on the topography of the bottom and the degree of rigidity of the structure: slab (raft), cylinder cells closed off by domes, etc.

VIII.3.2 Stability of structure with respect to slip

Slippage of the structure under horizontal loads can occur as a result of:

– either slip of the surface,
– or slip between the layers of the terrains.

The stability must be calculated making various assumptions as to the soils (sand, clay, alternating layers of sand and clay) in order to consider the most critical case.

Let us take the example of the structure already mentioned for numerical application of calculation of the stability (see Section VIII.1.5.3):

– soil area: A = about 8,000 m^2,
– static vertical load:
  – $F_{v\max} = 200{,}000$ t (tank filled with water),
  – $F_{v\min} = 150{,}000$ t (tank filled with crude),
– forces due to the action of the 100-year wave (30 m, 15 s):
  – vertical force $\Delta F_v = \pm\, 10{,}000$ t,
  – horizontal force $F_h = 50{,}000$ t,
  – overturning moment $M = 2.10^6$ t × m.

It is assumed that the structure rests on the soil on a foundation raft not equipped with skirts.

VIII.3.2.1 Slip of the structure on sand

The maximum friction preserving the stability of the structure and under the most unfavourable conditions is given by the equation [13]:

$$\operatorname{tg} \lambda = \frac{F_h}{F_{v\,\min}}$$

For a sand with an internal friction tg φ, the minimum friction:

– either at the sand-structure interface,
– or at the sand-clay interface,

is such that $\operatorname{tg} \delta < \operatorname{tg} \varphi$.

The recommendations of *Det Norsk Veritas (DNV)* advocate the adoption of:

– a $\dfrac{\operatorname{tg} \varphi}{\operatorname{tg} \delta}$ ratio of 1.2 for sands,
– a multiplying factor of 1.3 on the calculated horizontal force F_h.

This being so, the safety factor with respect to slip of the structure:

$$CS = \frac{\text{tg}\ \delta}{\text{tg}\ \lambda}$$

is:

$$CS \approx 1 \quad \text{for } \varphi = 28°$$

$$CS \approx 2 \quad \text{for } \varphi = 45°$$

VIII.3.2.2 Slip of the structure on stiff clay

The safety factor with respect to slip of a structure resting on clay is given by the equation:

$$CS = \frac{A\,\tau_f}{F_h}$$

where:

A = area of base of structure,
τ_f = shear strength of clay.

The recommendations of *DNV* advocate the adoption of:

– a ratio $\tau_f = \frac{c_u}{1.4}$ (c_u being the undrained cohesion),
– a multiplicant of 1.3 on the horizontal force F_h method.

For a foundation area of the structure $A = 8{,}000\ \text{m}^2$, the safety factor is (Fig. VIII.3.1):

$$CS \approx 1 \text{ for } c_u \approx 10\ \text{t/m}^2\ (100\ \text{kPa})$$

$$CS \approx 2 \text{ for } c_u \approx 22.5\ \text{t/m}^2\ (225\ \text{kPa})$$

As can therefore be seen, the undrained cohesion of the clay must be greater than 10 t/m² (100 kPa). However, the arrangement of skirts beneath the structure will reduce the danger of slip at soil level to a certain extent.

VIII.3.3 Stability of the structure with respect to overturning. Bearing capacity of the foundation soil and slip surface of the soil

The stability of the structure with respect to overturning can be calculated in three different ways:

– estimation of the bearing capacity of the foundation soils,
– determination of the slip surface of the soil on rupture,
– the method of finite elements.

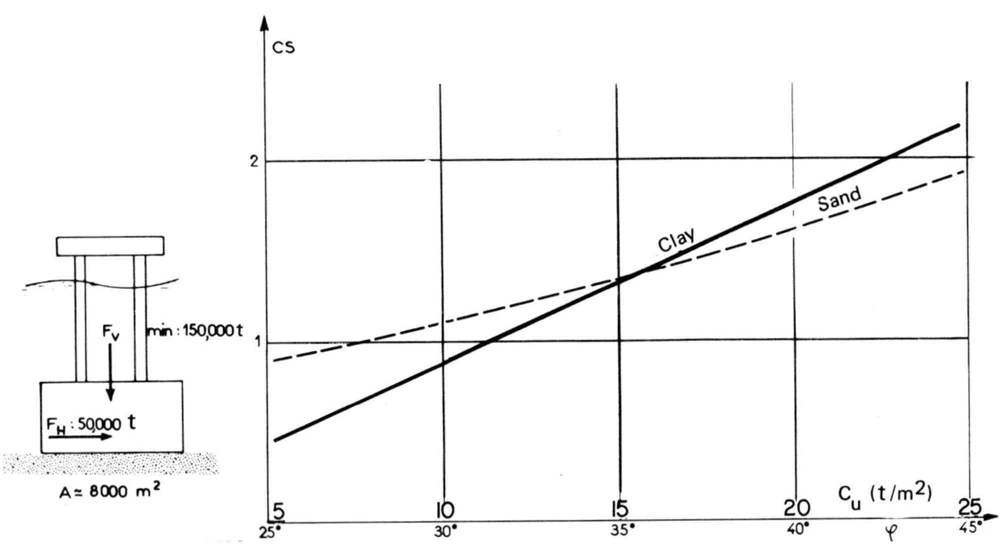

FIG. VIII.3.1. – Sliding stability of a gravity structure. Safety factor.

VIII.3.3.1 Bearing capacity of the foundation soil

The bearing capacity of a foundation soil can be calculated by the formulae of Hansen and Meyerhof. The general equation for the bearing capacity of the surface foundation is in the form [6] [15]:

$$q_u = \frac{Q}{BL} = \frac{1}{2}\,\gamma' BN_\gamma + c_u N_c + \gamma' D_f N_q$$

where:

q_u = bearing capacity of foundation soil,
Q = bearing capacity of soil,
B = width of foundation,
L = length of foundation,
D_f = depth to which foundation is buried,
γ' = submerged density of the soil ($\gamma' = \gamma - 1$),
c_u = undrained cohesion of soil,
N_γ, N_c, N_q = dimensionless factors depending on the angle of friction of the soil.

In the case of dense sands or stiff clays, the depth to which the structure is buried D_f will be small. The term $\gamma' d_f N_q$ will be neglected.

The previous equation can therefore be reduced to its first two terms:

$$q_u = \frac{Q}{BL} = \frac{1}{2} \gamma' BN_\gamma + c_u N_c$$

If the structure rests on a dense sand ($c_u = 0$), the previous equation can then be expressed as:

$$q_u = \frac{F_v}{BL} = \frac{1}{2} \gamma' BN_\gamma$$

It is assumed that the sand is drained in application of this formula.

This formula must comprise the various correcting factors allowing for the following:

– the eccentricity of the load on the foundation soil resulting from the action of the horizontal forces; the effective width of the foundation $B' < B$ is such that:

$$B' = B(1 - 2e) = B - 2E$$

where:

$$e = \frac{E}{B}$$

and:

$$E = \frac{M}{F_{v\,min}} \quad \frac{(M = \text{moment of overturning forces})}{(F_v = \text{minimum vertical load})}$$

– the shape of the structure:

$$s_\gamma = 1 - 0.4 \frac{B'}{L} i_\gamma$$

– the inclination of the forces acting on the structure:

$$i_\gamma = \left[1 - 0.7 \frac{F_h}{F_{v\ min}}\right]^5$$

The equation for the bearing capacity then becomes (Fig. VIII.3.2):

$$q_u = \frac{F_v}{BL} = \frac{1}{2} \gamma' BN_\gamma (1 - 2e)^2 s_\gamma i_\gamma$$

The factor N_γ which depends on the internal angle of friction φ:

– differs according to various authors by a factor of from 1 to 2,
– varies by 50% for a 2° variation of φ,
– depends on the dimensions of the foundation (scale effect).

For N_γ, the values given by B. Hansen, which would appear fairly close to values observed experimentaly, have been assumed (see Fig. VIII.3.3).

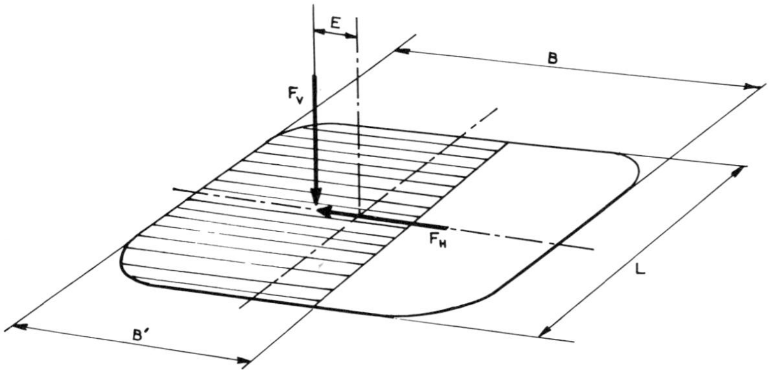

E

B

F_V

F_H

L

B′

$$q = \frac{1}{2}\gamma B N_\gamma (1-2e)^2 s_\gamma i_\gamma$$

Vertical bearing force for a centred load	$= \frac{1}{2}\gamma B N_\gamma$
Excentricity factor 1-2e	$= (B' = B\ (1 - 2\ e))$
Form factor s_γ	$= 1 - 0.4 \frac{B'}{L}$ $\quad i_\gamma$
Load inclination coefficient i_γ	$= \left[1 - 0.7 \frac{F_H}{F_V}\right]^5$

FIG. VIII.3.2. – Bearing force of a surface foundation resting on sand.

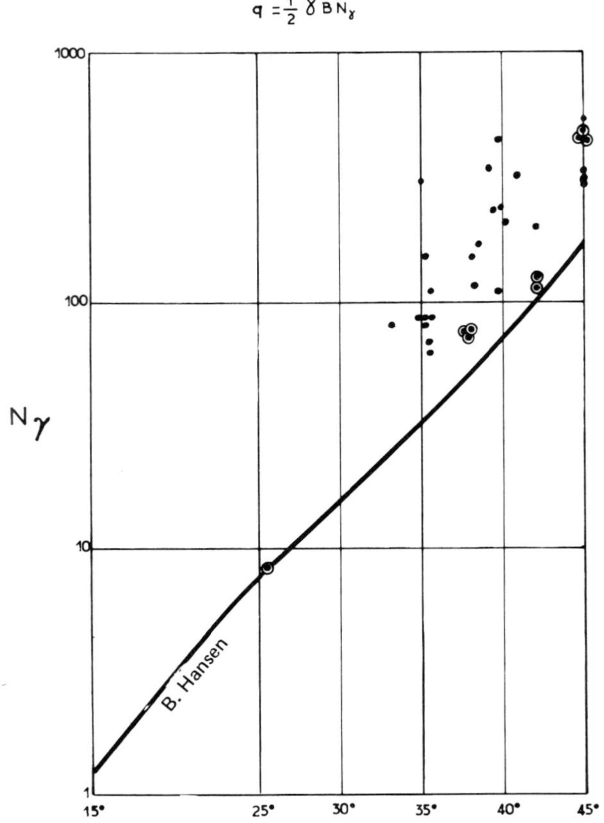

FIG. VIII.3.3. – Values of N_γ determined by experimentation

Numerical application:

– dimensions of the foundation: $B = L = 100$ m,
– forces acting on the foundation:

$$F_{v\,max} = 200{,}000 \text{ t},$$
$$F_{v\,min} = 150{,}000 \text{ t},$$
$$F_h = 50{,}000 \text{ t},$$

– moment of overturning forces $M = 2.10^6$ t x m,
– eccentricity of load:

$$E = \frac{M}{F_{v\,min}} = 13 \text{ m}$$

$$1 - 2e = 0.74$$

whence:

$$B' = B(1 - 2e) = 74 \text{ m}$$

– form factor $s_\gamma = 1 - 0.4\frac{B'}{L} i_\gamma \approx 0.9$,

– inclination factor $i_\gamma = \left[1 - 0.7\frac{F_h}{F_{v\,min}}\right]^5 \approx 0.27$.

Assuming that the density of the sand can vary from 1.5 to 2 t/m³, one obtains:

$$0.5 < \gamma' < 1 \text{ t/m}^3$$

The corresponding angle of friction φ varies:

– from 20 to 25° for $\gamma = 1.5$ t/m³,
– to 40-45° for $\gamma = 1.9$ to 2 t/m³.

The bearing capacity q_u of drained sand calculated under the above defined conditions will then theoretically vary from about 10 to 400 t/m² (100 to 4,000 kPa).

Under the assumption of uniformly distributed loading on the soil, the bearing capacity involved does not exceed about 20 to 30 t/m² (200 to 300 kPa). For a mean uniform strength of 25 t/m² (250 kPa), the theoretical safety factor is more than two if the angle of friction of the drained sand is greater than about 30° (Fig. VIII.3.4).

In actual practice, the sand is only partially drained during a storm, so the calculating assumption made is certainly highly optimistic (see Section VIII.3.3.2).

In addition, the stress distribution beneath the foundation raft is not uniform, either owing to local irregularities of the soil, or to the deformability of the soil and the slab (see Section VIII.3.1.2).

If the structure rests on a stiff clay ($\varphi = 0$), the equation expressing the bearing capacity becomes:

$$q_u = \frac{F_v}{BL} = c_u N_c$$

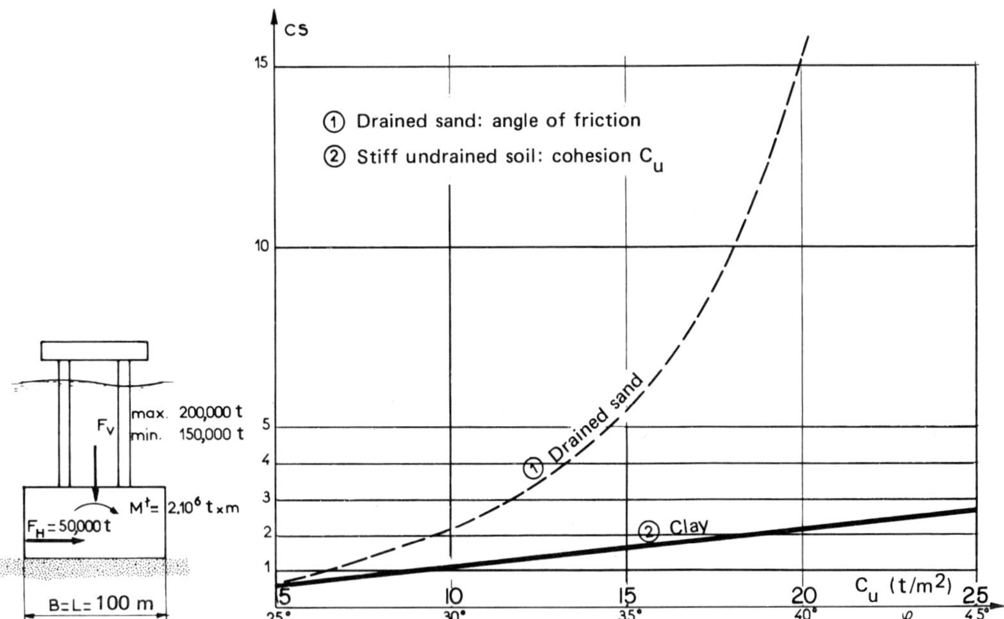

FIG. VIII.3.4. – Overturning stability of a gravity structure. Bearing capacity of foundation soil. Safety factor of soil on rupture.

As before, this equation must be modified by various correcting factors to allow for:

– the eccentricity of the load e, such that:

$$B' = B(1 - 2e) \quad \text{(effective width)}$$

– the shape of the structure:

$$s_c = 1 + 0.2 \frac{B'}{L}$$

– the inclination of the forces acting on the structure:

$$i_c = \left(1 - \frac{\alpha}{90}\right)^2$$

where:

$$\alpha = \text{arc tg} \frac{F_h}{F_{v\,\min}}$$

According to Meyerhof, the bearing capacity can then be expressed as:

$$q_u = \frac{F_v}{BL} = c_u N_c (1 - 2e)\, s_c i_c$$

The factor N_c on the surface equals $\pi + 2 \approx 5$.

Numerical application:

– dimensions of structure $B = L = 100$ m,
– force $F_{max} = 200{,}000$ t,
– force $F_{min} = 150{,}000$ t,
– force $F_h = 50{,}000$ t,
– moment $M = 2.10^6$ t × m,
– coefficient of eccentricity of load $= 1 - 2e = 0.74$,

whence:

$$B' = B(1 - 2e) = 74 \text{ m}$$

– form factor $s_c = 1 + 0.2\dfrac{B'}{L} = 1.15$,

– inclination factor $i_c = \left(1 - \dfrac{\alpha}{90}\right)^2 = 0.64$

Assuming that the cohesion of stiff clay varies from about 5 to 25 t/m² (50 to 250 kPa), the bearing capacity q_u ranges from approximately 13.5 to 68 t/m² (135 to 680 kPa).

Assuming also a uniform load beneath the structure, the bearing capacity actually involved is about 25 t/m² (250 kPa). It follows from that the safety factor varies (Fig. VIII.3.4):

– from 1 for $c_u = 9$ t/m² (90 kPa),
– to 2.7 for $c_u = 25$ t/m² (250 kPa).

VIII.3.3.2 Slip surface of the soil on rupture

In view of the great many correction factors introduced into the above equations, much caution should be exercised in considering the results of this method, which can be considered only as initial approach to the problem. Furthermore, the above method of calculating the bearing strength infers that:

– the interstitial pressure in the **sand** remains zero (sand drained), which would appear fairly remote from reality in fine and dense sands beneath large dimension foundations. Indeed, under the action of a wave with a period of about 15 s in the North Sea, the horizontal force rises from 0 to its maximum value in about 5 s, thus tending to increase the interstitial pressure,

– rupture of the foundation only occurs beneath the effective surface $B'L$ and not the total surface BL. This assumption would appear rather conservative, particularly in the case of clays.

This being so, it would appear necessary to check the conditions of stability obtained by means of another method.

The method of calculating stability which is commonly used therefore consists in determining the **critical slip surface** of the undrained foundation soil, by successive approximations. This method is applied [6]:

– **to the fine sands with low permeability** of the North Sea ($k \approx 10^{-3}$ cm/s, or 1 darcy), undrained or partially drained. The angular friction of the sands under undrained conditions is no more than 30-35° (instead of 40-43° drained). This problem would appear to have been solved for the first time in the case of sands by B. Hansen in calculating stability of the Ekofisk tank,

– **to undrained consolidated clays,**

– **to soils consisting of alternating layers** of sands and clay (where it would often appear very difficult to determine the mean shear strength).

In seeking to determine the critical stability, the following are considered:

– first, the effective stresses beneath the foundation resulting from stresses previously in the soil and those caused by the submerged weight of the structure,

– second, the variations in the water pressure at the sea bottom, near the structure.

This method of calculation, which is generally considered satisfactory, would lead to correct definition of the safety factor.

In the case of the Ekofisk tank (Fig. VIII.3.5), the solution given by B. Hansen, assuming:

– an undrained angular friction $\varphi_u = 38°$,

– perfect contact between the sand and the slab over the entire area of the foundation,

leads to a safety factor on tg φ_u of 1.36 to 1.61.

The method of calculation used for studying the stability of foundations installed on clayey soils has been gradually improved at the *Norwegian Geotechnical Institute (NGI)* for some years now. The numerical program [16] [17]:

– assumes that the vertical load is applied only to the effective foundation surface, whereas the horizontal force is applied to the total surface,

– enables the shear stresses on both sides of the slip interface to be determined.

Remark: Analysis by the finite elements method also enables the stress distribution beneath the structure and its displacement under the effect of the horizontal forces to be determined.

VIII.3.4 Stability of the foundation under cyclic loads

Under the effect of wave action on the structure, the foundation soil undergoes alternating shear loads, the positive and negative amplitudes of which are appreciably equivalent, resulting in "fatigue" of the soil, i.e. a drop in the shear strength [6] [13] [18] [19].

The average length of a storm in the North Sea is 5 to 6 h, which, for a wave with a period of 15 s, corresponds to about 1,500 loadings.

The evolution of the shear strength with the amplitude and number of applications of stress to the foundation soil is at present estimated on the basis of results of tests performed in various laboratories. However, it will be seen that this method must be considered with serious reservations.

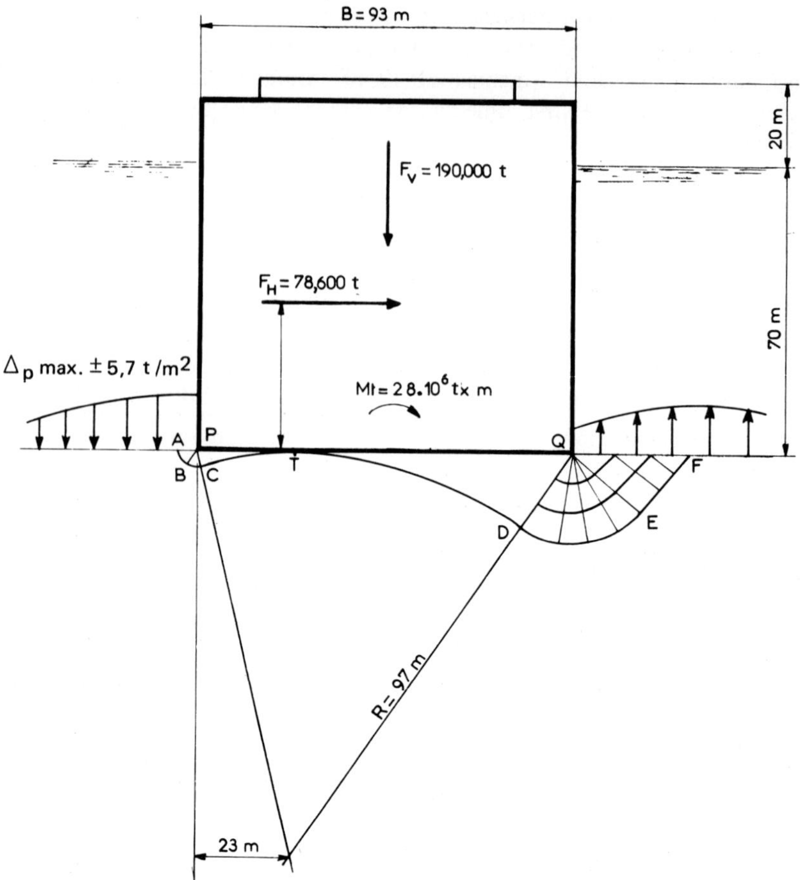

FIG. VIII.3.5. – Overturning stability of the Ekofisk tank. Slip surface of soil on rupture (B. Hansen's solution).

VIII.3.4.1 Cyclic loading of sands

Where there is no drainage, cyclic loading of fine sands in the laboratory is accompanied by a continuous rise in the interstitial pressure [20] [21]. Since the total stresses caused by the weight of the structure are kept constant, the effective stresses decrease uniformly and can cause the sand to liquefy:

$$\sigma_{eff} \searrow = \sigma_{tot(constant)} - p_i \nearrow$$

As an example, Fig. VIII.3.6 shows the behaviour of an undrained dense sand ($D_r > 95\%$) subjected to cyclic shear loads under a state of stress:

$$\tau_H / \sigma_1 = 0.13,$$

where:

τ_H = alternating shear stress,
σ_1 = constant vertical stress applied to the sand.

The interstitial pressure u increases steadily (until break-up) with the number of cycles N (Fig. VIII.3.6) in accordance with an approximately linear law such that:

$$\beta = \frac{\Delta u}{\sigma_1 N}$$

where β designates the mean straight line slope.

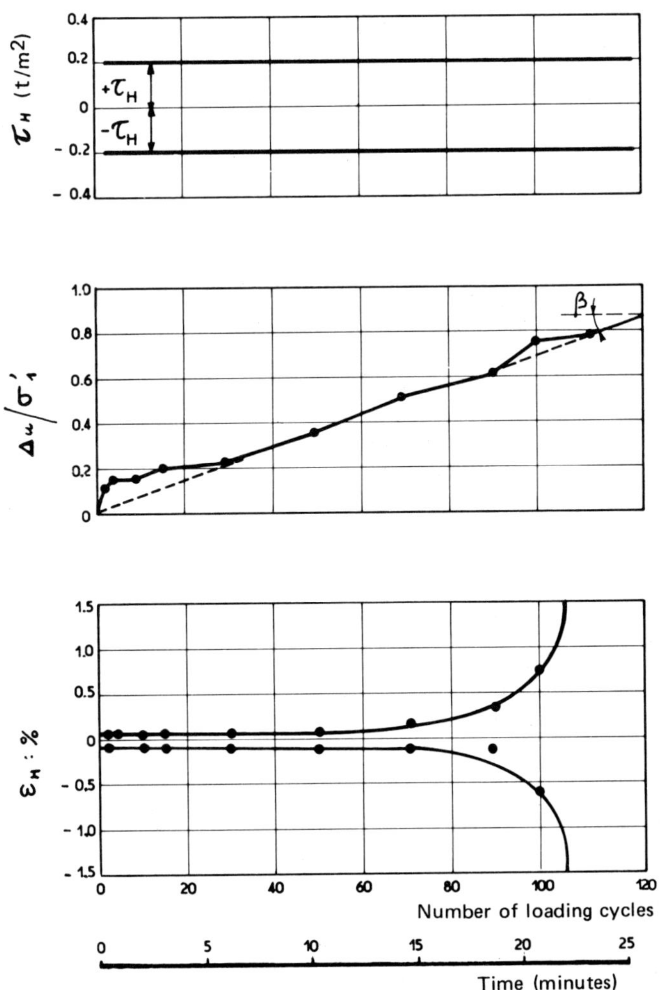

FIG. VIII.3.6. – Cyclic loading of a dense North Sea sand (NGI linear shear device).

Carrying out cyclic shear strength tests under various states of permanent stress τ_H/σ_1 gives a relationship between the relative increase in interstitial pressure and the shear stress applied to the sand (Fig. VIII.3.7).

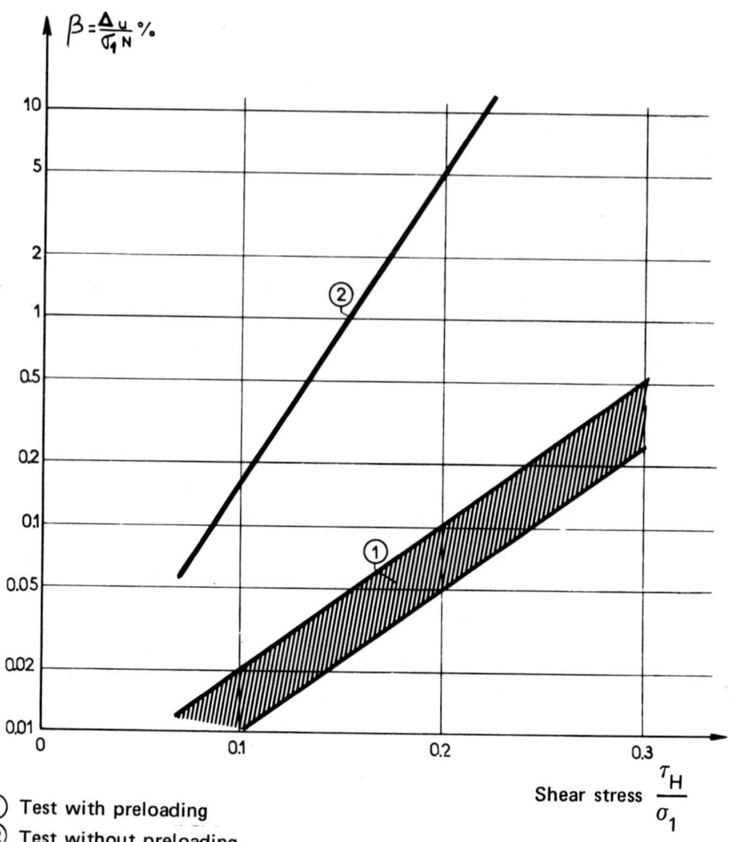

FIG. VIII.3.7. – Increase in interstitial pressure per loading cycle in undrained sand (relative density of sand $D_r = 90$ %)

In dense sands (relative density $D_I \geqslant 90\%$), it is assumed that:

– the 25 m wave amplitude gives rise to a shear stress in the soil of $\tau_H/\sigma_1 = 0.3$,

– the shear strength τ_H/σ_1 is proportional to the amplitude of the wave.

By means of the diagram in Fig. VIII.3.7, it is then possible to estimate a maximum foreseeable interstitial pressure u_{max} developed in the foundations and underneath the structure during a storm featuring a given wave amplitude spectrum as a function of time (about 6 h):

$$u_{max} = (\Sigma \beta_i N_i)\, \sigma_1$$

Numerical application:

The numerical data previously adopted are (see Section VIII.3.2.1):

$F_{v\,min} = 150{,}000$ t,
$F_h = 50{,}000$ t,
Augmentative factor applied to $F_h = 1.3$,
Angle of friction of sands $\varphi = 35°$,
Angle of friction between sand and raft δ, approximately 30°.

The slip safety factor depends on how u develops:

– in the case of **dissipation of the interstitial pressure** ($u = 0$), the slip safety factor CS = 1.34 (see Section VIII.3.2.1),

– in the case of **interstitial pressure which is continuously increasing** during a 6-h "100-year storm" including a 25 m maximum wave amplitude, calculation gives :

$$u_{max} = 0.30 \text{ to } 0.40\,\sigma_1$$

i.e. :

$$\sigma_{1\,eff} = \sigma_1 - u_{max} \approx 0.65\,\sigma_1$$

The slip safety factor falls to CS = 0.90.

It can therefore be observed that rupture of the foundation by slip occurs well before liquefaction of the sand.

The initial density of the sand and the previous stresses exerted play a major part in its final behaviour:

– the action of the low amplitude wave may result in a slight increase in the density of the sand beneath the structure and hence its subsequent shear strength under greater loads,

– the action of a large amplitude wave can result in a considerable drop in the static shear strength of the sand (fatigue phenomenon).

The execution of cyclic loading tests of undrained sand would not appear to be representative of reality, since, as indicated previously (Section VIII.1.2.4):

– drainage cannot take place instantaneously with each wave stress,

– however, the interstitial gradient created beneath the structure leads to at least partial drainage of the foundation sand during a storm.

In the absence of sufficiently representative experimentation and particularly of in situ measurements underneath structures, caution should be exercised in applying the above method of calculating the stability.

The execution of scale models of gravity structure foundations subjected to cyclic loadings is very probably a highly interesting method of investigation. The similarity is generally brought about by centrifugation [9] [22] [23] [24] [25].

Measurements made with piezometers located under the Ekofisk tank show maximum increases of 1 to 2 t/m² (10 to 20 kPa) for the interstitial pressure, for a wave amplitude of 20 m (during the November 1973 storm) [26].

VIII.3.4.2 Cyclic loading of clays

The behaviour of clays under the effect of cyclic loads, albeit very inadequately understood, would appear to be essentially a function of the initial consolidation and of the level of stresses applied [19]:

– sensitive clays of normal consolidation often display a tendency to "liquefy",
– slightly overconsolidated clays can display a degree of "adaptation" or "accommodation" to cyclic loads,
– highly overconsolidaded clays can first display a decrease of interstitial pressure, followed by an increase in volume and a drop in the shear strength ("fatigue" effect).

According to cyclic loading tests on overconsolidated clays (overconsolidation ratio of 4) conducted at the *NGI* [16], the loss of shear strength would be the following for over 1,000 cycles:

– about 5 to 8% for a cyclic deformation of 1.3%,
– about 25% for a cyclic deformation of 3%, corresponding in reality to the effect of a large amplitude wave.

The main difficulty in interpreting the cyclic loading tests of soils lies in the choice of the characteristic fatigue parameter, i.e. number of repeated stresses, cyclic deformation, etc. [27]. The previous loads exerted unquestionably affect the development of fatigue; in actual practice, can one apply a principle of superposition, such as the Miner principle for instance?

The stability of a gravity structure resting on a clay, subject to wave stresses for the duration of a storm, is at present calculated:

– by determining the cumulative deformation under the effect of cyclic shear stresses of various amplitudes during the storm,
– by relating the deformation to the critical shear strength after cyclic loadings.

VIII.3.5 Settlement of the soil beneath the structure

Settlement of the soil beneath the structure is the result:

– of settlement under the effect of the weight of the structure,
– of settlement under the cyclic loads.

VIII.3.5.1 Settlement of dense sands or stiff clays beneath a structure

An estimate of the settlement can be obtained by the method of finite elements, enabling the non-linear behaviour of the sand and/or clay on which the structure rests to be taken into account [6] [13].

The calculation makes it possible to determine:

– first, the total settlement of the soil beneath the structure,
– second, the differential settlement resulting from wave action.

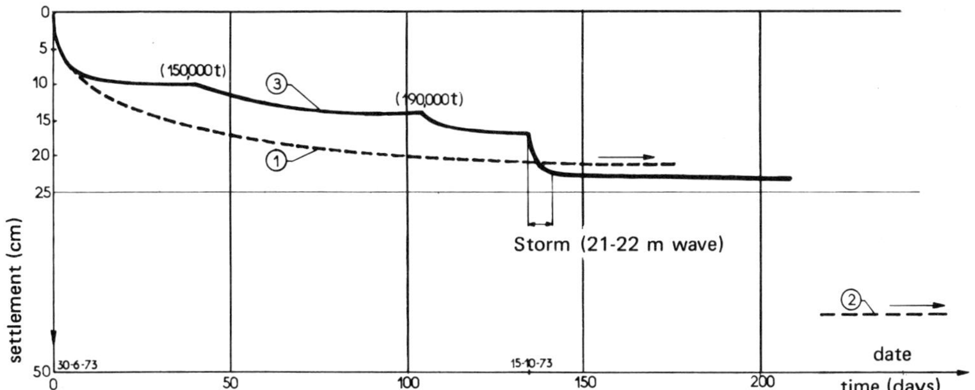

① Calculated settlements as a result of weight
② Calculated limit settlement as a result of weight + wave effect
③ Mean settlement observed (according to the various measurements made [25]).

FIG. VIII.3.8. – Calculated and observed settlements of the soil beneath Ekofisk tank.

The total settlement calculated for the Ekofisk tank, assuming that the layers are elastic, is of about 40 cm, resulting in appreciably equivalent proportions from (see Fig. VIII.3.8) :

– the weight of the structure,
– the action of the cyclic loading of the structure by wave action.

The settlement of about 20 cm under the effect of the weight of the structure is due:

– for 15% of the total, i.e. about 3 cm, to settlement of the top 26 m of highly dense sands,
– for the remaining 85%, i.e. about 17 cm, to the settlement of the underlying very stiff clay layer.

Figure VIII.3.8 shows the variations in the approximate settlements of the soil (calculated and observed) underneath the Ekofisk tank [26]. The reader may note the following:

– first, the relatively large influence of the November 1973 storm on the settlement of the soil,
– second, that the settlement of the soil less than one year after the structure was installed equals about half the total projected settlement ([10]).

Differential settlement of the soil beneath the structure may arise from:

– either the heterogeneity of the formation layers (frequent facies changes or possible backfilled valleys),
– or the action of waves.

In the case of the Ekofisk tank, the predicted effect of the 100-year wave (height 24 m) would result:

– first, in raising one side of the structure by about 15 cm,
– second, subsidence of the opposite side of the structure, also by about 15 cm.

The result would therefore be a predicted differential settlement of 30 cm.

(10) The settlement during 1974/1975 was no more than about 5 cm.

VIII.3.5.2 Settlement of soft soils underneath structures

Any contemplated emplacement of a gravity structure on a clayey soil of little consolidation is based on the fact that the layer of soft soil of relatively limited thickness (less than about 10 m) rests on a layer of draining sand.

Installation results in immediate sinking by a depth D such that the bearing capacity of the soil balances the weight of the structure [28] :

$$D = \frac{2R}{3\left(\frac{p}{c_u} - 5\right)}$$

where:

$2R$ = diameter of structure,
p = pressure on soil,
c_u = mean cohesion.

If:

$2R$ = 100 m,
p = 20 t/m² (200 kPa),
c_u = 1 t/m² (10 kPa),

the structure sinks immediately by a depth D = 2.2 m.

Settlement than continues beyond this point:

– either by slow consolidation, with lateral creep of the soft soils,
– or by fast consolidation, by applying the self-bending method (suction of the soft soil).

The total settlement can be expected to be little less than the initial thickness of the layer of soft soil.

VIII.3.6 Scour near the structure

As already pointed out in Paragraph VIII.1.2.5, scour at the base of a gravity structure may seriously undermine its stability [28] [29].

VIII.3.6.1 Causes of scour

Scour is the result of the action of currents and waves near the structure. Experience proves that most of the fine and uniform sand of the North Sea is subject to erosion.

The degree of scour that occurs is as yet difficult to predict:

– current knowledge of the problem is highly empirical, making no allowance for the effect of all the parameters related to hydraulic, geometry and grain size distribution,

– tests on models, with water particle speeds equivalent to those expected at the base of the structure, would not yet appear to permit satisfactory analysis of the problem, but major work is now going on at the Trondheim (Norway) *River and Harbour Laboratory*,

– in situ experiments on structures already set up are still very limited.

VIII.3.6.2 Techniques for preventing scour

Several techniques are currently recommended for reducing or eliminating scour at the base of the structure.

Laying of rip-rap around the base seems to be the solution envisaged for structures at present being built. However, current ideas reside on totally empirical data, the quantities recommended differing considerably with each project:

– about 10 m^3 per metre of perimeter (slope 1.5 to 2 m in height and 4 to 5 m in width), i.e. 4,000 m^3 for a 100 × 100 m square structure according to some,

– 20 to 30 m width and 1 to 2 m height, i.e. 10,000 to 20,000 m^3 for a 100 m diameter structure, according to others.

Plastic meshes (Filtrix, Seditech, etc.) laid around the structure in widths of 6 to 10 m and held in place by anchors and/or stones, appear to provide a satisfactory solution.

Equipping the base of the structure with skirts prevents scour from extending underneath the foundation raft.

The solution employed at Ekofisk:

– "Filtrix" rolled out on the soil before laying pebbles,

– about 10 m^3/m of pebbles of 10 cm (90 to 95%) + finer elements,

appears satisfactory according to an inspection of the base of the tank after the November 1973 storm. One can however wonder what would have been the effectiveness of a stone-gravel mixture were there to have been no Jarlan wall (alveolar). It should be noted that the mixture contains more fine elements than it originally did (addition of sand).

VIII.3.7 Instrumentation for verification of gravity structures

Calculations of the stability of gravity structure foundations appear highly elaborate considering the mediocrity of the data gathered on the geotechnical characteristics of the soils, whether by coring or in situ measurements. Only through measurements and checks made when the structure is being installed and afterwards will it be possible to assess the validity of the methods of calculation used and possibly to improve them. The instrumentation necessary for these various checks is now being developed. [30].

VIII.3.7.1 Verification when installing gravity structures

The purpose of verifications made when laying gravity structures is to determine [31] [32]:

– first, the relationships between the ballasted weight of the structure, the depth and the rate of penetration of the skirts into the soil,
– second, the stress distribution between the soil and the structure.

The purpose of retractable probes fitted into the walls of the skirts is to detect any abnormal resistance to penetration of the skirts and hence to adjust the ballasting of cells (alveolar cylinders).

In the case of the Condeep type structure, the following must be mounted beneath each cell:

– pressure sensors to follow the development of the pressure of the water captured inside the skirts (especially for clayey soils),
– systems for measuring the distance between the soil and the bottom of the cell in order to forecast the necessary quantities of grouting,
– total pressure sensors to check the contact stresses between the soil and the structure.

Installation of the structure will also be checked by means of:

– inclinometers enabling the position of the raft surface to be related to a horizontal plane of reference,
– a device to measure the vertical movement of the structure,
– various systems for checking the behaviour of the different parts of the structure.

VIII.3.7.2 Measurements after installing gravity structures

Measurements to be made after installing the structure essentially involve movements of the structure and the development of the interstitial pressure [30].

The settlement of a few tens of centimetres of the soil must be appraised to the nearest centimetre. The measurements are made in relation to a point which is assumed fixed, i.e. optical measurements from a neighbouring platform, device anchored about 100 m into the foundation soil, etc.

Alternating lateral movements under the effect of wind and wave action are checked by means of accelerometers.

Accurate measurement (to the nearest 1 or 2 cm) of the slow drift (slip) of the structure which may possibly occur is envisaged by means of devices equipped with inclinometers.

The total stresses underneath the structure are measured by means of sensors laid immediately underneath the foundation raft.

Knowledge of the development of the interstitial pressure in the foundation soil during a storm appears indispensable for appraising the validity of the experiments conducted and the method of calculation now being used.

The change in interstitial pressure is followed by means of piezometers placed at various depths in boreholes just beneath the structure.

BIBLIOGRAPHY

[1] *Data sheets* on different gravity structures.

[2] SNYDER (R.E.). – "Platform Technology Spurred by Deep-Water Developments", *World Oil,* July 1975, p. 79-84.

[3] EIDE (O.T.) and LARSEN (L.G.). – "Installation of the Shell/Esso Brent B Condeep Production Platform". *OTC,* Houston, 1976. Paper OTC 2434.

[4] WATT (B.J.). – "Gravity Structures. Installation and other Problems". *Offshore Soil Mechanics.* Published by Cambridge University, Engineering Department on Lloyd's Register of Shipping, September 1976, pp. 285-305.

[5] VAN OOSTRUM (W.H.A) and LUBBERS (A.). – "Deep Dredging for Offshore Purposes. A Method to Clear Stones and Boulders from the Seafloor"., *SPE European Spring Mtg.,* Amsterdam April 8-9, 1976, Paper SPE 5759.

[6] BJERRUM (L.). – "Geotechnical Problems Involved in Foundations of Structures in the North Sea", *Geotechnique,* No. 3, Sept. 1973.

[7] HOVE (K.) and FOSS (I.). – "Quality Assurance for Offshore Concrete Gravity Structures"., *OTC.* Houston, May 6-8, 1974, Paper OTC 2113.

[8] *DET NORSKE VERITAS* – "Rules for the Design, Construction and Inspection of Fixed Offshore Structures (1974)"

[9] YOUNG (A.G.), KRAFT (L.M. Jr.) and FOCHT (J.A. Jr.). – "Geotechnical Considerations in Foundation Design of Offshore Gravity Structures" *J. of Petrol. Techn.,* August 1976, p. 925-937.

[10] STUBBS (S.B.). – "Concrete Structures in Deep Water". *Offshore Scotland Conference,* Aberdeen 1973.

[11] CAQUOT (A.) and KERISEL (J.). – *Traité de Mécanique des sols,* Gauthier-Villars Paris, 1966.

[12] LEONARDS (G.A.). – *Foundation Engineering.* Mc Graw-Hill Book Co, New-York, 1962.

[13] EIDE (O.F.) – Marine Soil Mechanics. Applications to North Sea Offshore Structures". *Offshore North Sea Technology Conference,* Stavanger 3-6 Sept. 1974. Preprint NGI Publication, No 103.

[14] HITCHINGS (G.A.), BRADSHAW (H.) and LABIOSA (T.D.) – "The Planning and Execution of Offshore Site Investigations for a North Sea Gravity Platform" *OTC,* Houston, 1976. Paper OTC 2430.

[15] HANSEN (J.B.). – "A general Formula for Bearing Capacity", *Geotechnical Bull,* Vol. 5, No 11, 1973.

[16] SCHJETNE (K.). – "Foundation Engineering for Gravity Structures in the North Sea", *SPE European Spring Mtg.,* Amsterdam April 8-9, 1976. Paper SPE 5758.

[17] LAURITZSEN (R.A.) and SCHJETNE (K.). – "Stability Calculations of Offshore Gravity Structures". *OTC,* Houston, May 3-6, 1976. Paper OTC 2431.

[18] LEE (K.L.) and FOCHT Jr. (J.A.). – "Cyclic Testing of Soil for Ocean Wave Loading Problems". *OTC,* Houston, May 5-8, 1975.

[19] ANDERSEN (K.H.). – "Behaviour of Clay Subjected to Undrained Cyclic Loading". *Proceedings of BOSS' 76*, Vol. I, p. 392-403.

[20] LEE (K.L.). – "Fundamental Considerations for Cyclic Triaxial Tests on Saturated Sand". *Proceedings of BOSS' 76*, Vol. I, p. 355-373.

[21] SEED (H.B.). – "Some Aspects of Sand Liquefaction under Cyclic Loading". *Proceedings of BOSS' 76*, Vol. I, 374-391.

[22] ROWE (P.W.). – "Displacement and Rupture Modes of Model Offshore Gravity Platforms Founded on Clay" *Offshore Scotland Conference*, Aberdeen, 1975.

[23] ROWE (P.W.). – "Model Studies of Offshore Gravity Structures Founded on Clay". *Proceedings of BOSS' 76*, Vol. I, p. 439-448.

[24] GEORGE (P.J.) and SHAW (P.G.H.). – "Certification of Concrete Gravity Platforms in the North Sea". *OTC*, Houston, 1976. Paper OTC 2436.

[25] LE TIRANT (P.), LUONG (M.P.), HABIB (P.) and GARY (G.). – Simulation by Centrifugation of of Offshore Foundations. *Proc. of the IX Inst. Conf. on soil Mechanics*, Tokyo 1977, Vol. 2, p. 277-280.

[26] CLAUSEN (G.L.), DIBIAGIO (E.). DUNCAN (J.M.) and ANDERSEN (K.H.). – "Observed Behavior of the Ekofisk Oil Storage Tank Foundation", *OTC*, Houston, May 5-8, 1975. Paper No. 2373.

[27] VAN EEKELEN (H.A.M.). – "Fatigue Theory for Cyclic Loading of Soils" *SPE European Spring Mtg*, Amsterdam April 8-9, 1976. Paper SPE No 5773.

[28] HEIJNEN (W.J.). – "Some Soil Mechanical Problems Related to the Application of Gravity Type Structures". *Inter-Ocean 73*, Dusseldorf, 13-18 Nov. 1973.

[29] MARTINSEN (O.R.). – "Concrete Platforms" *Offshore North Sea Technology Conference*, Stavanger, 3-6 Sept. 1974.

[30] DIBIAGIC (E.), MYRVOLL (F.) and HANSEN (S.V.). – "Instrumentation of Gravity Platforms for Performance Observations". *Proceedings of BOSS' 76*, Vol. I, p. 516-527.

[31] BERGE (H.). – "Offshore Gravity Structures and Harbours in the North Sea Environment". *Offshore North Sea Technology Conference*, Stavanger, 3-6 Sept. 1974.

[32] LOW (E.). – "Gravity-Platform Foundations: Making the Best of Limited Knowledge", *New Civil Engineer Special Supplement*, 26 Sept. 1974.

CHAPTER IX

Jack-up Platforms

CONTENTS

INTRODUCTION

1. Jack-up platforms are mobile units resting on the sea bottom during operation and which have been used for drilling (or production) since 1953.

At present, there are about 120 jack-up platforms in service throughout the world operating in depths of from 20 or 30 m to about 100 m. About 10 units now being built or in the project stage are to operate in depths of water of 120-130 m and even 150 m (Fig. IX.1.1) [1] [2].

The building technology of jack-up platforms has been developing for 20 years and is essentially typified by the gradual replacement of cylindrical legs by lattice-work legs endowed with greater buckling strength and ability to withstand the elements (waves, wind).

2. There are two main types of jack-up platforms:

– platforms resting on piles ending in bases or caissons (known in the literature as spud tanks, spud cans or again doughnuts, depending on the geometry or structure), built by *Le Tourneau, Santa Fe, CFEM* (Figs. IX.1.2a and IX.1.3) [3] [4],

– "mat-supported" platforms resting on a single foundation raft of about the same dimensions as those of a platform, built in particular by *Bethlehem Steel-Beaumont* (Figs. IX.1.2b and IX.1.4) [5] [6].

The stability of jack-up platforms depends on the adaptation of the type of support (bases, caissons or mats) to the nature and the mechanical characteristics of the terrains. The shape and dimensions of the supports have developed along with the overall technology of the structure, but depend first and foremost on the constructor.

A distinction is drawn (see Section IX.1) between:

– bases, which are generally cylindrical,
– caissons (or spud cans) of widely varying geometries,
– the single or twin mat.

Jack-up platforms can also be characterized:

– either by the verticality of the piles,
– or their inclination, which varies from 0 to 10° (or even 15°) in order to increase the stability under lateral loads.

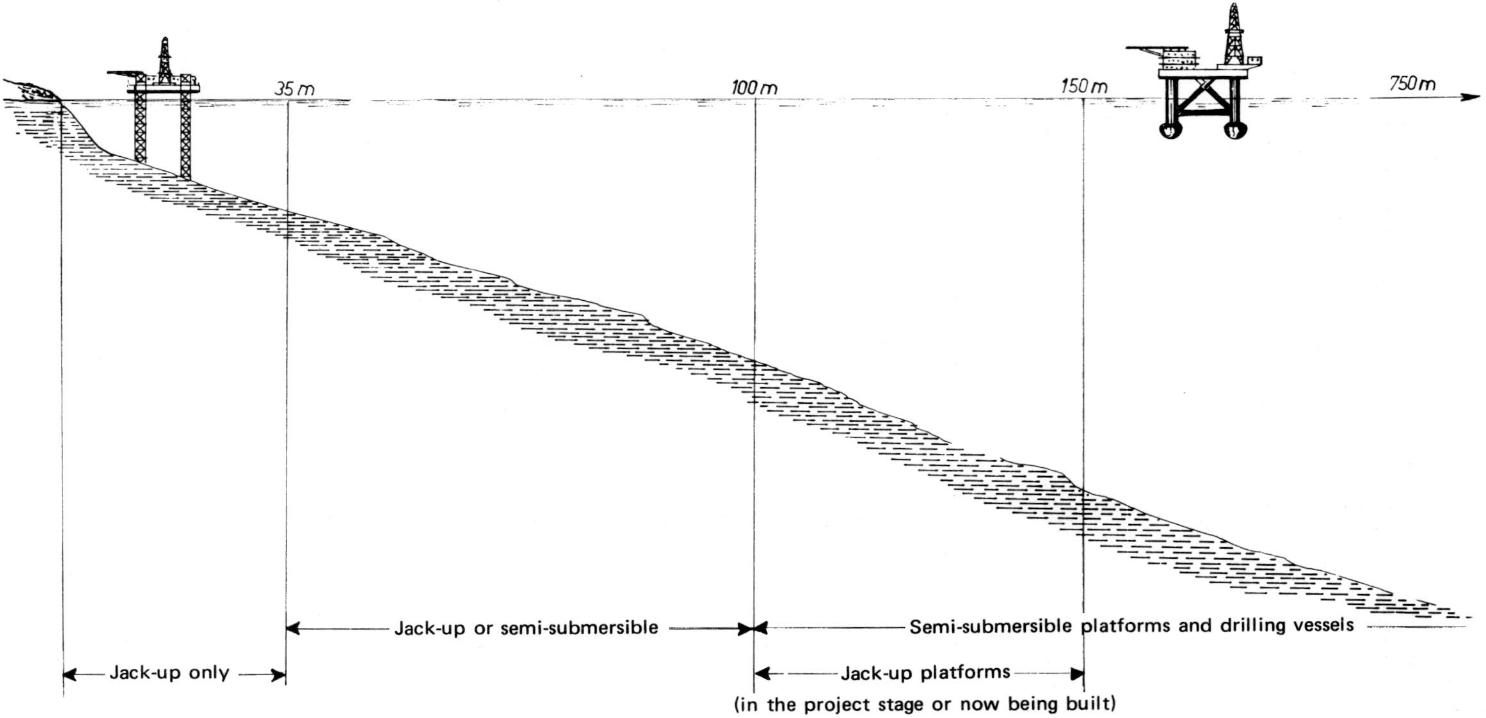

FIG. IX.1.1. – Diagram showing common depths of water and limits of use of jack-up platforms.

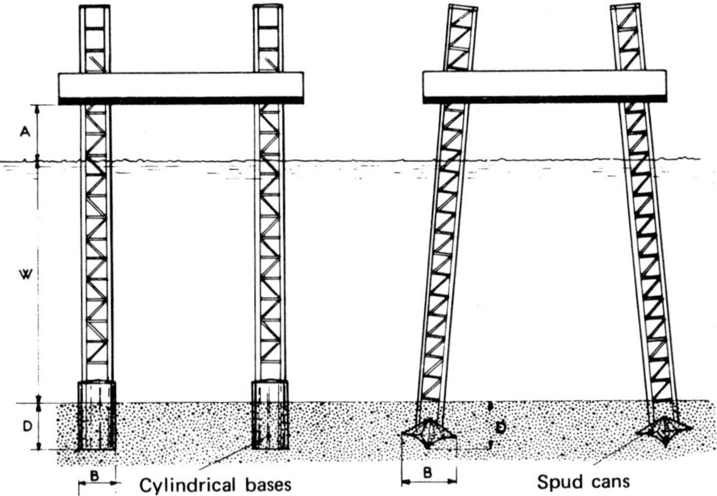

(a) Jack-up platforms with bases or spud cans

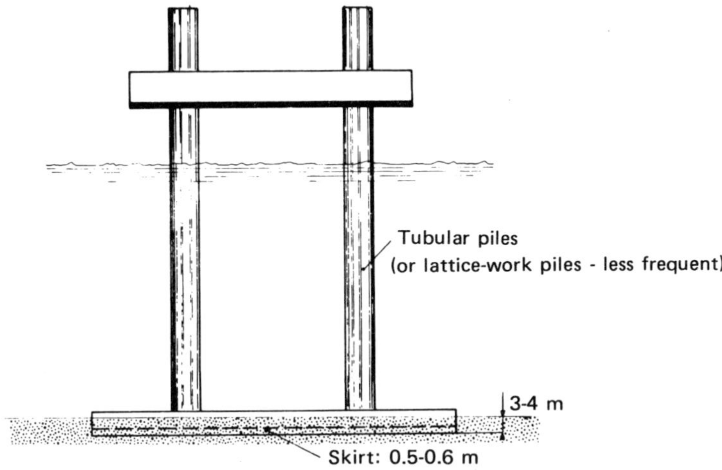

(b) Mat-supported jack-up platforms

FIG. IX.1.2. – Layout of jack-up platforms.

3. The soil reconnaissance needed before setting up a jack-up platform varies with the site and the legislation in force:

– geotechnical reconnaissance proper must invariably be preceded by geophysical reconnaissance,

– the depth of reconnaissance of the soils varies with the nature and consolidation of the formations encountered, the water depth, the weather and oceanographic conditions, the type of structure to be set up, etc.,

– prior knowledge of the site enables this soil reconnaissance work to be limited in certain cases before a new unit is set up. For instance, in the Gulf of Mexico it is considered

FIG. IX.1.3. – Tripod jack-up platform mounted on spud cans.

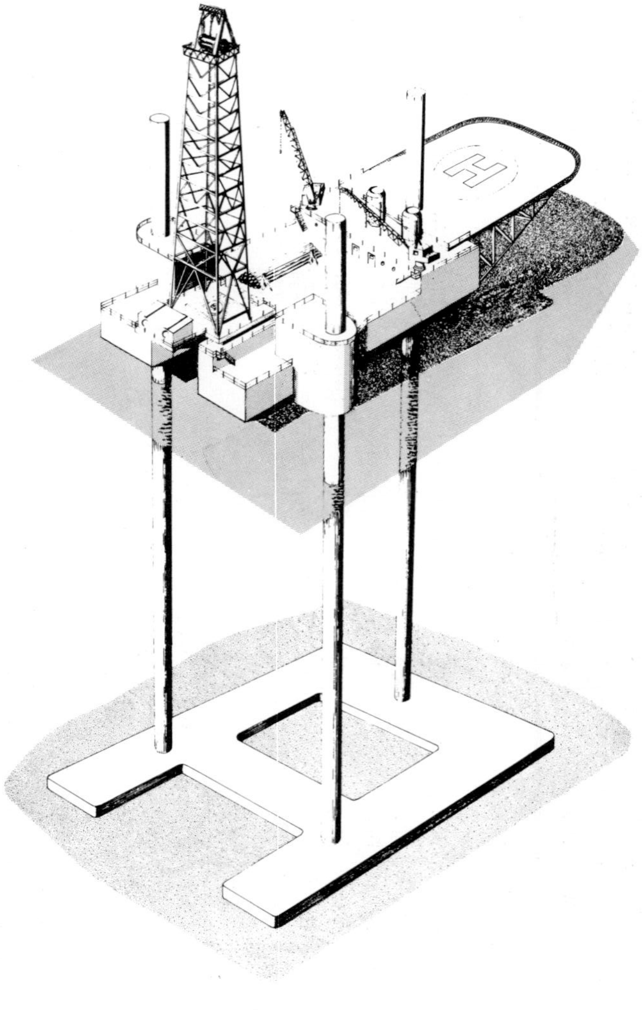

FIG. IX.1.4. – Mat-supported jack-up platform.

that the soils are sufficiently well known to render unnecessary any reconnaissance before erecting a new jack-up platform. On the other hand, in most of the other regions of the world, this reconnaissance is required by law.

4. The following will be examined in the present chapter:

– description of the supports of jack-up platforms,

– the geotechnical problems raised by the erection, stability during drilling and removal of jack-up platforms,

– the soil reconnaissance needed before installing jack-up platforms,

– application of the results of soil reconnaissance to calculation of the stability of jack-up platforms, together with the choice of the type of support best suited to the nature of the soils encountered.

IX.1 DESCRIPTION OF THE SUPPORTS OF JACK-UP PLATFORMS

The diversity of the geometry of jack-up platform supports would not appear to have any clear justification from the standpoint of the stability of the structure, depending on the nature and consolidation of the formations encountered.

In this section, we propose first of all to examine the possible advantages and disadvantages of certain types of supports, considering:

– first, platforms resting on piles ending in bases or spud cans,

– second, mat-supported platforms (twin or single mat).

Next, as an example, the characteristics of a number of jack-up platforms will be given, together with the approximate loads per pile.

Remark: We have deliberately confined ourselves to the two main types of jack-up platforms now in service, since new structures for depths of water of 100 to 150 m now being built or in the project stage such as the following:

– the "ETA" monopod jack-up platforms resting on the soil on 3 cylindrical gravity bases [7] [8],

– or the Sharp "two-stage" jack-up platform with conventional cylindrical bases [9],

do not appear to differ from conventional platforms from the standpoint of the stability of their foundations.

IX.1.1 Jack-up platforms with piles ending in bases or spud cans

The general architecture of jack-up platforms differs enormously with each individual unit:

– the number of piles varies from 3 to 14; however, structures with 3 or 4 piles are by far the most common,

– the diameter of the bases or spud cans varies from approximately 5 to 15 m,
– the piles may be vertical or inclined (up to 10-15°).

IX.1.1.1 Geometries of bases or spud cans

The origin of spud cans goes back to about 1955, after a number of accidents caused by the instability of jack-up platforms resting directly on the soil without supports.

Figures IX.1.5a and IX.1.5b show the configurations and main dimensions (height and diameter) of a number of jack-up platform supports in service in various regions of the globe: Gulf of Mexico, North Sea, Indonesia, Persian Gulf, etc. [2] [10].

The geometry of the bases or spud cans displays numerous particularities including:

– cylindrical bases with a conical extremity with a tip angle of from 60 to 120°,
– cylindrical bases with a concave or convex end,
– conical polyhedric or hyperbolic spud cans,
– spud cans with asymmetrical ends (asymmetrical cone or truncated cylinder).

IX.1.1.2 Dimensions of bases or spud cans and piles

Except for platforms built in about 1955, most bases or spud cans for jack-up platforms have a diameter of about 10 to 14 m, i.e. a cross-section of 75 to 150 m².

The $\dfrac{\text{height } h}{\text{diameter } B}$ ratio would appear to differ quite distinctly with the geometry:

– $\dfrac{h}{B}$ is about 1 for cylindrical bases with a diameter $B \approx 10$ to 12 m,

– $\dfrac{h}{B}$ is hardly more than 0.5 in the case of certain large diameter ($B = 13$ to 14 m) conical or hyperbolic spud cans.

The cross-section of the cylindrical base or spud can with a diameter B is invariably greater than that of the pile, which is being built more and more in the form of a metal lattice, often triangular in form.

After the sediments have caved in on top of it, the spud can therefore forms a veritable anchorage, and one can well imagine the difficulty in breaking out such an anchor (see Section IX.2.3.2).

IX.1.2 Large mat-supported jack-up platforms

Mat-supported jack-up platforms (one or several mats) have been designed with a particular view to setting up the platform in soft soils with a low bearing capacity.

These platforms are at present in service in various seas throughout the world:

– in the Gulf of Mexico, where they represent 30% of mobile drilling platforms,

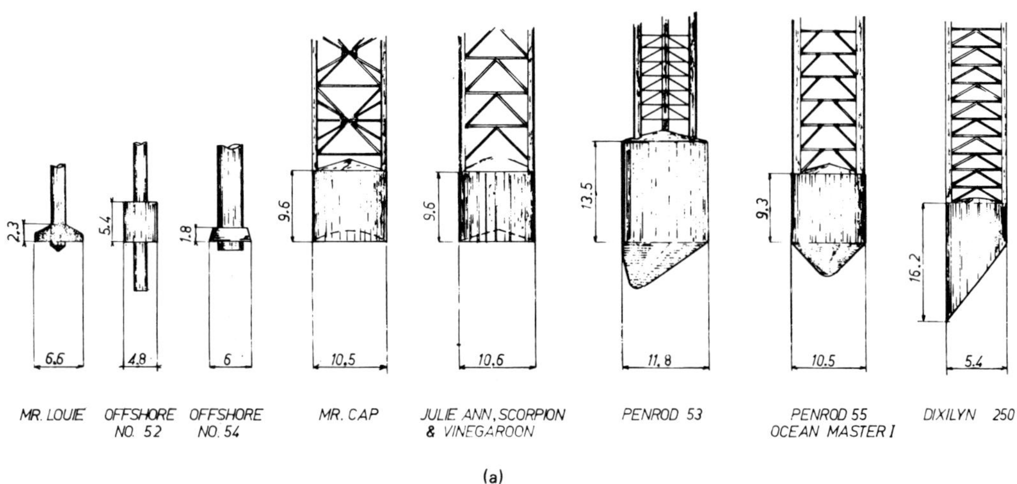

(a)

All dimensions in metres

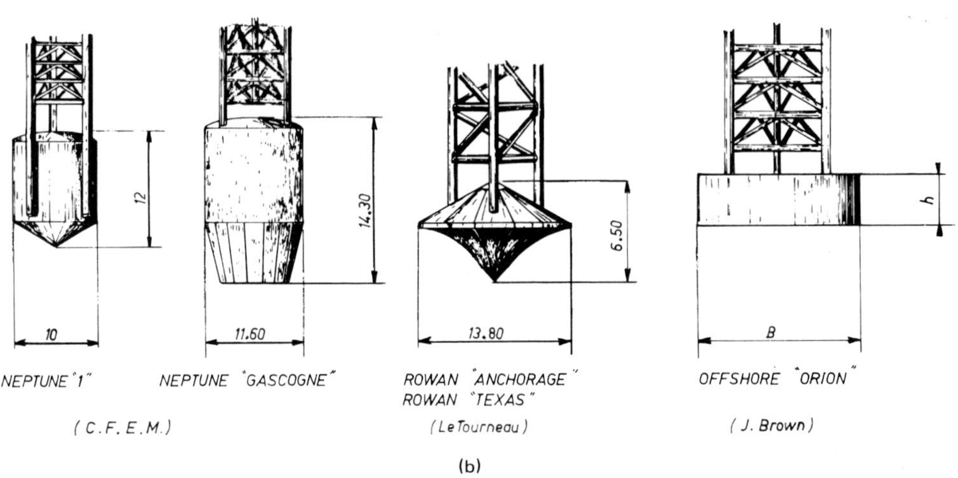

(b)

FIG. IX.1.5. – Bases and spud cans of various jack-up platforms.

– on the Pacific Coast of Latin America (Peru, Equador, etc.),
– in Indonesia and on the West African Coast.

IX.1.2.1 Geometries of mats

Figure IX.1.6 diagrammatically illustrates the various jack-up platform mats built by *Bethlehem-Beaumont* [6] [11]. Depending on the particular case, the mat consists of:

– a single rectangular mat with a central opening (A-shape),
– one or two mats (linked together by struts) with a hydrodynamic shape to make it easier to move them.

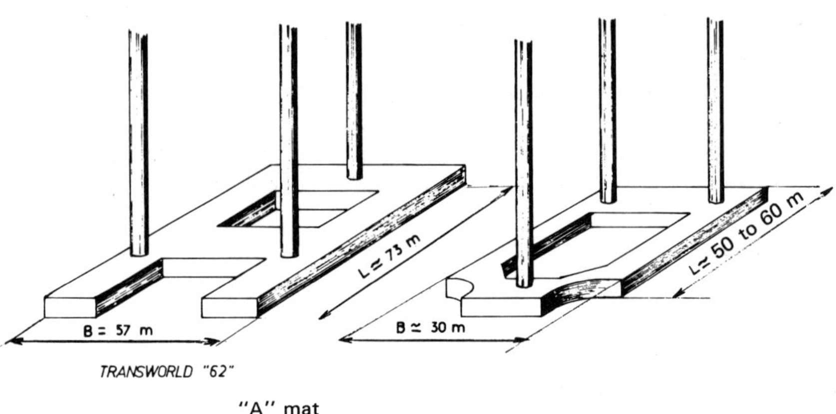

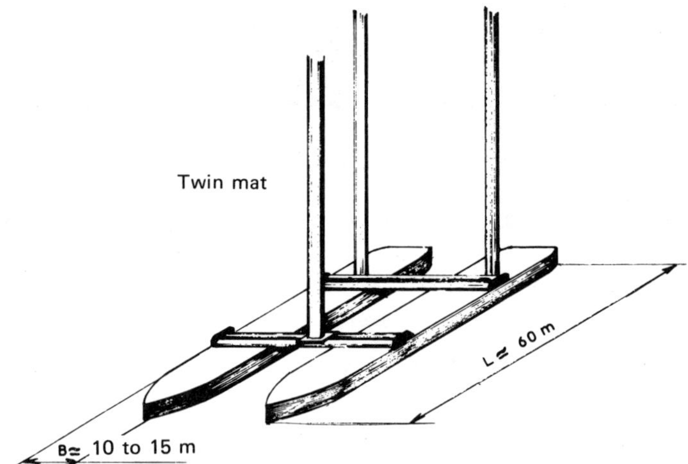

Bethlehem Steel Corporation jack-up platforms

FIG. IX.1.6. – A number of different types of mats of jack-up platforms.

The mat is connected to the platform by piles:

– generally tubular,
– less frequently in lattice-work

IX.1.2.2 Dimensions of mats

The overall dimensions (length and width) of the base are comparable to those of the platform:

– rectangular mat 50 to 60 m long with an area of 1,000 to 2,500 m^2
– hydrodynamic form mat(s) 50 to 60 m (or more) in length with an area of 1,000 to 2,000 m^2.

Generally, the mat is about 2 to 4 m thick.

IX.1.3 Characteristics of a number of jack-up platforms

The September 1976 issue of *Ocean Industry* gives the main characteristics of all the jack-up platforms in service or now being built up to the time that this publication went to press.

As an example, Table IX.1.1 gives the approximate characteristics of some jack-up platforms, enabling the reader to gain some idea of the loads per pile and the unit loads on the soil.

IX.1.3.1 Jack-up platforms on piles with bases or spud cans

The approximate loads per piles vary very widely with the dimensions and the number of piles on jack-up platforms:

– from 800 to 1,000 t per pile when there are no more than 8 piles,
– from 2,000 to 4,000 t per pile for most of the recent platforms with 3 or 4 piles.

The average unit loads on the soil beneath the mats or spud cans, calculated without allowing for their geometry (spud cans with conical or other type profiles), generally lie between 20 and 35 t/m^2 (200 and 350 kPa). As will be seen (see Section IX.4), these unit loads are relatively high and explain the difficulties experienced in installing platforms in soft or low consolidation soils.

IX.1.3.2 Mat-supported jack-up platforms

The weight of mat-supported platforms is about 8,000 to 10,000 t.

The unit loads on the soil are always well below 10 t/m^2 (100 kPa) and do not generally exceed 2 to 3 t/m^2 (20 to 30 kPa). A consequence is that the penetration of the mat and the stability of the structure depend first and foremost on the distribution of the loads and the homogeneity of the foundation soils.

TABLE IX.1.1

CHARACTERISTICS OF A NUMBER OF JACK-UP PLATFORMS

Designation of jack-up platform	Builder	Water depth (m)	Dimensions of platform (m)	Number of piles (or mat)	Approximate load per pile (t)	Dimensions of spud cans (or mat)	Unit load on soil (t/m^2)
Neptune 1 (1965)	*CFEM*	53	61 x 51	3	1,800	10 m Ø bases	23
Neptune-Gascogne (1965)	*CFEM*	89	49 x 53	3	2,700	11.6 m Ø bases	25
Ile-de-France (1966)	*IHC*	60	54 x 53	5		Bases	≈ 20
Rowan-Anchorage (1972)	*Le Tourneau*	75	61 x 50	3	2,400	13.8 m Ø spud cans	16
Ocean Master I (1966)	*Le Tourneau*	90	54 x 50	3	4,330	12 m Ø spud cans	38
Penrod No. 53 (1965)	*Le Tourneau*	90	54 x 50	3	3,160	12 m Ø spud cans	28
Penrod No. 55 (1966)	*Le Tourneau*	45	60 x 55	3	2,530	10.5 m Ø spud cans	29
Offshore No. 54 (1956)	*American Bridge*	45	60 x 31	8	880	6 m Ø base	31
R. Louie (1958)	*Le Tourneau*	54	73 x 40	12	1,000	6.6 m Ø base	30
Sagar Samrat (1972)	*Mitsubishi*	75	83 x 39	4	4,600	13.4 m Ø spud cans	33
Leading and Bats MacLean (1970)	*Bethlehem*	68	54 x 40	Mat	Overall load not specified 8,000 to 10,000	57 x 48 m mat = 2,700 m^2 72 x 54 m mat = 3.900 m^2	2 to 3
Stormdrill IV (1966)	*Bethlehem*	58	50 x 26	Mat			
Marlin No. 3	*Bethlehem*	75	50 x 26	Mat			

IX.2 DEFINITION OF THE GEOTECHNICAL PROBLEMS INVOLVED IN THE INSTALLATION, STABILITY AND REMOVAL OF JACK-UP PLATFORMS

The geotechnical problems raised by the use of jack-up platforms concern:

- the installation of the platforms,
- the stability of the platforms during drilling (or, where applicable, production),
- the removal of the platform on completion of drilling.

However, the method of installation or removal, just like the stability during operation, clearly varies with the type of structure used (mat or spud cans, three piles or more, etc.) and the nature of the soils encountered.

IX.2.1 Installation of a jack-up platform

The main problem involved in the installation of a jack-up platform concerns the penetration of the piles, which depends at one and the same time on :

- the nature and consolidation of the surface terrains,
- the geometry and the type of support: bases, spud cans or mats.

IX.2.1.1 Maximum acceptable penetration of the supports

Aside from the stability of the structure, the maximum penetration of the supports compatible with the installation of a jack-up platform on a given site is defined by the obvious equation (see Fig. IX.2.1) [7]:

$$D_{max} = H - (W + A_{min})$$

where:

D_{max} = maximum penetration of the supports,
H = total length of the piles (total height of platform),
W = depth of water + maximum wave amplitude,
A_{min} = total minimum height above waterline.

The maximum penetration D_{max} is the sum of:

- the penetration under the effect of the static loads (unballasted weight of the structure),
- penetration under the action of the horizontal forces (see Section IX.4.4).

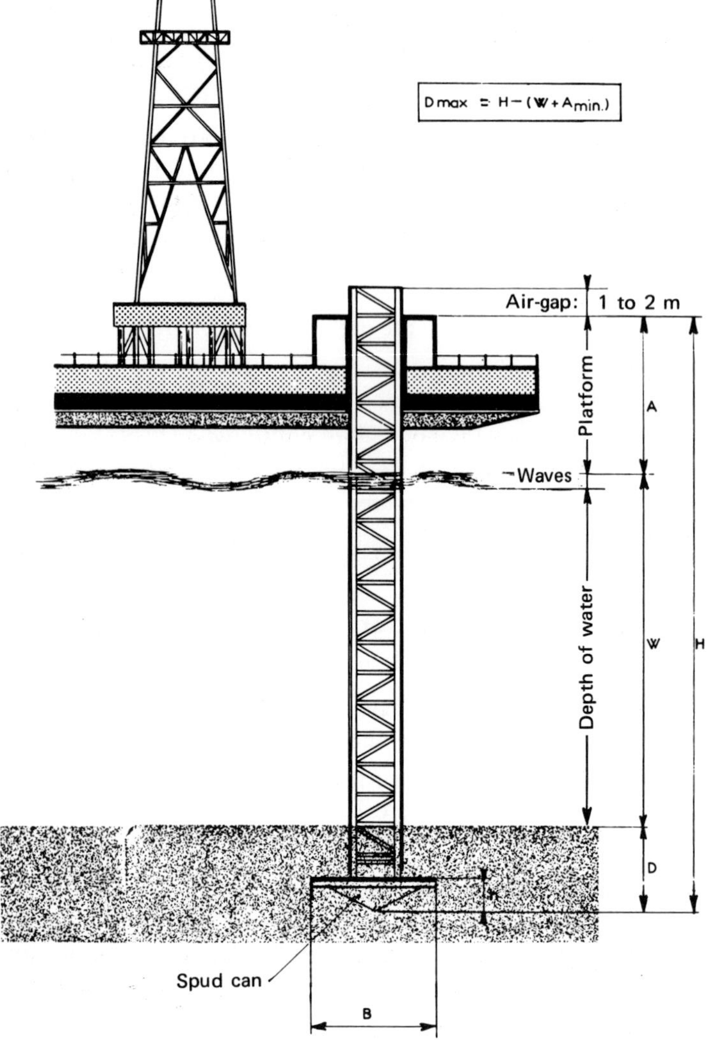

FIG. IX.2.1. – Jack-up platform installation conditions.

IX.2.1.2 Sequence of installation operations for a jack-up platform

Installation of a jack-up platform must take place following a well-defined sequence of operations.

Once the platform has been brought to the site, it will be positioned:

– either by means of tugs or supply vessels,
– or by means of its anchors (raised after the piles have penetrated into position).

The piles are then lowered as quickly as possible (about 0.30 to 1 m/min), particularly if the speed of the current is high (⩾ 1 knot).

The platform is preloaded evenly, verifying the stability by comparing the estimated and observed curves of penetration of the supports. The platform must be raised into the working position only provided the stability has been duly ensured.

When locating the barge, it is important if possible to check the depth of penetration of the piles by divers.

IX.2.1.3 Penetration into soft soils

In the case of supports consisting of bases or spud cans (Fig. IX.1.3), the penetration into the soft soils under the weight of the structure and an overload reaches the following values:

– commonly, 15 to 20 m in soils of low consolidation,
– sometimes, as much as 30 to 40 m in very soft soils.

In the case of large dimension **mats** (Fig. IX.1.4), which are particularly well suited to soft soils, penetration is generally no more than 3 to 4 m.

IX.2.1.4 Penetration into consolidated soils

Depending on the nature and degree of consolidation of the soil, in certain case penetration may :

– either be facilitated by jetting, though this technique has the serious drawback of considerably modifying the properties of the soil,
– or possibly be assisted by hammering the piles for mat type jack-up platforms (though it is true that this type of platform is poorly adapted to this type of soil).

The penetration of spud cans into consolidated soils is invariably shallow:

– about 1 to 2 m in dense sands,
– a few metres in medium consolidation clays.

IX.2.1.5 Preloading of piles

It is indispensable to preload the piles when setting up a jack-up structure in order to avoid (or reduce) the subsequent settlements of the soil during operation.

Regulations advocate that piles be preloaded to about one-third more than the normal service load. In actual fact, very few jack-up platforms can satisfy this requirements. Preloading is therefore generally fairly limited.

The arrangements needed for preloading differ with the type of structure:

– for a mat-supported platform resting on a soft soil, it is indispensable that the structure be uniformly ballasted,
– for a 3-piles platform (with bases or spud cans), the piles must be overloaded gradually and in turn, in order to avoid excessive penetration of any one pile,
– in the case of a platform with 4 or more piles, certain piles are overloaded while unballasting the remainder. This provides a certain safety margin when the forces exerted on the piles are distributed unequally (subsequently) owing to the eccentricity of the loads.

IX.2.2 Stability of jack-up platform during drilling

The stability of a jack-up platform depends on:

– the vertical loads,
– the horizontal stresses,
– the possibility of the soil liquefying,
– the risk of scour.

The problems of stability differ with the particular type of support (spud cans or mat) of the structure.

IX.2.2.1 Vertical loads

The stability of jack-up platforms under vertical loads is essentially governed by the **bearing capacity** of the soil.

The bearing capacity of the soil is defined by:

– shear strength measured in the laboratory or in situ,
– the limit pressure derived from the pressuremeter test.

The methods of calculating the bearing capacity of the supports were examined in Section IX.4.

Settlement of the soil, which is determined by its compressibility, will not be involved to any great extent in studying stability providing the piles have been subject to preloading when installing the platform.

The horizontal heterogeneity of the soil at the emplacement of the platform can cause:

– either different penetration depths of the piles, which is no great disadvantage within a certain limit compatible with their length and the water depth; however, an excessive differential penetration has caused some accidents,
– or inclination of the mat platform, which is difficult to correct.

IX.2.2.2 Horizontal loads

There are two sides to the problem of the stability of jack-up platforms under the effect of horizontal (dynamic) loads (1):

– first, the calculated penetration of the piles under static loads must be increased to allow for the effect of the dynamic loads,

– second, the overturning moment of the structure must be less than the reactive forces of the soil.

The stability of jack-up platforms under the effect of horizontal loads can be enhanced by various technological arrangements:

– lattice-work piles, which have more bending strength than cylindrical tubes, are now being used for all spud can jack-up platforms [2] [3] and for certain mat-supported jack-up platforms [11],

– the inclination of the legs of jack-up platforms very appreciably increases their stability by increasing the bending moment; this arrangement is advantageous, particularly, when the penetration of the spud cans is small (at the most a few metres), as is the case of dense sands where the horizontal reaction of the soil is obviously zero,

– the large dimensions of the mat, which are very commonly greater than those of the hull, generally endow the platform with good stability. In certain cases, the stability can be increased by means of skirts driven into the soil through cylindrical tubes.

IX.2.2.3 Liquefaction of the soil

The influence of shocks or vibrations when installing jack-up platforms or during drilling may cause loose soils to liquefy, namely:

– sands,

– muds or soft clays,

resulting in a considerable (temporary) reduction in shear strength and hence bearing capacity.

The difficulties encountered in setting up certain jack-up platforms in soft soils can be explained by the liquefaction of the soil caused by disturbance.

IX.2.2.4 Scour near supports

Scour near the supports is a highly complex function of several parameters, including:

– the nature of the sediments: fine sands,

– the speed of the currents at the bottom and the wave amplitude,

– the geometry of the supports: piles and spud cans or mats.

---

(1) The problem of estimating the lateral loads caused by the elements will not be dealt with here.

Scour can jeopardize the stability of a jack-up platform under the effect of lateral stresses owing to the relatively low penetration of the supports, particularly in layers of sands.

In actual fact, the period during which a jack-up platform is immobilized on a site is generally no more than a few months:

– in the case of soft soils, the considerable penetration of the spud cans or the cover over the mat avoids any scour effects,
– in the case of sands, the arrangement of stones (if necessary) around the spud cans would appear quite satisfactory.

IX.2.3 Removing a jack-up platform

When operations are completed, the jack-up platform is removed by the simultaneous effect of:

– jetting around the mats or spud cans,
– the pullout force of the raising devices mounted on the piles of the platform.

IX.2.3.1 Sequence of removal operations for a jack-up platform

The hull of the jack-up platform must be lowered before starting jetting. Several accidents have in fact occurred by jetting in the raised position.

The hull is lowered to a minimum height above the surface of the sea compatible with the amplitude of the waves.

Jetting is carried out in turn on each support (for instance for 1 h at a time). In the particular case of a tripod platform, it is important to prevent any single leg from settling through jetting for too long a period.

The duration of jetting needed to release the supports obviously varies considerably with the nature of the soil, the number of supports, their depth of penetration, etc. In muddy soils the duration may vary from a few hours to 1 or 2 days (or even more).

The piles are raised by means of hoisting devices, while the platform itself is gradually settled into the floating position.

As soon as the supports are extracted from their emplacement, they must be raised as quickly as possible, particularly if the currents are fast (> 1 to 2 knots); the piles of jack-up platforms have little strength against lateral loads. Generally speaking, raising the supports should not last more than a few hours.

IX.2.3.2 Pullout force

The necessary pullout force depends:

– on the penetration of the supports,

– on the geometry and dimensions of the supports,
– on the insertion techniques employed (static loading, driving, jetting, etc.),
– on the sensitivity of the soils: after recovery, the pullout force may be very much more than the initial bearing capacity.

In the case of spud can jack-up platforms, the total load to be raised equals approximately:

– the weight of the sediment filling each hole, created by lowering the supports (minus the volume of the lattice),
– the friction of the sediment on the racks and braces (lattice).

For example, the apparent weight of the sediments (density 1.5 t/m^3) accumulating on top of a 14 m diameter spud can sunk to 25 m into the soil is about 2,000 t. Allowing for friction, the load to be lifted (2) in the case of a tripod jack-up platform would probably be about 10,000 t, thus representing an extraordinarily strong anchor.

The load is generally much greater than the traction capacity of the hoisting devices mounted on the jack-up platforms, of about 3,000 to 4,000 t.

An improvement in the system would consist in the case of soft soils in enclosing the volume delineated by the pile racks to a height of about 20 to 30 m, in order to reduce the weight of the sediments which build up on the spud can.

In the case of mat-supported jack-up platforms resting on a soft soil (mud, soft clay), the weight to be raised is proportional to the force of suction between the mat and the soil. As will be seen (Section IX.4.6), the high level of this suction force has resulted in the breaking of the four cylindrical legs of a mat-supported jack-up platform during its extraction.

IX.2.3.3 Jetting action

A suitable jetting system near the spud cans makes for a considerable reduction in the load to be raised. The jetting systems on jack-up platforms generally appear satisfactory for most types of formations (Fig. IX.2.2):

– the number and layout of the jets is of considerable importance. According to certain specialists, the jets should be positioned beneath the mats or spud cans and not above them. This layout does in fact offer the advantage of reducing the suction force between the spud can and the soil,
– the initial jetting flow should be several m^3/min at a pressure of a few tens of kg/cm^2 (up to 50).

The difficulties encountered in extracting certain jack-up platforms, the spud cans of which have sunk 20 to 25 m into soft soils (the Green Dolphin in Indonesia for example),

(2) At the start of the pull on the piles.

would appear to have been caused by the inadequacy of the jetting system, with respect to both the number of jets and their layouts:

– the jetting action in the immediate vicinity of the spud cans is inadequate at high penetrations,

– the arrangement of jets along the pile driven into the soil should increase the effectiveness of the system, provided however that the jetting flow is sufficient.

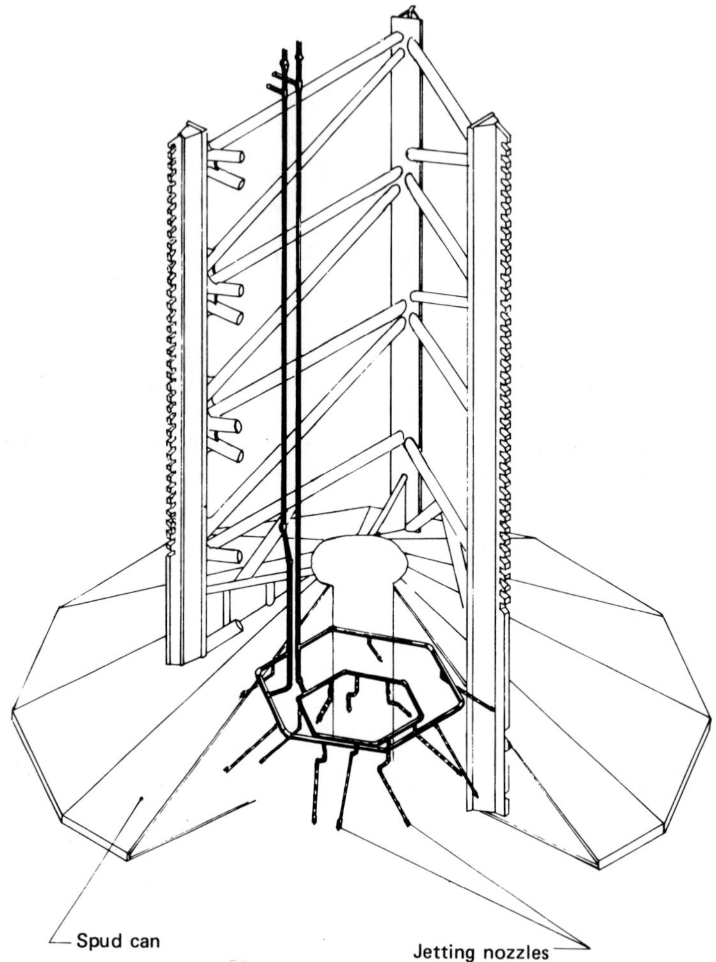

FIG. IX.2.2. – Diagram of jetting system (layout of nozzle pipes) of the spud can of a jack-up platform.

IX.2.3.4 Conditions of removal of jack-up platforms

Raising piles is invariably a critical stage in withdrawing a jack-up platform. Most of the accidents which have occurred with jack-up platforms (see Section IX.4.6) have in fact happened during this stage of the operation. It is therefore highly important to restrict the duration of the operation to no more than a few hours, particularly in the presence of fast currents exceeding 1 knot (i.e. 0.5 m/s). To give some idea, the maximum lifting speed of the piles of a jack-up platform (under load) is about 0.40 to 0.50 m/min.

The weather and oceanographic conditions needed for withdrawal of a jack-up platform are the following:

- wave amplitude < 1.50 m,
- wind speed 20 to 35 kph,
- fine weather period forecast = at least 12 h.

A jack-up platform must not be moved from one site to another should the wave amplitude exceed 3 to 4 m.

IX.3 SOIL RECONNAISSANCE FOR SETTING UP JACK-UP PLATFORMS

Although reconnaissance of the soils before positioning jack-up platforms is sometimes (wrongly) considered as superfluous in certain well-known zones (such as the Gulf of Mexico), it is required by most legislations and invariably calls for very careful examination, particularly in the case of soft soils.

There are two major stages to this reconnaissance:

- first, geophysical reconnaissance over the site as a whole,
- second, geotechnical reconnaissance on the actual expected emplacements for the structure.

IX.3.1 Geophysical reconnaissance of the site

Generally speaking, geophysical reconnaissance must precede any geological and geotechnical reconnaissance.

In the case of a jack-up platform, this principle is all the more applicable since in the light of the initial knowledge of the field, it is often impossible to forecast with any certainty the actual emplacements of wildcats wells.

IX.3.1.1 Coverage and meshing

On the basis of the information gathered during the geophysical reconnaissance it would be possible:

- to correctly decide the emplacement for the boreholes (core sampling or in situ measurements),
- to limit their number,
- if necessary, to extend the results of core sampling or in situ measurements.

The delineation of the field always requires several wildcats. A priori, therefore, geophysical reconnaissance of the site should be planned, extending well beyond the possible boundaries within which a jack-up platform may be installed in order:

– to avoid having to resume reconnaissance later on,

– to verify the continuity of the layers and to detect the presence of any possible geological accidents, which it is vitally important to determine in the event of changing the emplacement of the structure.

In the end, the coverage of the geophysical reconnaissance depends on the presumptive extent of the field to be delineated. Very frequently, reconnaissance will cover an area of 10 to 20 km^2 or more.

The meshing of the geophysical profiles may vary with the horizontal homogeneity of the layers from 100 to 500 m, depending on:

– the recordings made: bathymetry, seismics, etc.,

– the direction, in particular for side-scan sonar.

IX.3.1.2 The bathymetry and morphology of the bottom

The accuracy required for the bathymetry and morphology (roughness) of the bottom for installing jack-up platforms depends on the type of structure:

– in the case of spudcan platforms, the slope or moderated reliefs of the soil can easily be compensated for when setting up the structure,

– in the case of mat-supported platforms, their installation implies fairly precise knowledge of the slopes and reliefs on the bottom.

The installation of jack-up platforms implies that the water depth be less than 80 to 100 m. This being so, bathymetric recordings:

– are carried out by means of conventional echo sounders, the ultrasonic beam angle of which varies from about 10 to 20°,

– make it possible to obtain (after the usual corrections for tide, roll and pitch and calibration) an accuracy of about 1 m (i.e. $\leqslant 1\%$ of the water depth), which is quite acceptable for installing jack-up platforms.

The low penetration (from 0.50 to 1 m) of the ultrasonic signal gives an idea of the absorption of the bottom (muddy or sandy).

The recordings of the side-scan sonar make it possible:

– to detect the reliefs or obstacles involving differences in level of about 1 m,

– to supplement the data of the bathymetric profiles, which are generally set a few hundred metres apart.

IX.3.1.3 High resolution seismic prospecting

The depth of reconnaissance of the soils to be reached varies with both:

– the type of structure: spud can or mat supports,

– the nature of the formations encountered.

The following will be the objectives of the seismic reconnaissance stage:

- a penetration of 50 to 100 m,
- a vertical resolution of 2 m at a depth of 10 to 20 m.

These results can be reached particularly by means of the following:

- a sparker, the penetration and resolving power of which depends on the energy used,
- a boomer (Uniboom).

In the case of soft soils, where reconnaissance must be highly accurate, it would appear wise to make simultaneous recordings using:

- a sparker,
- a sediment sounder to improve the definition (of about 1 m) of the surface layers (10 to 20 m).

Table IX.3.1 summarizes the geophysical reconnaissance required before installing jack-up platforms, together with the performance of the devices that can be used.

TABLE IX.3.1

GEOPHYSICAL RECONNAISSANCE OF SEA BOTTOMS AND SOILS BEFORE INSTALLING JACK-UP PLATFORMS

<table>
<tr><th>Reconnaissance necessary</th><th>Techniques applicable</th><th>Coverage of zone to be reconnoitered</th><th>Meshing (m)</th><th>Emission frequency of device</th><th>Penetration</th><th>Accuracy or power of resolution</th></tr>
<tr><td>Bathymetry</td><td>Echo sounder</td><td rowspan="5">10 to 20 km^2 (geometry depends on the data obtained concerning the geological structure)</td><td>100 to 500</td><td>30 to 50 kHz</td><td>≈0</td><td>Accuracy < 1 m</td></tr>
<tr><td>Morphology</td><td>Side-scan sonar</td><td>ditto</td><td>37 or 100 kHz</td><td>0</td><td>Relief < 1 m</td></tr>
<tr><td rowspan="3">Seismic prospecting</td><td>Sparker</td><td rowspan="3">200 to 500</td><td>100 to 1,000 Hz</td><td>≈ 100 m</td><td>Resolution ≈2 m</td></tr>
<tr><td>Boomer (Uniboom)</td><td>500 Hz to 4 kHz</td><td>50 to 100 m</td><td>Resolution ≈1.5 to 2 m</td></tr>
<tr><td>Sediment sounder</td><td>3 to 9 kHz</td><td>10 to 20 m</td><td>Resolution ≈ 1 m</td></tr>
</table>

IX.3.2 Geotechnical reconnaissance of the actual emplacement(s) of jack-up platforms

While the data deduced from geophysical reconnaissance of the soils sometimes suffices to gain some idea as to the possibility or impossibility of installing a jack-up platform, the expected penetration of the supports of the platform can be obtained only from data gained by geotechnical reconnaissance.

IX.3.2.1 Purpose of geotechnical reconnaissance

The purpose of geotechnical reconnaissance is to determine the nature, alternation and thickness of the layers of soil: clay, sand, coral, etc. For instance:

– the presence of a layer of creeping clay a few metres thick between two layers of sand can result in rupture of the foundations by slip,
– the existence of a layer of soft soil beneath a bed of coral can jeopardize the stability of the jack-up platform.

In situ measurements and core samples must make it possible:

– to identify the soils: nature, density, water content,
– to determine their physical properties: sensitivity, risk of liquefication, possibilities of scour, etc.,
– to measure their mechanical characteristics: shear strength, compressibility, etc.

IX.3.2.2 Location, number and depth of boreholes

The choice of the emplacements for the jack-up platform on a site will be determined:

– first, in the light of knowledge of the geological structure,
– second, in the light of the data obtained as a result of the geophysical survey: bathymetry and high resolution seismics particularly.

After definition of the approximate emplacements of the jack-up platform, the following should be provided for each particular emplacement:

– **at least two boreholes** (to check the geological homogeneity of the layers) down to:
  – 20 to 30 m if the soils are relatively consolidated (undrained cohesion $c_u > 5$ t/m² (> 50 kPa)),
  – 40 to 50 m if the soils are soft ($c_u < 2 - 3$ t/m² (< 20-30 kPa)),
– **if possible, two or three in situ tests** down to 10-20 m,
– **a few surface core samples** down to a penetration of 5 to 10 m, in order to identify correctly the nature of the surface sediments and to predict any future risk of scour.

The depth of the reconnaissance boreholes should allow for:

– the type of structure to be installed,
– the foreseeable penetration D of the supports,
– the zone of rupture of the soil beneath the supports,
– the necessary knowledge of the underlying layers.

If the number of boreholes is limited to a strict minimum –for reasons of time or cost–, if possible, boreholes will be drilled with core sampling in preference to in situ measurements.

IX.3.2.3 Applicable coring techniques

Several coring and in situ measurement techniques are now operational for the reconnaissance of soils before installing jack-up platforms, allowing both for:

– depths of water from about 40 to 100 m,
– maximum penetrations of from 20 to 50 m.

Drilling with wireline coring by percussion or by push sampler from a surface support is the most common soil reconnaissance technique before installing jack-up platforms. The cores are sampled:

– practically continuously down to 10 to 12 m (one core per metre),
– then at intervals of 1 to 1.5 m down to a depth of 20 to 30 m, this being the depth generally concerned by installation of the jack-up platform.

The small diameter cores obtained (mostly 54 mm) are invariably subject to considerable disturbance, particularly in muds and low consolidated clays. It will be seen that the often important difference between the estimated and observed penetrations of the jack-up platform supports can frequently be explained by the non-representivity of the cores sampled.

Rotary wireline coring using a submerged device applied by divers is used particularly in the Persian Gulf, where the water depth is no more than 35 to 40 m.
This coring technique, which is theoretically continuous, results in a recovery rate which varies fairly widely with the nature of the soils.
In the absence of specific information concerning the mechanical characteristics of the soils, knowledge of the nature and lithology of the layers obtained by continuous coring will make it possible to estimate the penetration of the supports of a jack-up platform on the basis of experience gained on the same site or under similar soil conditions.

Surface core sampling by means of the Kullenberg type corer or the vibrocorer together with samples obtained by dredging is often done as part of the geophysical reconnaissance survey.

IX.3.2.4 Applicable in situ measuring techniques

It should be remembered that while in situ measuring techniques give highly useful information on the mechanical properties of the soils in place, they can in no event replace the core sampling necessary to ascertain the nature of these soils.

The Menard pressuremeter is often used for analysing soils prior to installing jack-up platforms.
The pressuremeter modulus E_p and the limit pressure p_l are affected by the method of insertion of the pressuremeter probe into the soil.

It should also be pointed out that:

– the limit pressure measured in soft soils is often only poorly representative,
– the risk of liquefaction of soft soils, which is commonly observed when positioning jack-up platforms, cannot be assessed by means of the pressuremeter,
– the pressuremeter modulus does not enter into the calculation of the stability of jack-up platforms.

As we shall see, estimation of the penetration of the supports of a jack-up platform on the basis of pressuremeter findings also implies a thorough understanding of the behaviour of the soils.

Remote-controlled modular penetrometers (of the Seacalf type) or penetrometers implemented by means of divers, the penetration of which can reach 10 to 20 m, are particularly well adapted to soil surveys before installing jack-up platforms.

The mechanical parameters of the soil:

– undrained cohesion c_u,
– and/or internal friction φ,

must be estimated on the basis of magnitudes measured with the penetrometer (cone resistance R_p and lateral friction f).

TABLE IX.3.2

GEOTECHNICAL RECONNAISSANCE OF SOILS BEFORE INSTALLING JACK-UP PLATFORMS

Operations to be carried out	**Extent of zone to be surveyed**	**Number of boreholes**	**Penetration necessary (m)**	**Techniques applicable**	**Performances**	**Implementation**
Boreholes + core sampling	To be defined on the basis of the results of high resolution seismic prospecting	2 boreholes + core sampling	Consolidated soils = 20-30 Soft soils = 40-50	Drilling and percussion wireline coring Nuclear logs	Average rate of advance = 2-3 m/h Continuous recording	From a vessel: wave amplitude < 2 m
Surface core sampling		10 to 20 corings	5 to 10 depending on consolidation of soils	Kullenberg type corer Vibrocorer	Low consolidation formations Consolidated formations	From the geophysical prospecting vessel
In situ measurements (if necessary)		2 to 3 in situ tests	10 to 20	Modular penetrometer (Seacalf type) TLM pressuremeter	Maximum penetration depends on consolidation of soils. Penetration varies with the method used	From a vessel provided wave amplitude < 1.5 m Varies with water depth

In the case of sands, the penetrometer is practically the only device by which an estimation of the relative density and hence of the shear strength can be made.

Tables IX.3.2 and IX.3.3 respectively summarize:

– the soil surveys to be made for each intended emplacement for installing a jack-up platform, together with the performance and conditions of implementation of the devices or techniques that can be used,

– the core and in situ measurements to be made before installing jack-up platforms.

TABLEAU IX.3.3

CORE AND/OR IN SITU MEASUREMENTS TO BE MADE BEFORE INSTALLING JACK-UP PLATFORMS

	Characteristics of the soils measured	Observations
Core analysis in the laboratory	**Identification of the soils**: Nature; Atterberg limits (W_L, W_P); Size distribution	In clayey soils
	Water contents w Density γ	Determined immediately aboard the vessel
	Shear strength Undrained shear strength (undrained cohesion c_u): – unconfined compression – vane test – "fall cone" – pocket penetrometer	Preferably carried out immediately aboard the vessel
	– triaxial	(possibly) in the laboratory
Penetrometer measurements	Cone resistance R_p Lateral friction f	These measurements are practically indispensable in sands (estimation of γ and φ)
Pressuremeter measurements	Limit pressure p_l	Method of calculating the stability from p_l

IX.4 CALCULATION OF THE STABILITY OF JACK-UP PLATFORMS

The sole purpose of this paragraph is to indicate the principle for calculating the stability of jack-up platforms. Indeed, there can be no question of going into detailed examination of the specific problems relevant to each structure, for instance to allow for the individual geometries of the supports, the eccentricity of the loads, etc.

On installing a jack-up platform, the supports continue to penetrate until the bearing capacity of the soils balances the unit loading caused by the weight:

– if the structure can be preloaded, in principle the maximum penetration is attained,
– if the structure cannot be preloaded, an additional penetration may occur during the operation.

However, the penetration calculated for the static load must be increased to allow for the effect of the dynamic loads.

The pullout force of the supports is invariably difficult to estimate in view of the importance of jetting in withdrawing the structure (See section IX.2).

The choice of the type of support to suit the nature of the terrains is often merely academic in view of the actual availability of the structures.

A number of accidents (or incidents) that have occurred on jack-up platforms may be attributed to the instability of the foundations.

IX.4.1 Conventional method of calculating the penetration of the supports under the effect of the static loads

This method uses conventional soil mechanics formulas for designing foundations resting on the surface or penetrating slightly into the soil.

IX.4.1.1 Penetration of the supports into clays or muds

The general equation for the bearing capacity is [10]:

$$q_u = c_u N_c + \gamma' D$$

where:

q_u = maximum bearing capacity of the soils,
c_u = undrained cohesion of the soil under the base or spud can,
N_c = dimensionless factor,
γ' = submerged density of soil ($\gamma' = \gamma - 1$),
D = penetration of supports.

For bases or cylindrical (or square) spud cans, with a diameter B, buried to a depth of D, the dimensionless coefficient N_c is expressed by Skempton's empirical formula:

$$N_c = 6\left(1 + 0.2\frac{D}{B}\right)$$

The coefficient N_c varies from about 6 to 9 (for a maximum ratio $\frac{D}{B} = 2.5$).

For rafts with a width of B and a length of L buried only little, the coefficient N_c is given approximately by the equation:

$$N_c = 5\left(1 + 0.2\frac{B}{L}\right)$$

In practice, N_c can be taken to be about ≈ 5.5.

The expression for friction:

$$f = kc_u$$

(where $f \leqslant 1$ depending on the value of c_u)

is negligible compared to q_u for a value of $\frac{D}{B} \leqslant 2.5$, which is rarely reached when installing jack-up platforms.

As an example, Fig. IX.4.1 shows the variations in the maximum bearing capacity Q of a soft clayey soil in terms of the penetration D of cylindrical bases with a diameter B:

$B = 12$ m

Clayey soil $\begin{cases} \gamma = 1.5 \text{ t/m}^3 \ (\gamma' = 0.5 \text{ t/m}^3) \\ c_u = \begin{cases} 1 \text{ t/m}^2 \text{ (10 kPa) on the surface} \\ \text{to } 4 \text{ t/m}^2 \text{ (40 kPa) at a depth of 18 m} \end{cases} \end{cases}$

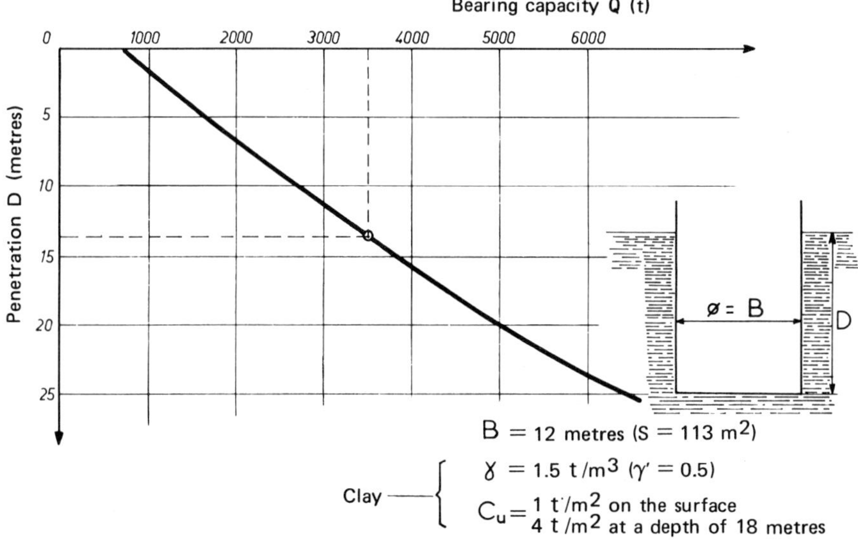

FIG. IX.4.1. – Bearing capacity Q of a cylindrical base in terms of penetration into the clay D.

A bearing capacity $Q = 3{,}500$ t is theoretically reached at a penetration of the bases D of 12.2 m.

With the same soil conditions prevailing, the stability of a jack-up platform mounted on foundation mats:

– weight = 10,500 t,
– area of mats = 1,500 m^2,

is obtained for a penetration D = 3 m (about the same order of magnitude as the thickness of the slabs).

These theoretical estimates are essentially based on the measured cohesions c_u. Now, particularly in soft soils, the measured cohesion is certainly only poorly representative of the true cohesion of the soil in place, in view:

– of the considerable disturbance of the cores (or the non-representativity of the in situ measurements),
– of the possible liquefaction of the soft soils on installing the jack-up platform.

Uncertainty therefore invariably ensues concerning the true penetration of the bases, all the more so the less consolidated the soil.

IX.4.1.2 Penetration of supports into sands

The general equation for bearing capacity can be written:

$$q_u = \frac{1}{2}\gamma' B N_\gamma + \gamma' D N_q$$

where:

q_u = maximum bearing capacity of the sand,
γ' = submerged density of the sand ($\gamma' \doteq \gamma - 1$),
B = diameter (or width) of supports,
D = penetration of supports,
N_q, N_γ = dimensionless coefficients depending on the angle of friction of the sand φ.

The magnitudes of N_q and N_γ differ considerably depending on the assumptions made and the particular writer. It is proposed to adopt:

– Terzaghi's values for N_q (for $\varphi = 30°$, $N_q \approx 20$),
– B. Hansen's values for N_γ (for $\varphi = 30°$, $N_\gamma \approx 16$).

The lateral friction term in the equation is always negligible or zero owing to the low penetration of the piles.

It is verified that the stability of a jack-up platform with:

– a load per pile Q = 3,500 t,
– a base diameter B = 10 m,

set on a sand with the following characteristics:

– γ = 1.8 t/m^3 (γ' = 0.8 t/m^3),
– φ = 30°,

is theoretically certain, without the supports penetrating into the terrain at all.

Although the penetrations of jack-up platform supports into the sand are invariably low, it should be pointed out that estimates arrived at by calculation essentially depend on the value of the angle of friction φ. Now, in actual fact:

– the angle of friction is difficult to appraise without any penetrometer tests,
– the coefficients N_φ and N_q vary very sharply with a change in φ.

An obvious degree of uncertainty therefore prevails with regard to the estimation of the penetration D.

IX.4.2 Method of calculating penetration of supports deduced from pressuremeter measurements

This method, used by *Techniques Louis Menard (TLM)* allows for the maximum pressure as measured on a pressuremeter.

IX.4.2.1 Principle of calculation of bearing capacity of soil

The method of calculation deduced from pressuremeter measurements makes no distinction between sands and clayey soils. Only the value of the coefficient k differs.

The maximum bearing capacity of the soil is given by the equation:

$$q_u = q_0 + k(p_l - p_0)$$

where:

q_u = bearing capacity of the soil on rupture,
q_0 = effective vertical stress in the soil at the support,
p_l = limit pressure measured with the pressuremeter,
p_0 = effective horizontal pressure of the soil at rest,
k = bearing coefficient depending on the type of soil and the relative insertion of the support.

The pre-existing vertical stress q_0 equals a stress due to the weight of the submerged soil at the level in question:

$$q_0 = \gamma' D$$

where:

$\gamma' = \gamma - 1$ (γ being the density of the soil),
D = penetration in question.

The limit pressure p_l considered is an equivalent value equal to the geometric mean of the values measured at levels: D, $D + B$ and $D - B$ (B being the diameter of the support in the case of a cylindrical base or spud can).

The horizontal pressure p_0 of the soil at rest is a fraction of q_0. For highly compressible soils or soils undergoing consolidation, p_0 is taken as equal to q_0.

The difficulty in the calculating method lies in determining the coefficient of bearing strength k which depends:

– on the nature of the soil,
– on its consolidation (value of the limit pressure p_l),
– on the depth of the foundation D,
– and particularly on the type of insertion: foundations inserted by driving or not, covered or not by sheared soil, etc.

The values of the coefficient k given in Table IX.4.1 (merely as examples) have been grouped together in accordance with various TLM calculating rules and some numerical applications (Table IX.4.1).

TABLE IX.4.1

Type or state of soil	Limit pressure p_l (t/m²)	Coefficient k
Liquefied soft soils		0.8
Non-liquefied soft soils though covering the support		1.25 to 1.45
Clayey or silty soils of low consolidation, not covering the supports	5 to 60	≈ 1.5
Soils of the low compaction (low penetration of the support)	40 to 80	1.5
Compacted soils (low penetration of the support)	> 100	2

The friction term f, related to the maximum pressure by an experimental relationship:

$$f \approx \frac{p_l}{5} \text{ to } \frac{p_l}{10} \quad (\text{for } p_l < 40 \text{ t/m}^2 \ (< 400 \text{ kPa}))$$

is practically negligible compared with the tip term q_u for the supports of a jack-up platform.

IX.4.2.2 Application of the TLM calculating method

As an example, let us consider the case of a jack-up platform resting through conical spud cans on a soft soil that may liquefy:

– conical spud cans $B = 14$ m,
– soft soil $\begin{cases} \gamma = 1.5 \text{ t/m}^3 \ (\gamma' = 0.5 \text{ t/m}^3), \\ p_l - p_0 \text{ depending on the depth, as shown in Fig. IX.4.2,} \end{cases}$
– necessary bearing capacity: $Q = 4{,}500$ t.

The bearing capacity Q of a soil differs considerably depending on the calculation assumptions, which are in actual fact very imprecise and difficult to verify. In order to obtain a theoretical value of the bearing capacity $Q = 4{,}500$ t, the penetration D of the spud cans needed would, according to Fig. IX.4.2, lie between 16 and 35 m, depending on the calculation assumptions:

– soil does not cover the spud can ($k = 1.5$):

$$D = 16 \text{ m}$$

– soil covers the spud can ($k = 1.5$, but the stress q_0 need not be allowed for):

$$D = 22 \text{ m}$$

– the soil flows over the spud can, but is not at an interstitial overpressure ($k = 1.25$, $q_0 = 0$):

$$D = 25.5 \text{ m}$$

– the soil flows over the spud can and is liquefied ($k = 0.8, q_0 = 0$):

$$D = 34.5 \text{ m}$$

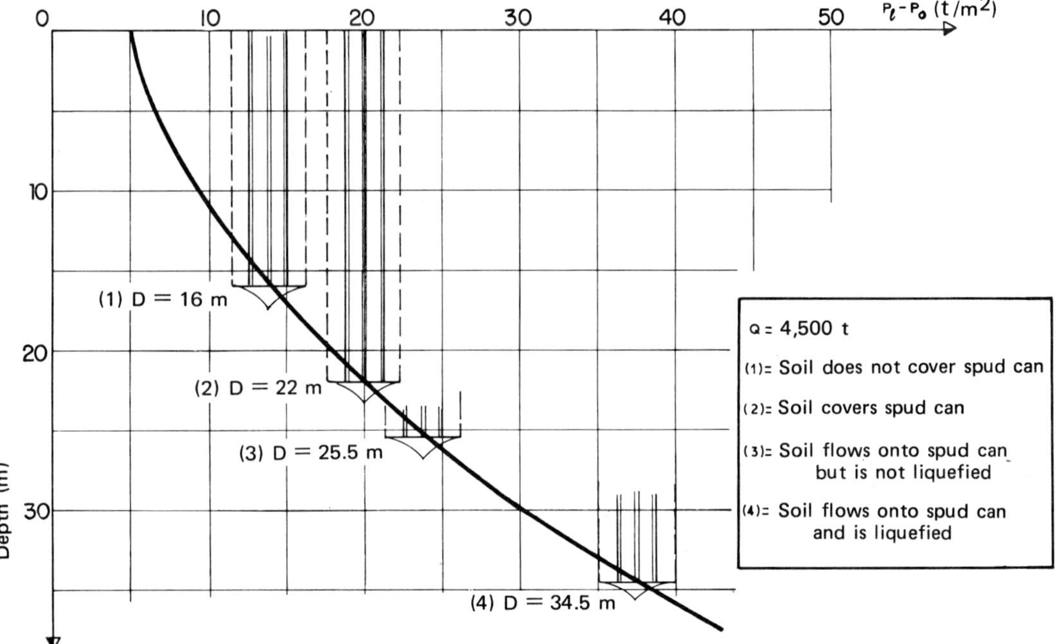

FIG. IX.4.2. – Penetration of spud cans of a jack-up platform into the soil in accordance with different calculation assumptions (TLM calculation method).

The values of coefficient k and the various calculation assumptions made may appear arbitrary and often lead to widely differing results for the stability of the jack-up platform.

It should however be pointed out that the uncertainty involved in the choice of k and the calculation assumptions stem at one and the same time from:

– the reconnaissance of the soils which in the present state of the arc does not render possible measurements representative of the characteristics of soft soils,
– the poor knowledge of the true state of the soil caused by disturbance when installing the jack-up platform.

Finally, the arbitrary factor in the method deduced from pressuremeter measurements probably remains comparable to that of the conventional methods set forth earlier (see Section IX.4.1).

IX.4.3. Remarks concerning methods of calculation and calculated and observed results

In the previous paragraphs, we have simply set forth the principle of methods of calculating the stability of jack-up platforms. For each particular case, it is clearly necessary to make greater allowance for the various factors concerning the heterogeneity of the soil and the geometry of the supports.

IX.4.3.1 Uncertainties involved in calculating the penetration of supports

Alternating thin layers of sands and clays complicate the appraisal of the bearing capacity of the soil and hence calculation of the penetration of the supports.
It should be noted that the existence of thin layers (about 10 cm) can be detected with certainty only by means of penetrometer recordings.

Possible scour near the supports, depending on the nature of the seabeds, can reduce the bearing capacity of the soils and may result in:

– either additional penetration of the supports, which is difficult to estimate (but should be considered in the light of the length of the piles),
– or slip of the supports, should for instance the penetration of the sand be insufficient ($\leqslant$ 1 m) (whence the advantage of conical bases with a tip angle $< 90°$ for this type of soil).

The considerable height h of the spud cans or bases compared to the penetration D introduces a difficulty in calculating the stability, particularly in heterogeneous soils. Experience would appear to show that the following have to be considered (Fig. IX.4.3a):

– first, the shear strength of the soil at the level of the cylindrical base or the cylindrical part of the spud can,
– second, the total penetration D (including the conical tip) [10].

The conventional theory of rupture beneath a cylindrical base (the Prandtl theory) shows that the soil is still sheared beneath this base down to a depth equivalent to its diameter B (Fig. IX.4.3b). It is therefore indispensable, in studying the stability of the

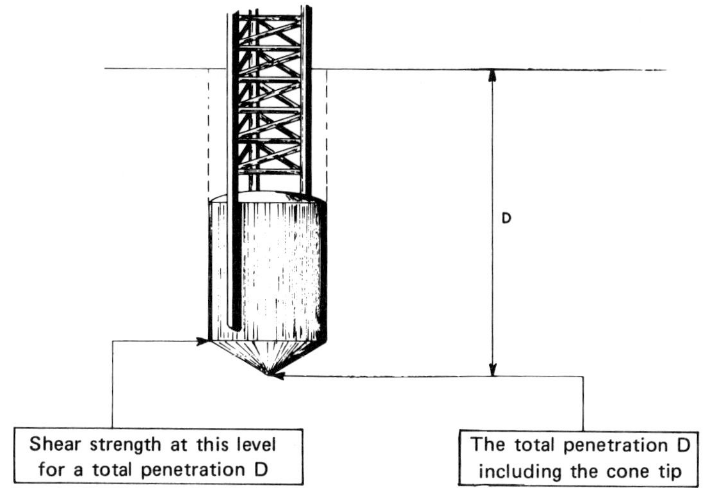

(a) Calculation assumptions adopted

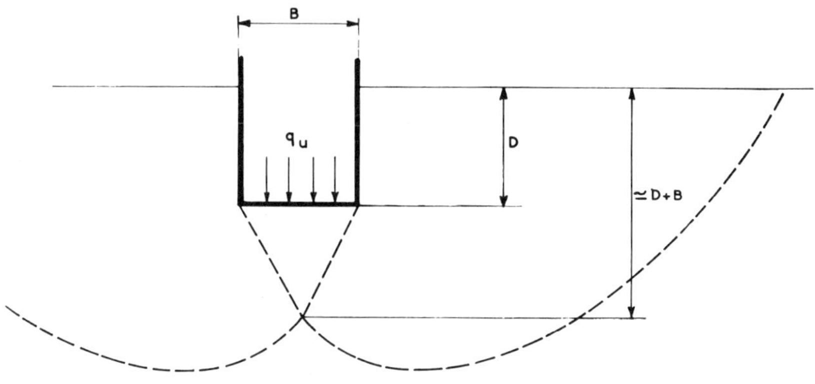

(b) Rupture diagram of soil beneath a cylindrical base

FIG. IX.4.3. – Conditions of calculation of the penetration of bases and spud cans of jack-up platforms.

supports of a jack-up problem, to examine the nature and characteristics at a minimum depth of $D + B$ (where D is the penetration and B the diameter) [10].

IX.4.3.2 Inclined piles

The piles of certain jack-up platforms can be inclined to 10 or 15° from the vertical, particularly in water depths of over about 50 m.

According to studies by Meyerhof and B. Hansen, it would appear that the bearing capacity of spud cans decreases theoretically in the case of inclined loads. It follows that the penetration of the spud cans increases slightly with the inclination of the piles.

The stability of a jack-up platform under horizontal stresses is increased:

- first, by the inclination of the piles,
- second, owing to the greater penetration of the spud cans into the soil.

IX.4.3.3 Comparaison of calculated and observed values of supports penetration

Confining themselves to the Gulf of Mexico, the engineers of *McClelland Eng.* have compared the calculated and the observed penetrations of several jack-up platforms on 120 different sites (Fig. IX.4.4). It would appear that the difference rarely exceeds ± 25%. This good correlativity can be explained at one and the same time by [10]:

- the perfect knowledge of the soils in this zone,
- the experience gained thanks to a great number of observations.

On the other hand, during initial installation of a jack-up platform on the given site, considerable differences often arise between the estimated penetrations and those actually observed, as is demonstrated by the values given in Table IX.4.2.

TABLE IX.4.2

Type of jack-up platform support	Type of soil	Estimated penetration (m)		Penetration observed (m)
		Conventional method	TLM method	
(Site A) Cylindrical bases (diameter 6.10 m)	Alternating layers of sand and soft muds	< 10	10	18
(Site B) Conical spud cans (diameter 13.80 m)	Soft soil ($p_l < 25$ t/m^2)	≈ 9	12.5	21
(Site C) Rectangular section spud cans 15.5 x 8.5 x 2.9 m	Soft soil $c_u \approx 1$ t/m^2; $z < 12\text{m} \rightarrow p_l \leqslant 10$ t/m^2; $z > 12\text{m} \rightarrow p_l > 20$ t/m^2	≈ 21	23 (Assumption of lique-faction)	It was not possible to install this jack-up platform on this site

These facts therefore reveal obvious shortcomings in the reconnaissance of soft soils with a cohesion of less than 5 t/m^2 (50 kPa).

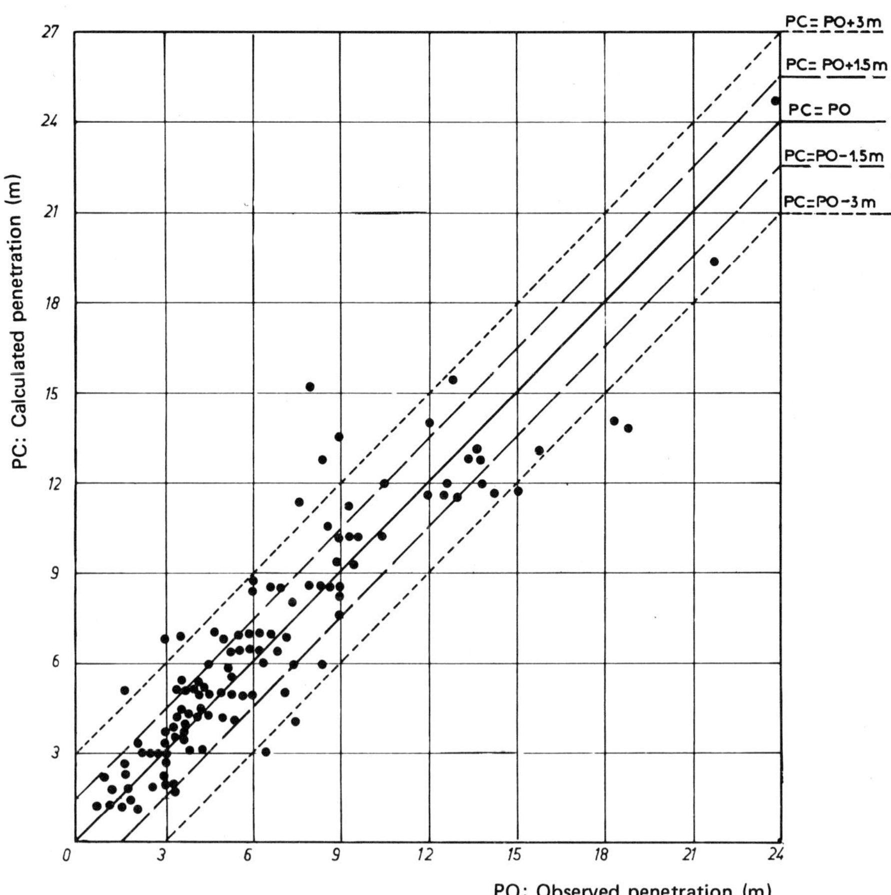

FIG. IX.4.4. – Relationship between the calculated and observed penetrations of the supports (cylindrical bases and spud cans) of jack-up platforms in the Gulf of Mexico (according to J.P. Gemenhardt and J.A. Focht).

IX.4.4 Stability of jack-up platforms under the effects of horizontal loads

As indicated earlier (see Section IX.2.2.2), verification of the stability under horizontal forces comprises two factors:

– estimation of the increased penetration of the piles as calculated under the action of the static loads,
– estimation of the overturning moment of the structure.

IX.4.4.1 Increased penetration of the supports under horizontal loads

A very appreciable increase in the penetration (particularly in soft soils) of the supports of jack-up platforms as a result of storms is very frequently observed.

This additional penetration can probably be explained by the liquefaction of the soils caused by the action of cyclic loads of relatively high amplitude. An upper limit to the penetration D could therefore be estimated by making the calculation assuming that the soil liquefies.

In the absence of precise data, we shall confine ourselves to a reminder of the operational criteria adopted for using new jack-up platforms with a total pile height of $H = 127$ m in various regions of the world (see Table IX.4.3).

TABLE IX.4.3

	North Sea		Gulf of Mexico (hurricane)	South East Asia
	Winter	Summer		
Wave amplitude (m)	20	11.5	17	9
Wind velocity (km/h)	165	120	200	130
Penetration of spud cans D (m)	21.5	11.5	22	12
Maximum acceptable water depths (m)	75	90	75	90

It will be noted in particular that the penetration D accepted for the North Sea will vary by a factor of 2 with the weather and oceanographic conditions.

IX.4.4.2 Overturning moment of the structure under the action of horizontal forces

It will be assumed that the supports rest directly on the ground if the penetration is less than 3 to 4 m. Such is notably the case for:

– mat-supported jack-up platforms,

– jack-up platforms with piles ended in bases or spud cans resting on a compact soil (dense sand, stiff clay, coral, etc.).

We shall confine ourselves here to setting forth the principle for calculating the stability, without going into details relative to each type of structure.

The stability of the structure is theoretically certain if the overturning moment resulting from horizontal loads is less than the moment of the vertical forces applying the structure to the soil (Fig. IX.4.5):

$$F_H h < Pa$$

where:

P = weight of structure,
F_H = horizontal force due to the action of the elements,
a = distance from the point of application of P to the edge of the mat (width $= 2a$) or the side of the pile-supporting polygon,
h = height of application of F_H above the seabed.

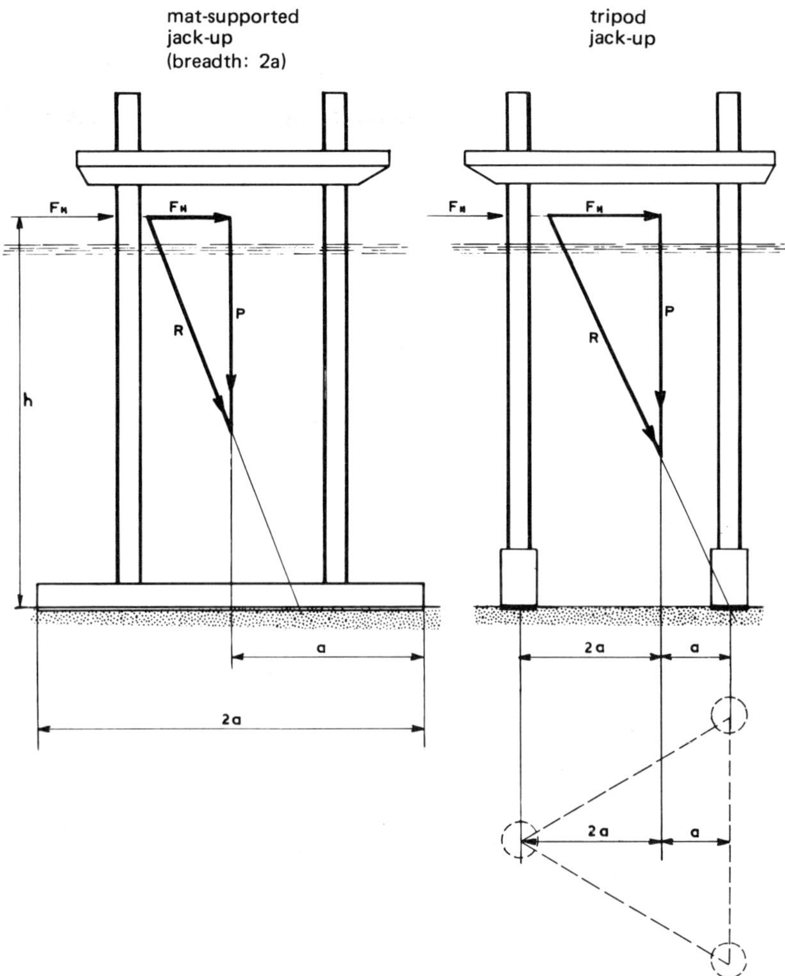

FIG. IX.4.5. – Stability of jack-up platforms under horizontal loading.

The action of F_H modifies the stress distribution beneath the mat (or spud cans). It is therefore also important to verify the risks of punching the soils beneath the supports.

As an example of the numerical application, let us consider:

– weight of structure $P = 10{,}000$ t, i.e. a load per pile of about 3,300 t in the case of a tripod jack-up platform,

– level of application of a horizontal force $h = 80$ m.

In the case of a mat-supported platform $2\,a = 50$ m wide, stability would imply:

$$F_H < 3{,}000 \text{ t}$$

In the case of a spud-can tripod platform, with a distance between piles of 45 m, resting directly on the bottom, stability implies a horizontal force per pile such that:

$$F_{H\text{pile}} < 500 \text{ t}$$

where:

$$F_{H\text{ total}} < 1{,}500 \text{ t}$$

If the penetration of the piles of a jack-up platform into soft soil reaches 20 to 30 m or more, it must be verified that the influence of the horizontal reaction of the soil on the stability of the structure subject to horizontal forces is still low, i.e. that the moment resulting from insertion of the piles into the soil does not generally exceed 10% of the moment of application on the soil of the vertical forces acting on the structure.

IX.4.5 Choice of supports for jack-up platforms in the light of the nature of soils

Appraisals of the possible influence of the geometry and dimensions of the supports on the stability of the structure can be of only a relative nature, depending on the characteristics of the formations (Fig. IX.4.6).

Furthermore, the choice of a jack-up platform on a given site very often depends on the availability of the structure and only little on the nature of the sands. However, strictly speaking, it would now appear possible to make certain recommendations as to the choice of the supports in the light of the results of the reconnaissance of the bottom.

IX.4.5.1 Influence of geometry of bases or spud cans on the stability of jack-up platforms

In the sands in the southern part of the North Sea, for instance:

– concave bases fitted with a skirt around the edge display instability in the event of scour,

– slightly conical bases (tip angle 120°), penetrating into dense sands to a depth of about 1 m (or less) may slip as a result of scour,

– highly conical bases (tip angle 60 or 90°) or pointed, generally display better stability owing to their greater penetration into the sands.

In soft soils, the stability of the jack-up platform depends:

– essentially on the master torque of the bases or spud cans,

– probably only little on the geometry of the spud cans. Bases with asymmetrical ends possibly offer certain advantages for structures with inclined piles.

IX.4.5.2 Influence of mats on the stability of jack-up platforms

In the case **of soft soils**, penetration will invariably remain low with the result that:

– both installation and withdrawal raise no great difficulties, although the platform may lean as a result of unevenly distributed loads or heterogeneities in the soil, without generally being possible to be corrected,

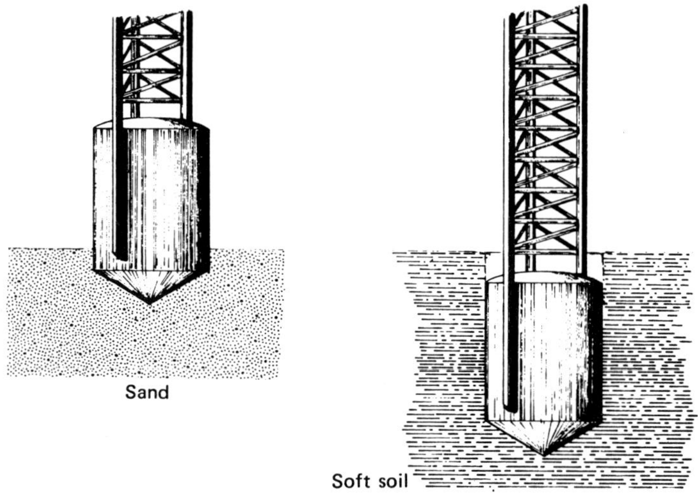

Cylindrical base

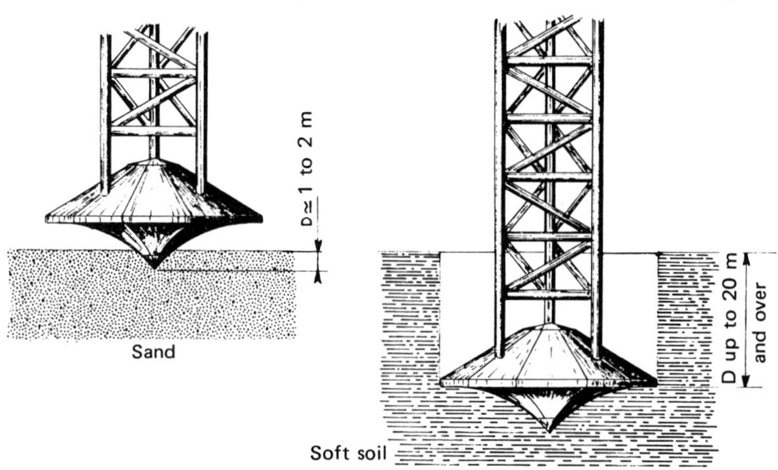

Large diameter spud can (13 to 14 metres)

FIG. IX.4.6. – Penetration of cylindrical bases or spud cans into the soil.

– thanks to the dimensions of the foundation mat cover, it generally appears possible to render the platform stable under the effect of the horizontal forces (wind, waves, current).

In the case **of hard or very hard soils**, it would hardly appear possible to recommend these jack-up platforms since :

– their installation presupposes a flat horizontal soil,

– their stability under the effect of horizontal loads appears less certain than before owing to the danger of slip.

The danger of scour around the mat often raises certain difficulties in the use of these structures. Depending on the nature of the sediments and the expected length of time the platform is to be positioned, it may sometimes be necessary to lay rip-rap around the mat.

IX.4.5.3 Conclusions concerning the choice of jack-up platforms to suit terrains

A distinction should be drawn between:

– on the one hand, sands and clays with a medium consolidation (cohesion $c_u > 5$ t/m² ($c_u < 50$ kPa),

– on the other, soft soils (clays and muds) with a cohesion of $c_u < 5$ t/m² ($c_u < 50$ kPa).

In **medium consolidation sands and clays**, the cylindrical bases:

– with a diameter of 8 to 12 m,

– and a conical extremity (60 to 90°),

penetrating a few metres into the soil, would appear to give satisfaction.

On the other hand:

– cylindrical bases with concave ends bearing on the skirts,

– cylindrical bases with conical ends and which are convex (tip angle 120°) the penetration of which is too small in dense sands,

– large diameter (12 to 14 m) conical spud cans with only slight penetration,

may slip in the event of scour under horizontal loads (Fig. IX.4.6).

In **soft soils**:

– large dimensions (12 to 14 m diameter) spud cans of various geometrical forms may penetrate to depths of 20 m or more. Withdrawal thanks to jetting may be facilitated by enclosing the pile over a height of 20 to 30 m in order to reduce the collapse of sediment onto the spud cans,

– mats generally prove satisfactory provided the loads on the structure are uniformly distributed.

IX.4.6 Jack-up platform accidents imputed to foundation problems

Information provided in the literature is generally highly fragmentary and makes it difficult to determine the exact causes of incidents or accidents that have occurred on jack-up platforms which may be imputed to the instability of the foundations. An attempt has been made in Table IX.4.4 to compile information concerning a number of jack-up platforms which have been lost or damaged.

TABLE IX.4.4

A NUMBER OF ACCIDENTS WHICH HAVE OCCURRED ON JACK-UP PLATFORMS IMPUTED TO THE INSTABILITY OF SOILS

Year	Designation of jack-up platform	Type of jack-up platform (maximum water depth)	Conditions and site of accidents
1957	Qatar Rig. No. 1	8 square section piles Water depth = 30 m	Destroyed by a storm when preparing to move in the Persian Gulf. Not recovered.
1957	Mr. Gus 1	Mat-supported (4 cylindrical piles) Water depth = 30 m	Overturned in the Gulf of Mexico when preparing to move. The high level on the suction forces prevented the mat from being withdrawn, causing the legs to break. Structure partially recovered in 1957. Mat recovered in June 1974.
1957	Deepwater No.2	3 triangular piles Water depth = 20 m	Subsided during drilling in the Gulf of Mexico. Recovered, but not brought back into service.
1959	Unit No. 10	12 cylindrical piles Water depth = 26 m	Overturned in the Gulf of Mexico when preparing to move. Not recovered.
1965	Sea Gem	10 cylindrical piles Water depth = 30 m	Collapsed in the North Sea when preparing to move. Not recovered.
1968	Dresser II	Mat -supported (4 cylindrical piles) Water depth = 20 m	Overturned in position on the Gulf of Mexico. Not recovered.
1973	Rowan Anchorage	3 triangular piles Water depth = 75 m	Collapsed on 1 pile during withdrawal under the effect of jetting in Indonesia (depth 37 m). Recovered.

IX.4.6.1 Accidents during installation or removal of jack-up platforms

The majority of accidents occurred as jack-up platforms were preparing to leave the site (during pullout of the supports) [12]. Overturning generally occurred:

– either under the effect of the lateral loads applied to the structure,

– or that of preferential jetting on one support,

– or again owing to the amount of the suction forces between the foundation mat and the soil.

It will be noted that the jack-up platforms where accidents have occurred are all of widely differing designs:

– structures resting on spud cans or mat-supported,

– number of piles,
– total height of structure (and water depths).

IX.4.6.2 Accidents during drilling

It will appear that the number of accidents during drilling are very low. On the basis of information obtained, one cannot say whether they were caused by:

– rupture of the soil (inadequate bearing strength) beneath the load,
– or overturning under the effect of lateral forces.

Subsidence during drilling may be explained:

– either by the unsymmetrical distribution of the loads on the platforms,
– or by differential settlement of the soil beneath one of the supports.

BIBLIOGRAPHY

[1] LOVIE (P.M.) and LOWERY (E.L.) – "Jack-up for 400 ft Water Depths Feasible". *Oil and Gas Journal,* Jan. 10, 1972.

[2] TUBB (M.) – "Jack-ups". *Ocean Industry,* Sept. 1976, p. 69-126.

[3] HOWE (R.J.) – "Offshore Mobile Drilling Units". *Ocean Industry,* July 1968.

[4] MACY (R.H.) – "Mobile Drilling Platforms". *Symposium on Offshore Technology.* New Orleans, 23-24 May, 1966, Paper SPE 1410.

[5] LACY (R.) and ESTES (J.C.) – "Evolution of Mat Supported Marine Oilfield Structures". *Offshore,* April 1966.

[6] CROOKE (J.O.) and LACY (R.) – "Advantages of Mat Support for Marine Structures". *Ocean Industry,* March 1970.

[7] LOVIE (P.M.) – "Jack-ups, a Future in the North Sea". *Northern Offshore,* No. 4, 1974.

[8] "New Innovative Designs for Jack-ups for the North Sea Oil Patch". *Offshore,* April 1974.

[9] GIBSON (R.P.) and MINORSKY (V.V.) – "Mobile Jack-up Platform Will Drill in 500 ft Water". *Ocean Industry,* Sept. 1973.

[10] GEMENHARDT (J.P.) and FOCHT (J.A.) – "Theoretical and Observed Performance of Mobile Rig Footings on Clay". *2nd Ann. OTC,* Houston, April 22-24, 1970, Paper OTC 1201.

[11] "Transworld 62 begins Drilling in Gulf". *Ocean Industry,* June 1974.

[12] GAUCHER (P.O.) – "Moving Drill Barges Safely". *The Drilling Contractor,* March-April 1967.

CHAPTER X

Anchoring Floating Structures

CONTENTS

INTRODUCTION

1. The various techniques for anchoring floating structures fall into several groups:

– anchors proper, with geometric shapes which vary widely with the constructors, used particularly for the temporary mooring of mobile structures (ships, platforms, semi-submersibles, etc.), but also for permanently securing loading buoys,

– drilled and cemented anchoring piles constructed differently depending on the nature of the formations used for securing loading buoys, floating tanks (now being developed), semi-submersible platforms in the case of rocky bottoms, future floating structures for deep-sea production, etc.,

– gravity anchors at present envisaged for anchoring tension-leg platforms and different structures.

In certain cases, the choice of the anchoring system may be dictated by the nature of the soils. For instance, in the case of bedrock, anchoring will be achieved first and foremost by piles regardless of the type of structure and period during which the object is to be immobilized.

2. Anchors oppose breakout in the horizontal direction with considerable force, but can be lifted without difficulty by vertical traction.

Permanent anchoring systems can withstand the action of horizontal and vertical forces of up to several hundred tons.

By convention, we shall define:

– the holding capacity of an anchor or anchoring system as the force that can be exerted horizontally (or vertically) on the anchoring line,

– the breakout force of the anchor as the vertical (or subvertical) force for raising the anchor.

3. The following will be examined in this chapter:

– the description and laying of anchors, anchoring piles and gravity anchors,
– the soil reconnaissance to be carried out before laying down various anchoring systems,
– the methods of estimating the capacity of anchors, anchoring piles and gravity anchors.

X.1 DESCRIPTION AND LAYING OF ANCHORING SYSTEMS

As was stated in the introduction, a distinction will be made between:

- anchors proper, with a wide variety of geometric shapes,
- anchoring piles,
- gravity anchors, which are as yet used only very little.

X.1.1 Anchors

The following will be described in turn:

- the most conventional type anchors, namely fluke anchors [1],
- special anchors: spade anchors, variable geometry anchors, etc.

For the record only, mention will be made of surface anchors [1] laid on the bottom and buried only slightly into the ground, generally small in size and used for securing positioning, marking-out, measuring, etc. buoys.

X.1.1.1 General description of the fluke anchor. Various types of fluke anchors

A fluke anchor consists essentially of (Fig. X.1.1) :

– the stem or shank, this being the main element of weight, transmitting the strains on the anchoring line to the flukes,

– the crown, providing a hinge-point and ensuring the relative orientation between the shank and the flukes,

– the fluke(s), ensuring the anchoring action proper, by causing the anchor to bury and hold in the soil as the anchor is pulled along,

– the stock, representing a balancing element, the purpose of which is to orient the flukes correctly in the soil and prevent the anchor from overturning.

There are many types of anchors, differing the geometry of the flukes, crown, etc. in common use:

– stockless anchors (the most conventional) with which most of today's vessels are equipped,

– LWT (light weight type) and Danforth anchors weighing 30,000 to 60,000 lbs (13.5 to 27 t) by which a great many semi-submersible platforms are anchored,

– Stato anchors, the opening angle of the flukes of which can vary from 34° (for sands) to 50° (for clays and muds),

– Boss anchors, with a large fluke area, giving good results in sandy and muddy bottoms,

– Offdrill anchors, with large area flukes, also suitable for sands and muds,

The Bruce anchor ensures a correct attitude of the single fluke, despite the fact that there is no stock (Fig. X.1.2). The specific shape of the fluke ensures absolute stability, avoiding any rotation or lifting in very soft soils. The constructor of this anchor claims:

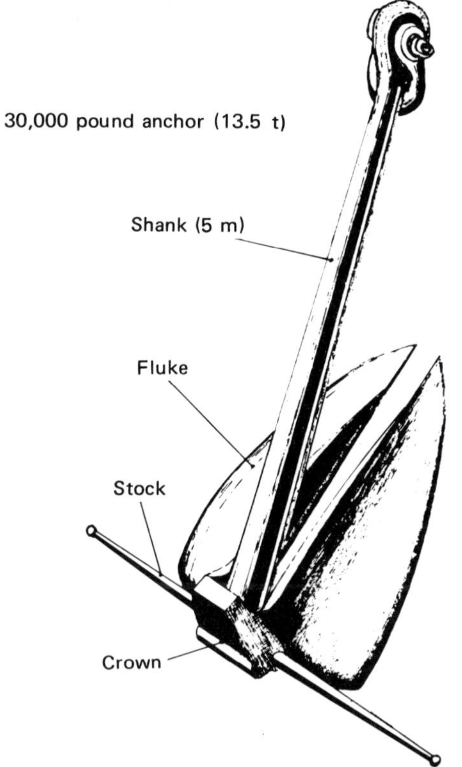

FIG. X.1.1. – Fluke anchor (conventional), type LWT.

– that it should be usable at angles of attack as low as 30% without any appreciable loss in holding power,
– that it can be used on highly consolidated bottoms.

The ease with which this anchor can be handled compared with conventional stock anchors should also be noted.

The Stevin anchors offer a relatively greater fluke area than most other type anchors and a great stability. The "STEV FIX" model should bite well into the consolidated clay.
The Flipper-Delta anchors present a great stability and are very effective in all soils: sands, clays and soft muds.
Stevin and Flipper-Delta anchors of similar size have almost identical holding power in sands and muds.

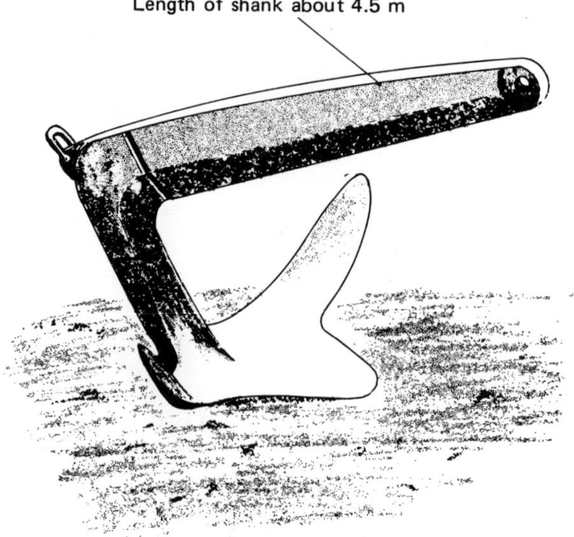

FIG. X.1.2. – 14,000 lbs (6.3 t) Bruce anchor.

Table X.1.1 gives examples of dimensions of certain conventional anchors together with the 6-ton Bruce anchor (see Section X.1.1.4).

TABLE X.1.1

Type of anchor	Weight (lbs)	Approximate length of shank (m)	Approximate area of flukes (m)
Danforth	30,000 (13.5 t)	≈ 5	≈ 4
Stato	25,000 (11.3 t)	5.80	≈ 7
Stevin	22,000 (10 t)	4.60	6
	55,000 (25 t)	6.25	12
Boss	40,000 (18 t)	7.50	
Bruce	14,000 (6.35 t)	4.50	≈ 5
Flipper-Delta	30,000 (13.5 t)	4	≈ 9.4

X.1.1.2 Holding action of fluke anchors

Fluke anchors are designed to bury into the ground when pulled along the bottom, thanks to the angle of attack of the flukes.

For some anchors, this angle of attack can be adapted to suit the type of soil:

– 30-35° for sands,
– 45-55° for muds and soft clays.

The behaviour of fluke anchors depends on their penetration into the soil. Use of these anchors therefore implies that the bedrock is covered with a sufficient layer of sediments. The burying depth, which depends on the type and dimensions of the anchor, will on the average be:

– 6 m for a 4.5 t anchor (10,000 lbs),
– 10 m for a 13.5 t anchor (30,000 lbs).

Fluke anchors can be used in various different types of sediments :

– consolidated sands to muddy sands,
– consolidated clays to soft clays.

However, the breakout force differs considerably with the nature of the soils.

Fluke anchors display high holding power against breakout in a horizontal direction only. It is therefore important to ensure that the length of the anchor lines satisfies this requirement. Should the angle between the anchor line and the surface of the soil be different from 0°, a coefficient of reduction as follows should be applied to the horizontal holding capacity of the anchor:

12 to 15 %	for an angle of 3°
20 to 25 %	for an angle of 6°
35 to 40 %	for an angle of 12°

Should the holding power of an anchor be insufficient, pack-up anchors (piggy backing) can be used, consisting in placing a series of two or more anchors separated by strands of cable a few tens of metres in length, each anchor then contributing to the holding power of the anchor line as a whole.

Unsatisfactory behaviour of a fluke anchor can be explained (Fig. X.1.3a) [2]:

– by the absence of penetration because the flukes have not opened owing to the ineffectiveness of the crown in muds (Fig. X.1.3b),
– by the absence of penetration owing to too high an angle of attack of the flukes in dense sands in particular (Fig. X.1.3c),
– by pulling of the anchor (Fig. X.1.3d),
– because the anchor has rotated where there is no stock,
– because of the unduly low shear strength of the soil (negligible cohesion in certain muds).

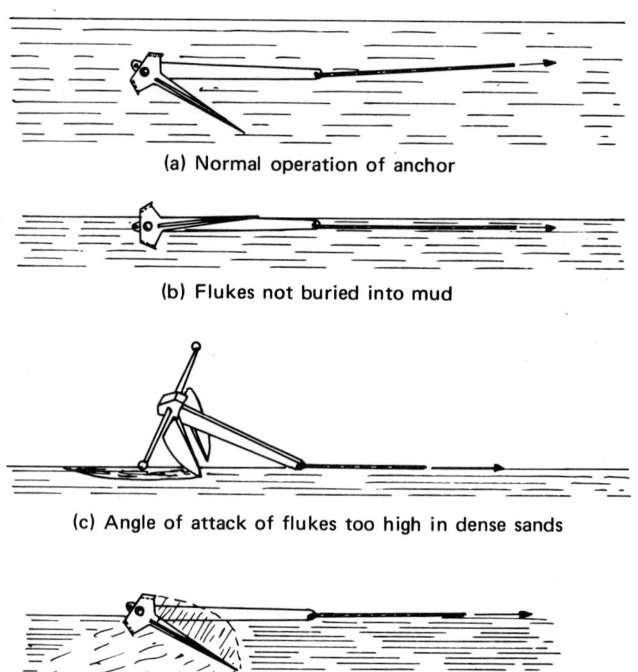

(a) Normal operation of anchor

(b) Flukes not buried into mud

(c) Angle of attack of flukes too high in dense sands

(d) Build-up of mud compressed by the pull of the anchor onto the flukes

FIG. X.1.3. – Various types of behaviour of fluke anchors.

X.1.1.3 Laying of anchors

The anchors of a floating support are laid:

– by the vessel itself, for 3 or 4 anchors, and sometimes even 5 to 6,
– by a tender (supply vessel or other) for semi-submersible platforms, very large tonnage vessels, etc.

Two anchor-laying methods are used:

– sometimes, pulling of the anchor along the bottom,
– more often, towing of the anchor line, carrying the anchor on the tender.

The pulling method is the fastest. It presupposes that there are no obstacles on the bottom. The anchor is lowered to the bottom and pulled by means of a trip line (generally reinforced) from the positioning buoy, while paying out the anchor line from the support to be moored.

By maintaining the anchor on the surface, the anchor line towing method offers the advantage of limiting the length of line dragged along the bottom and hence reducing the power required on the tender vessel. It does however involve the risk of not holding the anchor line sufficiently taut and thus allowing tangles to form which may result in snapping of the line under high loads.

X.1.1.4 Special anchors (DORIS, TLM, OSE)

Various special anchors are proposed either to counteract the inefficiency of fluke anchors, for instance in muds, or to reduce the weight and thus facilitate laying. Only a few examples of anchors at present being used will be mentioned.

Spade anchors (DORIS or Security type) consist of large blades below the vertical, the shape of which is similar to that of the blade of a bulldozer (Fig. X.1.4).

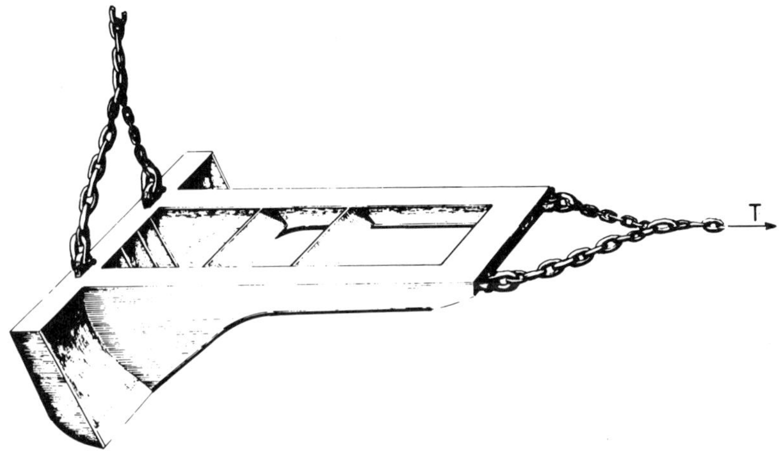

FIG. X.1.4. – DORIS anchor.

DORIS anchors are normally designed for muddy formations with very low cohesion (for example $c_u < 1$ t/m^2 (10 kPa), where there is much doubt as to the efficiency of fluke anchors. Figure X.1.5 shows how a spade anchor is laid and broken out.

Security anchors, equipped with two fixed feet enabling them to "catch" into a layer of compact sediments, can also be used on hard or clayey bottoms.

The variable geometry TLM anchors consist essentially of two plates welded to a central tube (Fig. X.1.6). The securing arm for the anchor line lies offcentre from the plates, so that the anchor buries vertically and then adopts a horizontal position under the effect of horizontal traction.

The burying depth of the anchor (by driving) depends on the quality of the formation, the holding power sought and its dimensions. For the 1.60 × 1.20 m anchor, the necessary depth would be:

- 3 to 5 m for a holding power (theoretical) of 100 t,
- 9 to 10 m for a holding power (theoretical) of 400 t.

This type of anchor would be usable in all sandy or clayey soils.

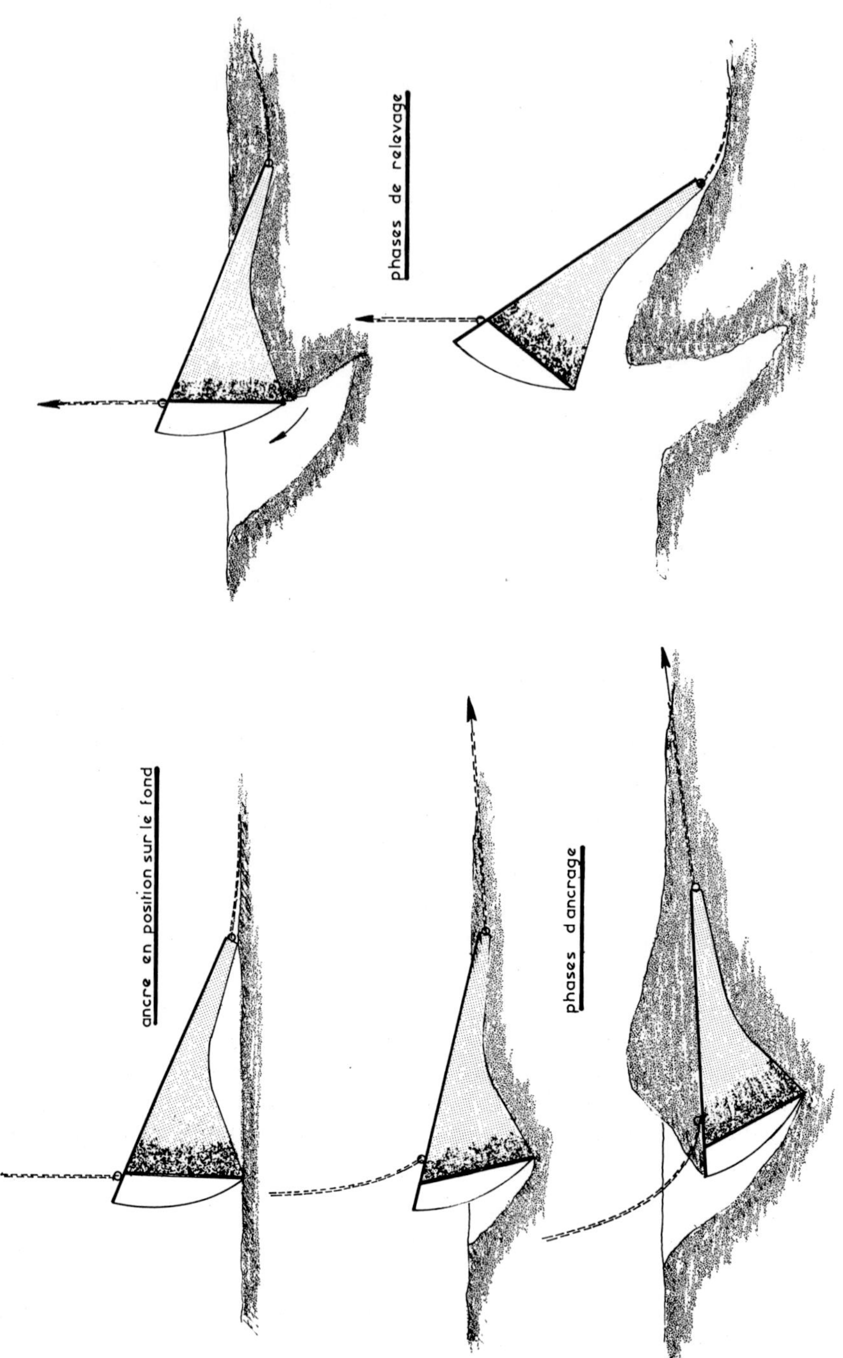

FIG. X.1.5. – Laying and lifting of a DORIS anchor.

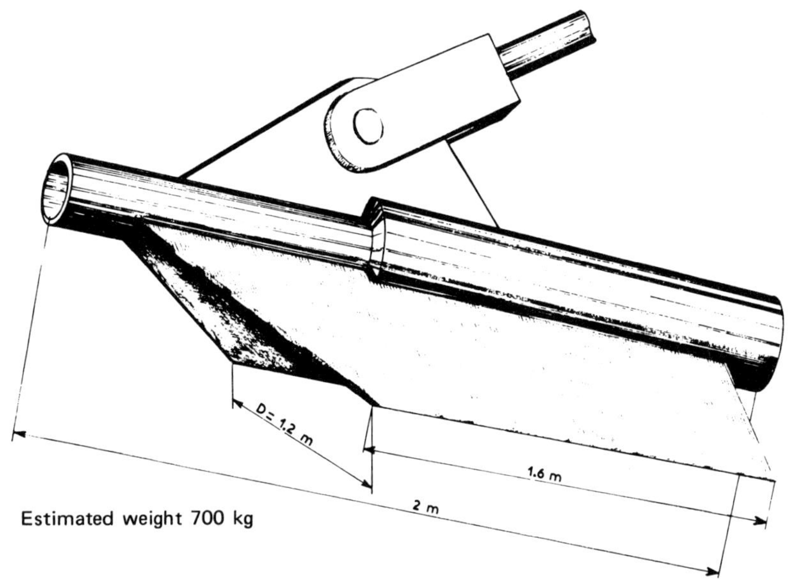

FIG. X.1.6. – TLM variable geometry anchor.

Ocean Science and Engineering (OSE) build a pile anchor consisting of a swivel plate mounted on the end of a rod. Offcentring between the centreline of the plate and the stem enables the plate to block against the stem during penetration, by pivoting when the anchor comes under load. The device is placed in the soil by vibrodriving.

Large dimensions (5 t and more) mushroom anchors are as yet still only little used, despite their advantages, particularly in very soft soils.

X.1.2 Anchoring piles

Permanent anchoring systems, used first and foremost so far for securing loading buoys [3] are unquestionably slated for considerable expansion with the development of deep sea petroleum operations, for anchoring floating production and storage structures.

Regardless of the implementation technology, these anchoring systems consist essentially of a drilled and cemented pile. A number of indications on the main techniques now proposed will be given.

X.1.2.1 AZ anchoring piles

Drilling is carried out through the pile which is lowered with the drilling string. The drill equipment is recovered after the pile has been cemented.

The drilling operation is carried out from a floating support which varies with the depths of water: drilling platforms, dynamic positioning vessel, etc.

The average time needed to install a pile is normally no more than about 12 h.

AZ piles, 13 to 26 inches in diameter and 10 to 20 m long, connected to the anchor lines by a swivel head are used or proposed :

– for anchoring semi-submersible platforms in the case of rocky bottoms (for instance the sea of Iroise or loading buoys,
– for securing floating storage tanks (Fig. X.1.7).

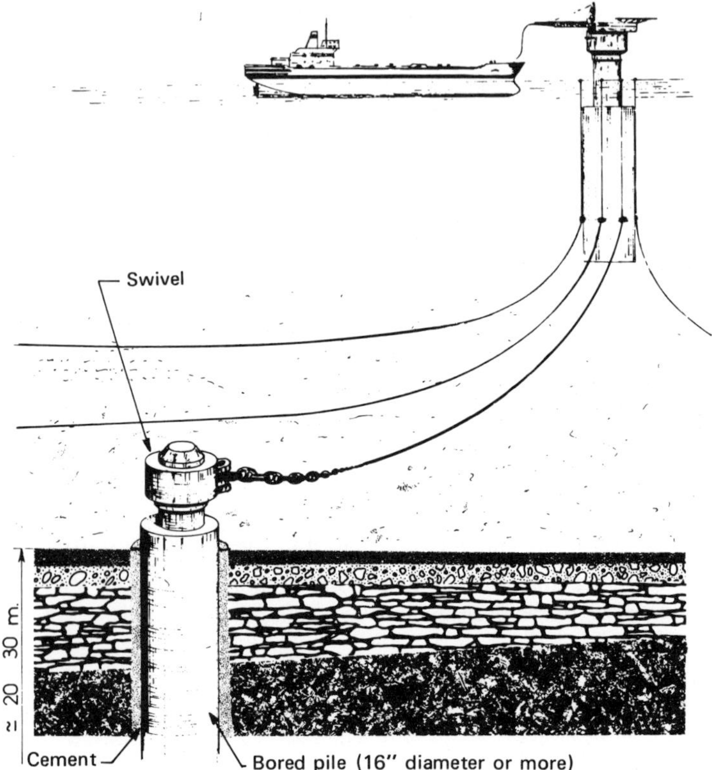

FIG. X.1.7. – Anchoring pile bored into hard soil (AZ technique).

The horizontal forces in an anchor line are taken up by the rigidity of the "thick pile-formation" combination. Clearly, this requirement can only be satisfied provided the surface terrains are relatively well consolidated, thus excluding the use of this technique in muddy bottoms.

X.1.2.2 DORIS anchoring piles

The drilled pile technique topped with a deformable chain tube is proposed by CG DORIS for the realization of anchoring systems where the surface soils are soft (Fig. X.1.8).

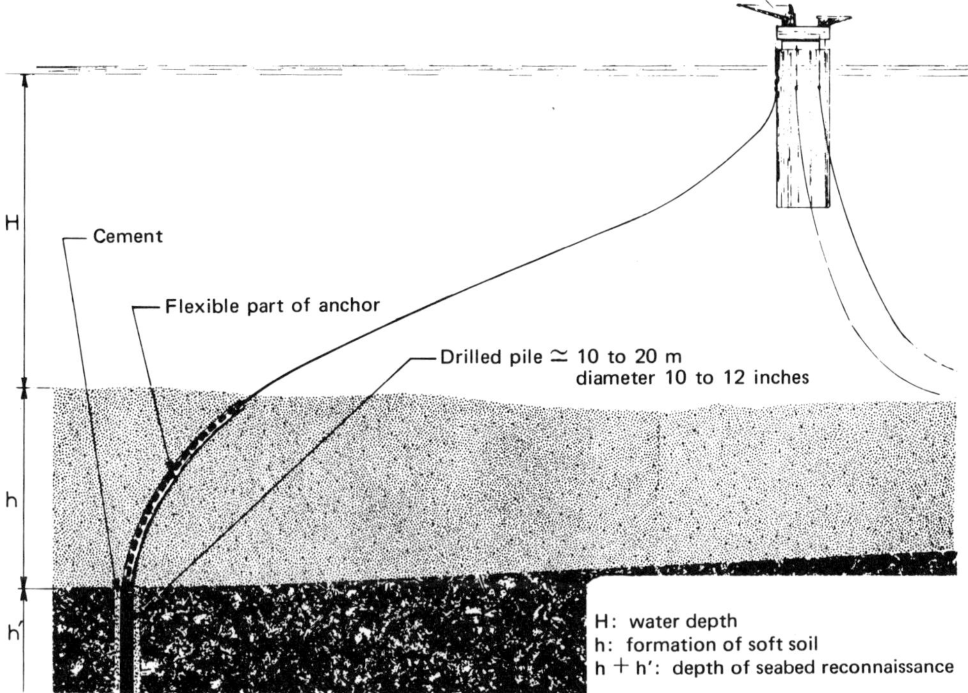

FIG. X.1.8. – Anchoring pile drilled into a hard soil lying beneath a soft soil (CG DORIS technique).

The anchoring system comprises:

– a pile consisting of a thick tube or drill collar, drilled and cemented in a sufficiently consolidated layer of soil at a depth which varies with the thickness of the loose surface sediments,

– a thin tube (chain type) which is bendable, crossing the loose layers and providing a link between the head of the pile and the anchor line,

– an articulated upper element to provide the link with the anchor line.

Normally, drilling is done from a floating support.

The bendable thin tube is used only for drilling and mud circulation.

The cementing, filling the annular space up to the surface of the soil makes it possible for a diver to check that there are no losses beneath the pile. The rock bit remains in position with the pile.

The dimensions of the piles inserted for securing SBM buoys are generally:

– diameter = 10 to 14″ (25 to 35 cm) and even 18″ (45 cm),
– length = 6 to 9 m.

Examples of this anchoring system have been laid so far in the following water depths:

– average = 30 to 40 m,
– maximum = 80 m.

The implementation of this anchoring system in great depths calls for certain technological modifications (the link between the head of the pile and the anchor line).

The number of piles laid for anchoring a structure is always an even number, so as to make it possible to carry out fraction tests on two opposite piles.

Under the horizontal forces in the anchor line, the thin tube bends (or self destructs) and the chain adopts a shape of equilibrium depending on the characteristics of the loose sediments. The result is that the drill pile is stressed in particular in the vertical direction, particularly if the thickness of the loose layers is considerable (over 20 or 30 m).

X.1.2.3 The Baker and Global Marine anchoring piles

With the Baker technique, the rock bit remains in position, using the solid rod as the pile. Cementing through the rock bit often results in irregular cementing in the ring.

Diameters of piles built by the Baker technique are 7 3/4″, 9″ and 13 3/8″.

In the Global Marine technique, the pile is inserted into a previously drilled hole and then cemented. This method requires work by divers.

X.1.2.4 Anchoring piles built by "insert-pile" technique

The "insert-pile" techniques have been briefly described already in the context of pile-mounted platforms (see Chapter VII).

The conventional "insert-pile" consists of a pile drilled through a main pile and cemented by normal circulation of the cement [4].

The approximate characteristics of an "insert-pile" of this type are:

- anchoring depth = 50 m,
- diameter = 30″ (75 cm),
- holding strength in an average clay = 600 t.

The Soletanche "insert-pile" method under pressure [5] distinctly improves the characteristics obtained by the conventional technique.

Thanks to successive cement grountings under pressure, the 30″ pile with an anchoring depth of 50 m is theoretically capable of a holding capacity of about 1,000 t in an average clay.

This type of pile would be particularly advantageous in hard or medium consolidation soils where a slight increase in the volume of the cement grouting would bring a considerable gain in strength. It could be an alternative solution to the conventional insert-pile for anchoring down to depths of 200 to 300 m.

X.1.2.5 The TLM expanding anchor pile

The expanding anchoring system proposed by TLM for consolidated soils or bedrock consists of (Fig. X.1.9):

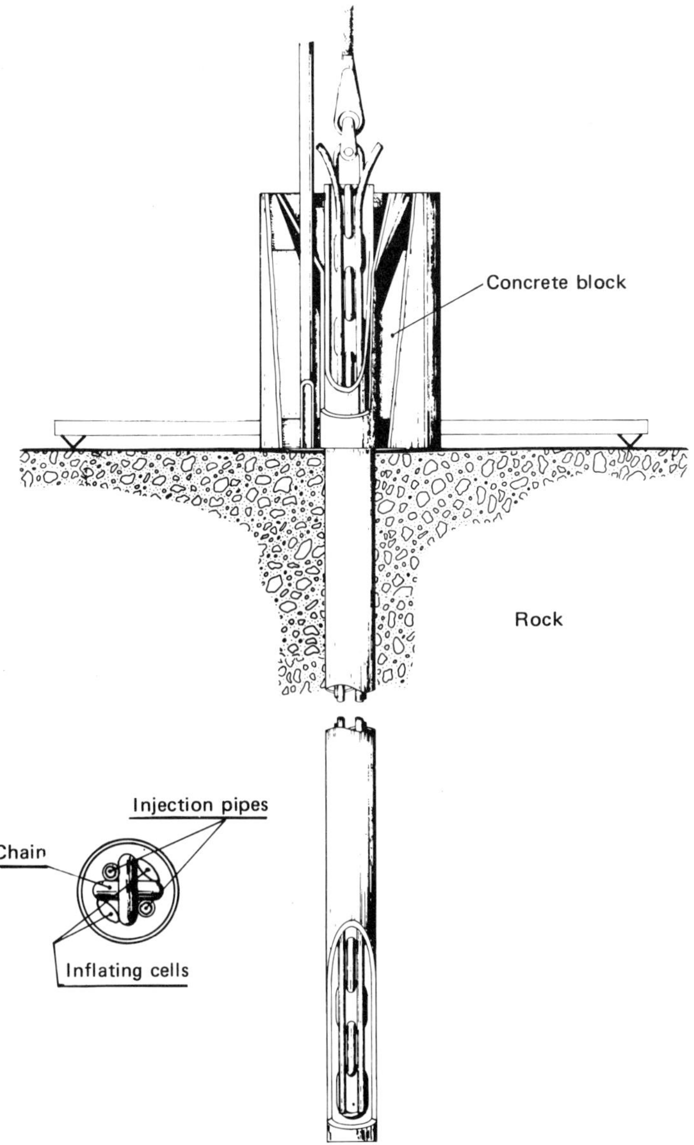

FIG. X.1.9. – Expandable anchorage system applicable for rocks (*Menard technique*).

– a drilled casing shorter than the depth of the borehole,
– a chain lowered inside the tube,
– two expandable cells.

The expansion of the cells enables the tube to be deformed or even locally burst. Pressurized cementing then increases the friction of the rock-cement-chain-tube combination.

X.1.2.6 Groups of piles

Laying anchoring systems in great water depths and capable of taking vertical or sub-vertical loads of 1,000 t or more can at present be envisaged by means of groups of piles connected together by a base.

On the other hand, the execution of anchoring systems consisting of groups of inclined piles to withstand horizontal loads of several hundred tons would involve considerable technological difficulties.

Remark.

The technology of the link between the head of the pile and the anchor line differs with the method concerned:

– cable fixed by a loop (around the pile) help in place by welded lugs,
– base plate with shackle welded to the head of the pile,
– chain cemented into the pile,
– rotating joint, with shackle, etc.

X.1.3 Gravity anchors

Large dimension gravity anchors, at present built or envisaged for anchoring marine structures, comprise the following:

– gravity anchors for securing structures with funicular anchoring,
– the bases of certain manifolds or production risers, with a submerged weight of a few hundred tons,
– the bases of articulated columns of SALM type buoys with a submerged weight of about 1,000 t,
– gravity anchors for securing tension-leg platforms, with a foreseeable submerged weight of 2,000 to 5,000 t.

X.1.3.1 Gravity anchors for securing structures with funicular anchoring

Gravity anchors are not yet used very much for securing structures with funicular anchoring. However, the SPAR storage buoy on the Brent field in the North Sea is held by six funicular anchoring lines, fixed by rectangular-shaped gravity anchors, weighing 1,000 t, with the possibility of ballasting (provided with 0.45 m long teeth) and arranged in a 1,000 m radius circle.

X.1.3.2 Manifold or riser bases

Certain bases for production manifolds or risers consist of concrete gravity anchors.

In the case of Argyl (North Sea) where the production support is a semi-submersible platform, the riser is kept under tension by (Fig. X.1.10):

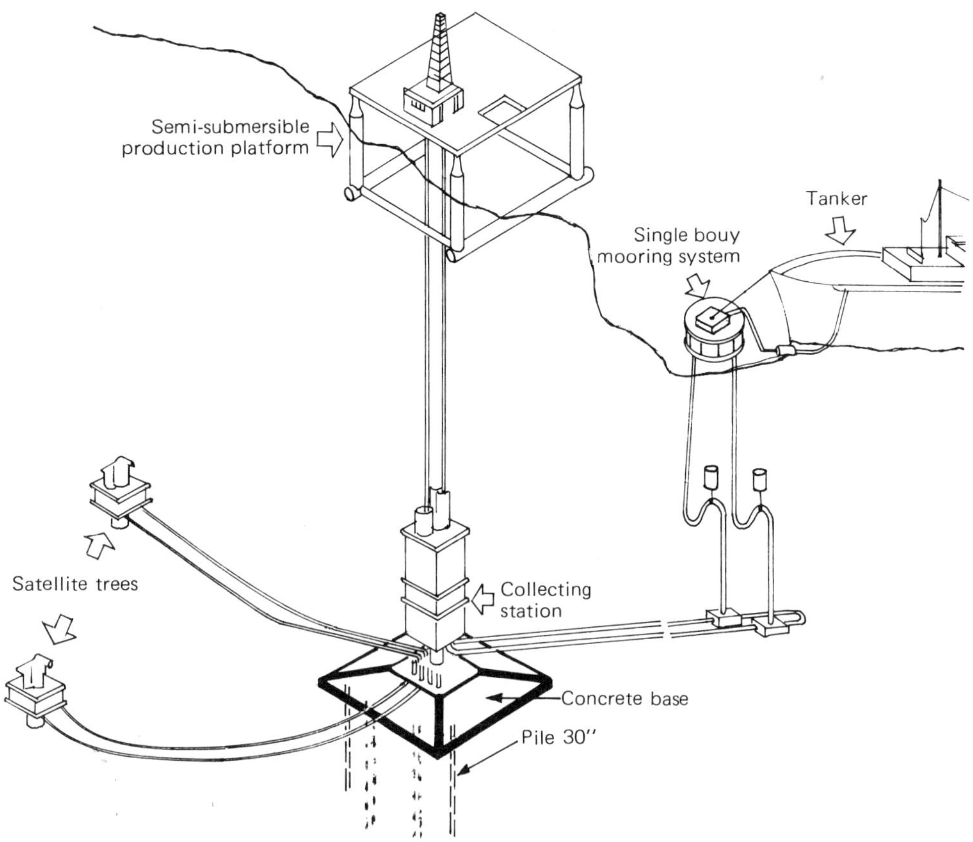

FIG. X.1.10. – Production riser of Argyl (North sea).

– 30″ piles (75 cm),
– a concrete base with sides measuring 6 to 8 m and weighing (in air) 350 t.

The Argyl concrete base was prefabricated ashore and laid by a "derrick-barge".

In the case of gravity anchors weighing a few hundred tons, one could also consider laying them by immersion in a metal casing followed by cement grout ballasting.

X.1.3.3 Articulated column or buoy bases

1,000 to 2,000 t bases:

– are used for anchoring terminals or flares of the articulated column type (Frigg, Brent, Beryl in the North Sea) or single point mooring buoys of the SALM type (Temburgo field in the China Sea),

– are envisaged as a possibility for the funicular anchoring systems of large semi-submersible platforms now in the project stage (of the Condrill or Conprod type) where the horizontal loads in the lines may attain 400 t.

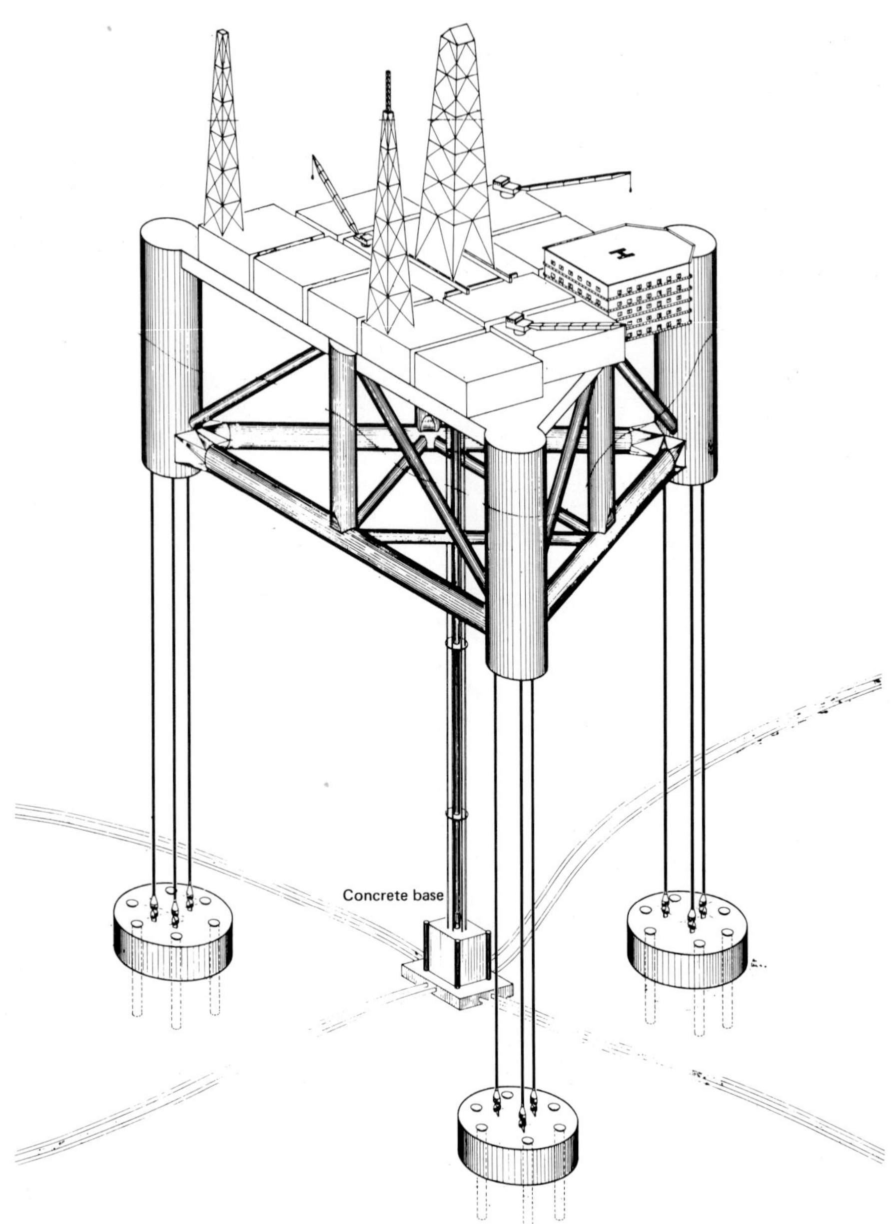

FIG. X.1.11. – Tension leg platform, VAP 200 *(CFEM–IFP)*.

At present, the base structure is invariably built of steel and then ballasted.
The weight of steel is:

– about 250 t for articulated column bases,
– about 300 t for the SALM buoy base.

For the articulated column, the base is positioned by the column itself.
For the SALM type buoy, the envelope of the base is carried on a barge and then submerged by ballasting floats. After being laid on the bottom, the base is ballasted by filling the envelope with granulates lowered by gravity through a 20″ tube (moved by guidelines in order to ensure uniform filling).

X.1.3.4 Gravity anchors for anchoring tension leg platforms

The large dimensions of tension leg platforms [6] now in the project stage (Fig. X.1.11) (*DOT, ACKER, EXBOY, CFEM,* etc.) will have to be held by three or four anchorage points with a capacity of about 2,000 to 5,000 t (or more). These may be achieved:

– either by means of groups of piles connected together by a mat,
– or large dimension gravity anchors,
– or again gravity anchors fixed by piles.

The technology for laying large dimensions gravity anchors would not yet appear to have been finally fixed:

– immersion of an envelope capable of withstanding the pressure, or with balanced pressures,
– ballasting with concrete, barite, heavy residues, etc.

For example, a base with a submerged weight of 2,000 t:

– may consist of an envelope weighing 900 to 1,000 t (in air),
– has to be supplied with about 2,000 t of ballasting material, i.e. 600 to 700 m^3 of barite or heavy residues.

X.2 SOIL RECONNAISSANCE FOR LAYING ANCHORS OR INSERTING ANCHORING PILES

A preliminary reconnaissance, albeit scanty, of the bottoms and soils will make it possible to reach an informed decision as to which anchoring method to use: fluke or spade anchors, piles, or gravity anchors.
However, apart from this preliminary reconnaissance of the nature of the surface soils, a distinction will be made between:

– on the one hand, the soil surveys which would be desirable for laying the anchors themselves,
– on the other, the surveys needed before inserting the anchoring piles.

X.2.1 Reconnaissance of soils before laying anchors

At present, soil reconnaissance before laying conventional anchors is in most cases very rough or even inexistent. The holding power of the anchors is determined only by anchoring tests when the structure is moored.

The surveys advocated here are therefore of a theoretical and maximum nature, but by making them, one unquestionably has some prior idea of how the anchors will behave, particularly by reference to other sites.

The importance of the surveys to be carried out will also vary with the type of structure to be anchored, or in other words how long it is to remain on the site:

- only a few months for semi-submersible platforms,
- 20 years or more, for instance, for loading buoys.

X.2.1.1 Geophysical reconnaissance

The coverage of the zones to be surveyed depends on the depths of water.

The $\dfrac{\text{length of anchor lines}}{\text{water depth}}$ ratio varies:

- from 3 to 7 for chains,
- from 6 to 14 for cables.

For water depths of 100 m, mean values of 5 to 10 will be adopted, depending on the particular case.

In view of the frequent imprecision as to the location of the structure, the geophysical survey will cover a zone measuring about 3 km along the sides if the depth is about 100 m.

A meshing of from 200 to 500 m will appear quite satisfactory for obtaining approximate information concerning the site.

The depth of investigation needed for laying the anchors will not be more than about 20 m.

The geometry of the layers will therefore be determined:

- either by a sediment sounder (or penetrator), particularly in soft soils,
- or by a boomer (Uniboom),

with a resolution power of better than 2 m.

The choice of the method of laying the anchors (by dredging or surface towing, see Section X.1.1.3) could possibly be governed by the presence or absence of obstacles (wrecks etc.) on the bottom.

Any possible obstacles will be sought by:

- a side-scan sonar,
- a magnetometer.

Provided the bottoms are found to be sufficiently flat (according to general marine charts), exact knowledge of the bathymetry is generally of little interest for laying anchors.

X.2.1.2 Geotechnical reconnaissance

The depth to which anchors are buried depends on their weight and the nature of the surface soils. This depth does not exceed:

– about 10 m in sediments of average consolidation,
– 15 to 20 m at the most in soft soils.

Allowing at one and the same time for:

– the maximum penetration to be reached: 10 to 15 m,
– the fairly rough approximation of the results sought,

only light weight means need be used for the survey which will therefore normally be conducted from the vessel used for geophysical prospecting.

Surface coring or sampling can be done by means of:

– a Kullenberg type corer in soft soils or soils with low consolidation. The maximum penetration reached will often provide an indication as to the shear strength of the soil and hence the probable depth of burial of the anchors,
– a vibrocorer (Zenkovitch type or other) in the case of sands or soils of medium consolidation. Penetration is generally no more than a few metres,
– systems for sampling by dredging (for example a grab sampler).

The number of cores or samples will depend mostly on the heterogeneity of the layers, deduced from an examination of the seismic profiles. In principle, at least 10 to 20 samples will be taken on the site.

Owing to the extent of the zone for laying the anchors and the frequent heterogeneities in the surface soils, it can be expected that the anchors will not all be placed in the same type of soil (sands, clays, muds, etc.)

Most in situ measuring devices can be used (in depths of 100 m and more) from the vessels used for geophysical propecting.

Adaptation of the TLM pressuremeter on a Kullenberg type jettisoning device (instead of a core barrel) would however be a simple method of rapidly characterizing soft soils in situ.

An adaptation of the same type is now being developed for a dynamic penetrometer (see Section VI.2). The difficulties, however, reside in the interpretation of the results and their correlation with the mechanical characteristics of the soil.

The in situ measurements and core samples advocated above at best will render possible a qualitative characterization of the soils (nature, consolidation, etc.), without providing representative values of their shear strengths. However, the search for a correlation between the quality of the soils and the holding power of anchors would already be of the highest interest (compared with the rough approximation possible using existing manufacturers' charts).

Table X.2.1 summarizes the bottom and soil surveys that would be desirable before laying anchors.

TABLE X.2.1

RECONNAISSANCE OF SOILS AND BOTTOMS BEFORE LAYING ANCHORS FOR SECURING SEMI-SUBMERSIBLE PLATFORMS, LOADING BUOYS, ETC.

	Coverage		**Meshing or number of soundings**	**Applicable devices and techniques**	**Penetration sought after (m)**
Geophysical reconnaissance	Zone with sides measuring about 3 km	Seismic prospecting	Meshing 200 to 500 m	Sediment sounder Boomer (Uniboom) Sparker	≈20
		Morphology of bottoms Search for wrecks		Side-scan sonar Magnetometer	
Geotechnical reconnaissance		Cores and samples	10 to 20 cores or more A few samples made by dredging	Kullenberg type corer Vibrocorer Grab sampler	5 to 10
		In situ tests		Pressuremeter jettisoned by Kullenberg device	A few metres

TABLE X.2.2

MEASUREMENTS ON CORES AND/OR "IN SITU" NECESSARY BEFORE LAYING ANCHORS

	Characteristics of soils measured		**Observations**
Laboratory core analysis (cores sampled using a Kullenberg corer)	**Identification of soils**	Nature Atterberg limits (W_L, W_p) Size distribution	On clay soils
	Water contents w Density γ		Preferably aboard the vessel
	Shear strength Undrained shear strength	Unconfined compression Vane test Pocket penetrometer	Preferably aboard the vessel
Pressuremeter (occasionally)	Limit pressure p_l		Pressuremeter possibly jettisoned by means of a Kullenberg device

Table X.2.2 gives the main measurement to be made on cores (and in situ) for identifying and determining the shear strength of the soils before laying anchors.

X.2.2 Reconnaissance of soils before inserting anchoring piles

Soil reconnaissance before inserting various anchoring systems using piles is simpler than that made for installing pile-mounted platforms, while nonetheless different in certain aspects: bathymetry, roughness of the bottom, number and penetration of the boreholes. etc.

X.2.2.1 Geophysical reconnaissance

It is not proposed to return to a consideration of the coverage of the zone to be surveyed as defined in Paragraph X.2.1.1. The meshing will be from about 100 to 200 m.

The necessary depth of investigation essentially depends on the nature and the consolidation of the soils :

- 20 to 30 m at the most if the soils are consolidated,
- several tens of metres if the surface layers are soft.

In the end, a seismic survey will identify the geometry of the layers:

- down to a depth of 50 to 100 m,
- with a definition of about 2 m.

These results will be achieved using a sparker or boomer (Uniboom).

A side-scan sonar and possibly a magnetometer will make it possible:

- to check the roughness of the bottoms,
- to verify the presence or absence of obstacles (wrecks, etc.).

Approximate knowledge of the bathymetry will be obtained at the same time as the other profiles are determined (seismic prospecting and sonar).

X.2.2.2 Geotechnical reconnaissance

Depending on the nature and consolidation of the soils, the reconnaissance depth will vary:

- from at least 20 to 30 m (consolidated surface soils),
- to 50 to 100 m (soft soils covering more consolidated formations).

In the case **of a layer of relatively consolidated soil** lying very close to the surface, the survey will comprise:

- boreholes with wireline core sampling,
- penetrometer tests using a modular penetrometer laid on the bottom (Seacalf type),
- pressuremeter tests (possibly).

The number of boreholes and cores samples will essentially depend on the results of the seismic survey:

– at least two or three boreholes if the layers are relatively homogeneous,

– preferably a borehole near the planned location for each pile, if the formations are heterogeneous (it should be noted that the piles often lie about 1 km apart).

A number of penetrometer tests down to refusal (at 10 to 20 m, depending on the nature of the soils) will be at least the same as the number of boreholes.

In the case of **soft soils** covering a more consolidated layer, drilling with wireline coring represents the only method of obtaining the necessary penetrations (50 m or more).

As before, the number of boreholes will depend on the horizontal heterogeneity of the layers as revealed by the seismic survey.

The penetrometer tests, limited to a penetration of about 20 m, will not generally provide any indications as to the characteristics of the deep consolidated layer in which the pile will normaly be anchored.

Table X.2.3 summarizes the bottom and soil surveys to be carried out before installing anchoring piles.

The measurements to be made on cores are naturally the same as those indicated for pile-mounted platforms (see Section VII.2).

TABLE X.2.3

RECONNAISSANCE OF THE BOTTOM AND SOILS BEFORE INSTALLING ANCHORING PILES FOR SECURING FLOATING STRUCTURES

	Coverage		**Meshing or number of boreholes**	**Devices and techniques applicable**	**Penetrations sought (m)**
Geophysical reconnaissance	Zone with sides measuring 2 to 3 km	Seismic	Meshing: 100 to 200 m	Sparker Boomer (Uniboom)	50 to 100
		Morphology of the bottoms Search for wrecks		Side-scan sonar Magnetometer	
		Bathymetry (if necessary)		Echo sounder	
Geotechnical reconnaissance		Drillings + corings	Minimum = 2 to 3 Preferably 1 borehole per pile	Boreholes and wireline core sampling	Hard soils = 20 to 30 Soft soils = > 50
		In situ tests	idem	Modular penetrometer Pressuremeter	≈ 20

X.3 ESTIMATION OF THE HOLDING POWER OF ANCHORING SYSTEMS

The holding power of anchoring systems can be estimated only to a rough approximation. For a given type of anchor and a given location, the holding power can vary by a factor of 2 depending on:

- the depth to which the anchor is buried into the soils,
- the type of anchor line (cable or chain),
- the angle of attack,
- the "cemented pile-soil" rigidity, etc.

The choice of the anchoring system to be laid depends at one and the same time on the nature of the soil, the type of structures to be secured, the cost price (for laying and maintenance), etc.

X.3.1 Holding power of anchors

The holding power of an anchor (line plus anchor) is the sum total of:

- the holding power of an anchor proper,
- the friction of the anchor line on the bottom.

Indeed, the length of the anchor line is generally calculated so that a portion of the line rests on the bottom.

X.3.1.1 Friction of the anchor line on the bottom and holding power of the anchor

The coefficients of friction of chains on the bottom are very much greater than those of cables. The anchor lines therefore mostly consist of:

- a cable,
- 100 to 300 m of chain placed between the anchor and the cable.

The holding power due to the friction of the chain on the bottom is obtained by the equation [7]:

$$HP_{(\text{chain})} = f \cdot \Delta l \cdot p_{a,l}$$

where:

f = coefficient of friction of chain on bottom sediments,
Δl = length of chain resting on bottom,
$p_{a,l}$ = submerged (apparent) linear weight of chain.

Table X.3.1 gives a few values of the coefficient f as measured on different soils when testing anchor lines [8].

TABLE X.3.1

Nature of surface soils	Coefficient of friction f	
	Chain resting on bottom	**Chain pulled on bottom (during laying)**
Sands	≈ 1	0.75
Muddy sand	≈ 1	0.75
Mud	≈ 1	0.6
Soft clay		0.5 to 0.6
Medium consolidation clay		0.1 to 0.2
Santa Barbara Channel clay	≈ 1	0.35

The holding power of an anchor is governed by:

– first, the type and weight of the anchor, the form and surface conditions of the flukes, the angle of attack of the flukes, etc.,

– second, the nature and shear strength of the soil, the depth to which the anchor is buried, etc.

It is generally considered that the ultimate holding power of an anchor is the sum:

– of its submerged weight,

– of the shear strength of the soil on the rupture surface, which is a function of the cohesion and/or of the internal friction angle of the soil, the area and geometry of the flukes, etc.,

– of the submerged weight of the displaced soil, depending on its apparent density γ', its cohesion and/or its internal friction.

The difficulty lies in the impossibility of predetermining the rupture area and volume of soil displaced by an anchor.

The reader can therefore comprehend the difficulty in making a theoretical calculation:

– only correlationships or experimental formulae allowing for the characteristics of the soil will provide approximate results,

– the development of numerical models would perhaps provide an approach to the problem on the basis of knowledge of the shear strength of the soil.

X.3.1.2 Estimation of the holding power of anchors on the basis of experimental formulae

In the case of conventional fluke anchors, the holding power would be obtained approximately by the formula:

$$HP_{(\text{fluke anchors})} = cP^b$$

where:

c and b = dimensionless coefficients typical of the characteristics of soil for a given anchor,

P = weight of anchor.

The values obtained by these formulae would represent the ultimate maximum capacity.

Table X.3.2 reproduces a number of values for c and b deduced from the testing of anchors on various sites [8].

TABLE X.3.2

Nature of soils and testing zone	c	b
Compact sand	65	0.82
Dense sand of average grain size distribution (Sabine Pass)	110	0.76
Mud (Sabine Pass)	37	0.91
Clay (South Timbalier)	98	0.82
Soft mud (Santa Barbara Channel)	2.6	1.15

These values would hardly appear applicable to sites other than those for which they have been determined.

In the case **of flat buried anchors,** *TLM* advocates the following formula for calculating the holding:

$$HP_{(\text{flat anchors})} = k s p_l$$

where:

s = surface area of anchor perpendicular to direction of pull,

p_l = limit pressure deduced from the pressuremeter tests (integrated according to a decreasing law starting from the anchor point),

k = bearing capacity coefficiency depending on the geometry of the anchor and the nature of the terrain.

The values of k to be adopted in this formula are given in Table X.3.3.

TABLE X.3.3

Type of soil	Limit Pressure P_l (t/m²)	k
Soft clay	$20 < P_l < 60$	1.5
Stiff clay	$60 < P_l < 200$	2
Loose sand and silt	$10 < P_l < 50$	2.5
Dense sand	$50 < P_l < 200$	3

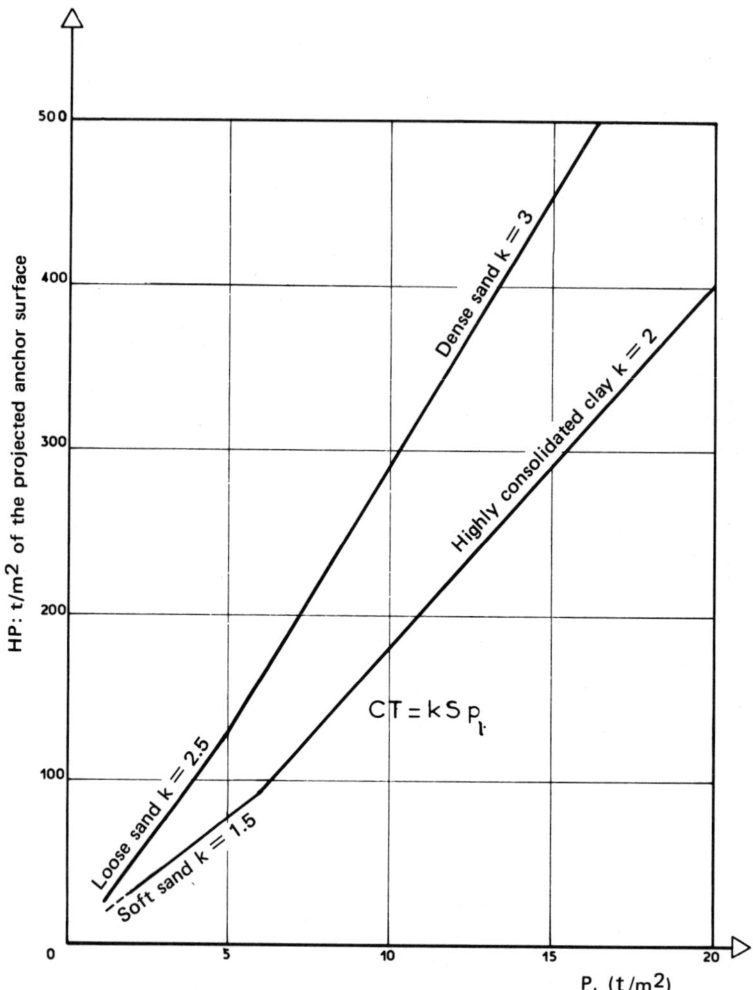

FIG. X.3.1. – Holding power (breakout force) of TLM variable geometry anchors in terms of the limit pressure of the terrain.

The holding power (or breakout force) expressed in tons per square metre of anchor area (perpendicular to the load) is shown in Fig. X.3.1 for various types of soil. A safety margin of two has to be added to these values. It should be noted that the "break out force/anchor insertion force" ratio is about 2 to 6, depending on the nature of the soils.

X.3.1.3 Information from manufacturers on the holding power of their anchors

Manufacturers of anchors give very approximate curves on the holding power of various types of anchors in terms of their weight and dimensions.

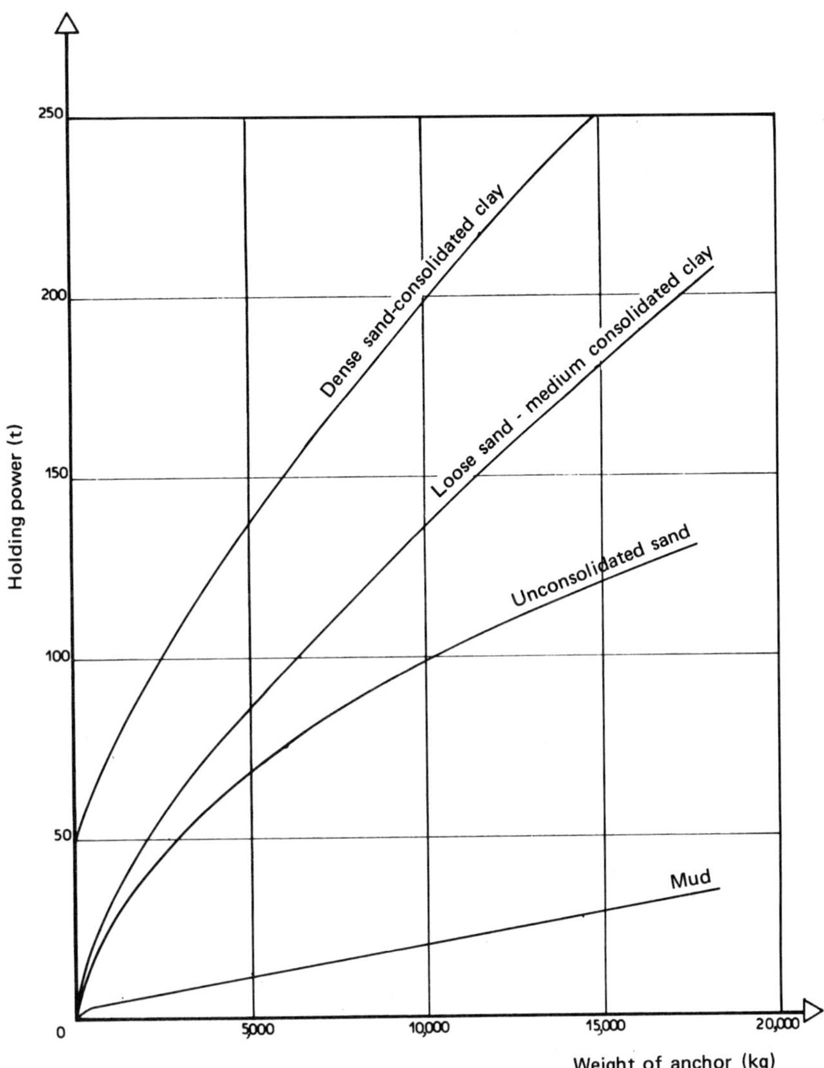

FIG. X.3.2. – Holding power of LWT anchors.

These curves generally do not allow for the nature of the sea bottoms. It would therefore always be wise to apply a high safety factor to these values, which are often greatly overestimated.

The holding power **of LWT anchors** in terms of their weight in various types of soils is shown in Fig. X.3.2. In dense sands, this power would reach 10 to 15 times the weight of the anchor.

The values of the holding power of **Stato** (built by *Baldt-Chester*) are shown in

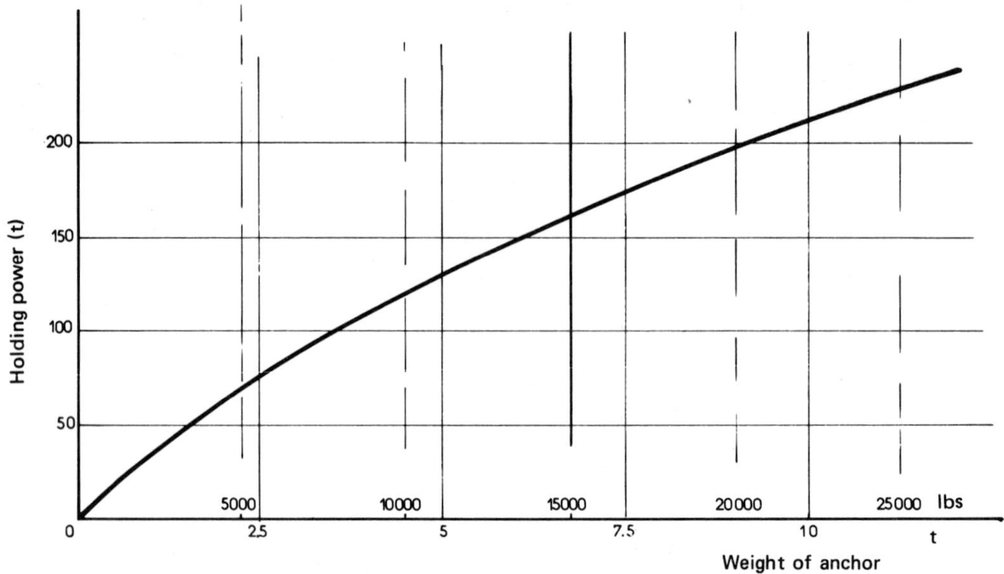

FIG. X.3.3. – Holding power of Stato (according to contructor *Baldt-Chester)*.

Fig. X.3.3. The $\frac{HP_{(Stato)}}{\text{weight}}$ ratio would decrease from 30 to 20 as the weight increases from 1.35 t (3,000 lbs) to 11.2 t (25,000 lbs).

The high holding power of **Stevin anchors** can be explained by the relatively large area of their flukes. A 9 t Stevin anchor would offer a holding power equivalent to that of a 15 t Danforth anchor.

Experience in the North Sea would appear to show that the 6.3 t (14,000 lbs) **Bruce anchor** offers performances comparable to those of conventional 13.5 t fluke anchors, both in hard soils (dense sands, overconsolidated clays) and soft soils (soft clays and muds).

Figure X.3.4 gives the performance of Bruce anchors as claimed by the maker.

Figure X.3.5 represents the average holding power of **DORIS spade anchors** in terms of their weight, in a mud with a cohesion of about 1 t/m^2 (10 kPa). For the same weight, the holding power of LWT anchors in muds is considerably less – not much more than the weight of the anchor.

X.3.1.4 Estimation of the holding capacity of anchors on the basis of in situ tests

In the face of difficulties in estimating the holding capacity of an anchor for a given site, some contractors conduct in situ tests on small dimension anchors before setting up the structure on the site.

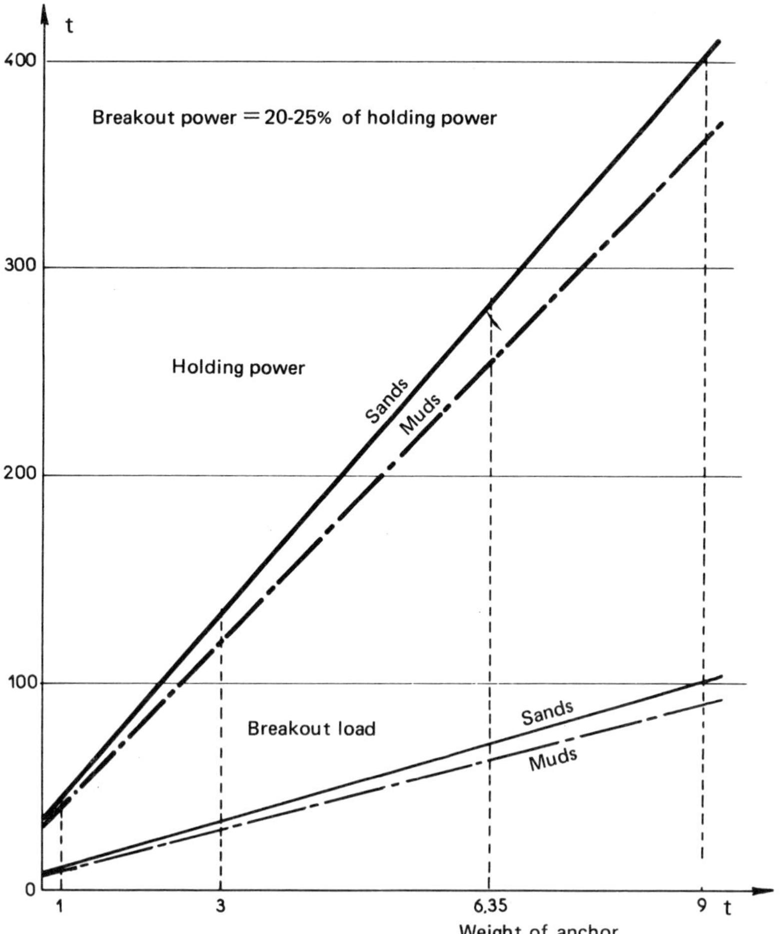

FIG. X.3.4. – Holding power and lifting force of Bruce anchors (Mark 2).

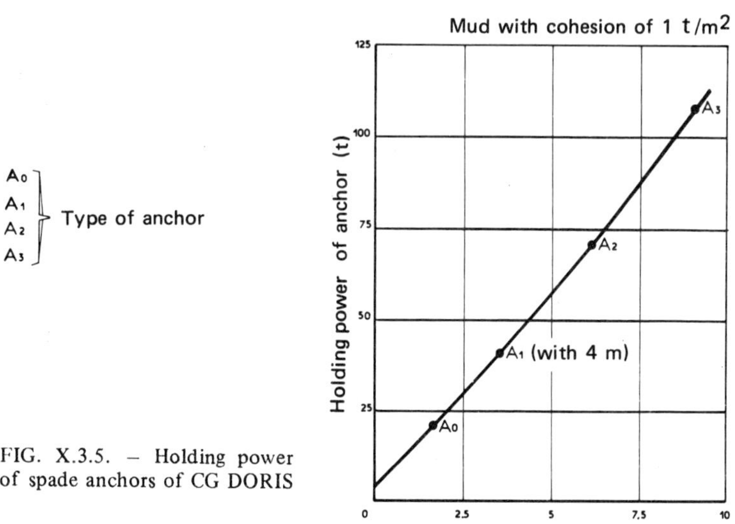

FIG. X.3.5. – Holding power of spade anchors of CG DORIS

Execution of these tests requires:

– a supply boat or tug,
– 100 to 3,000-lb anchors (45 to 1,350 kg), of the same type as those envisaged for anchoring the structure on the site,
– equipment for measuring the tension in the anchor lines.

It is necessary:

– to ensure that the anchor line forms a 0° angle with the surface of the soil (horizontal force),
– to know the length of the anchor line resting on the soil in order to deduce the inherent frictional holding power (see Section X.3.1.1).

The results obtained are scaled-up graphically to larger dimension anchors by means of the following formula:

$$HP_{(\text{anchors})} = c\ P^b$$

represented by a straight line in a bilogarithmic diagram (Fig. X.3.6). The value of b (gradient of the straight line) generally lies between 0.7 and 0.9.

Remark. It is evident that cyclic loadings have a significant influence, still poorly understood, on the holding power of anchors, particularly in clays.

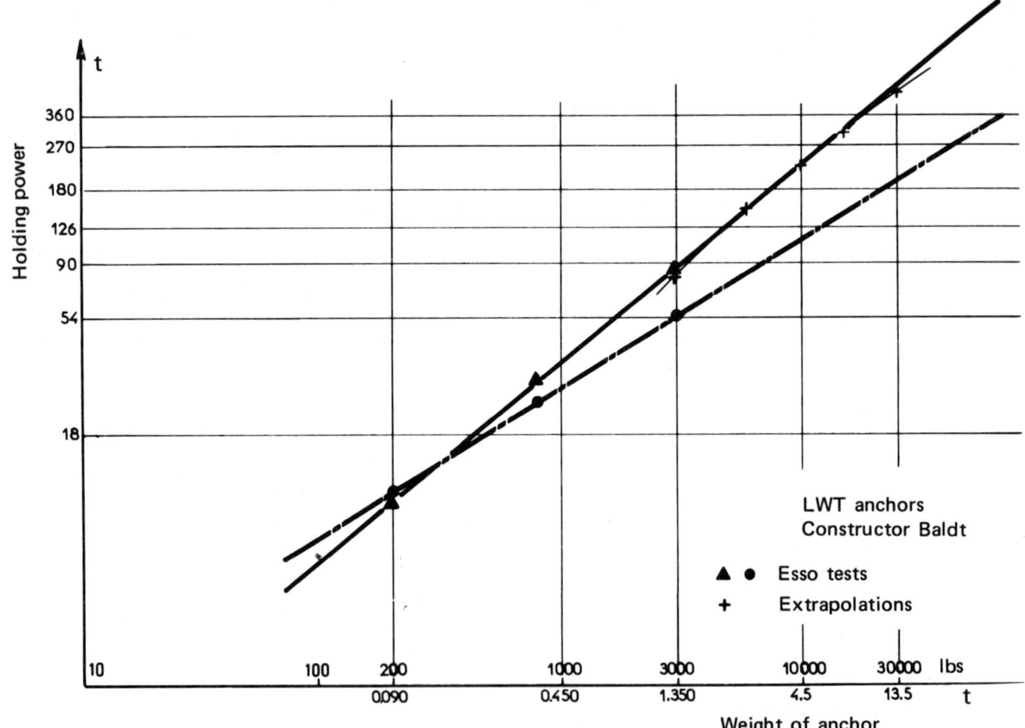

FIG. X.3.6. – Esso test method for testing the holding power of anchors.

X.3.1.5 Present trends for estimation of the holding power of anchors

The rational calculation of an anchoring system involves:

– choosing a type of anchor especially suited to the problem,

– making an accurate estimation of the holding of the type selected as calculated from soil data.

The joint research project undertaken in 1976 by the *Institut Français du Pétrole (IFP)* and the *Centre National pour l'Exploitation des Océans (CNEXO)* had the following two main objectives:

– gaining better understanding of the phenomena characterizing anchor behavior and kinematics,

– improving ways of predicting anchor holding power as a function of the soil encountered [9].

The in situ test program involved two phases:

– testing of simple configuration models weighing about 100 kg, with the possibility of independently varying several parameters,

– comparative tests on small-size commercial anchors (about 100 kg).

TABLE X.3.4

Fundamental Parameters	Main Functions	Opening	Penetration	Burial	Stability	Holding power
GEOMETRICS	Distribution of masses-rotation axis position	X			X	
	Trimming palm area	X				
	L/F ratio or fluke spacing				X	
	Fluke surface area			X		X
	Bearing surface of shank			X		
	Shank length S		X			
	Stocks, stabilizers				X	
	Fluke-shank angle α		X	X		
	Pulling angle β					X
	Burial depth D					X
MECHANICS	Mechanical properties of soil: φ, c_u	X	X	X	X	X
	Fluke roughness δ		X	X		X

Table X.3.4 (according to A. PUECH and al.) lists the basic parameters brought out as a result of experiments and shows their direct and preponderant effect on the different functions of the anchor.

Figures X.3.7 and X.3.8 show comparative obtained results with different commercial anchors:

– in calcareous sand (friction angle $\varphi \simeq 40°$),
– in soft mud (cohesion $c_u \simeq 0{,}5\ \text{t/m}^2$ (5 kPa)) and cohesion gradient of about $0{,}3\ \text{t/m}^2/\text{m}$ (3 kPa/m)).

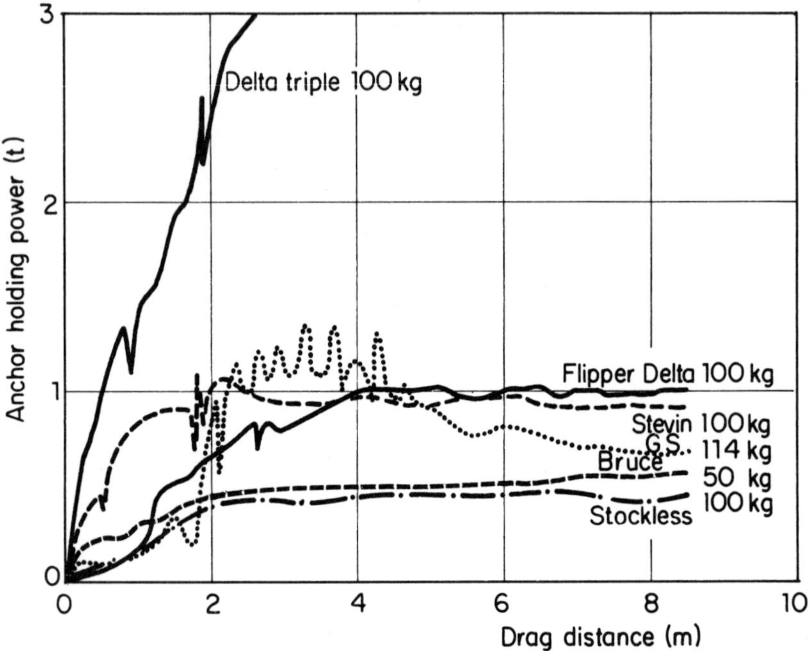

FIG. X.3.7. – Comparative tests in sand: typical holding power-distahce of drag records (according to A. PUECH, J. MEUNIER and M. PAILLARD).

X.3.2 Holding power of anchoring piles

The performances of anchoring piles must be calculated under vertical and horizontal loads.

X.3.2.1 Holding power of anchoring piles subject to vertical pullout forces

The calculation is identical to that given for the piles of fixed platforms on piles (see Section VII.3).

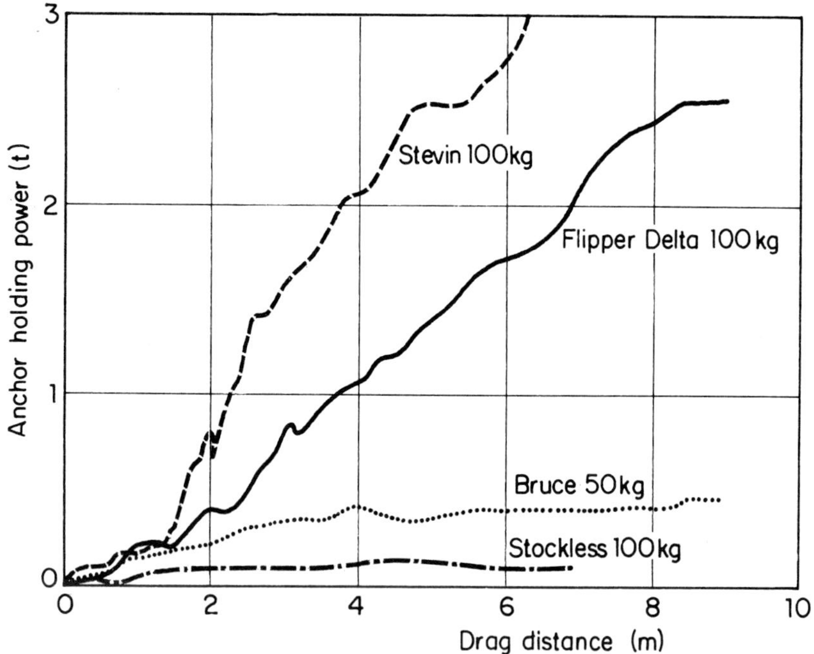

FIG. X.3.8. – Comparative tests in mud: typical holding power-distance of drag records (according to A. PUECH, J. MEUNIER and P. PAILLARD).

The vertical pullout force of a pile is given by:

$$HP_{\text{vertical}} = f \cdot A_s$$

where:

f = lateral frictional resistance,
A_s = surface area of pile, including that of the cement bulb for cemented piles.

In the case of piles cemented into sands, the lateral frictional resistance is given by the equation:

$$f = Kp_0 \text{ tg } \delta$$

where:

K = coefficient of lateral pressure of the soil, about 0.5 to 0.6 for piles cemented under pressure,
δ = angle of friction between pile and sand, approximately equal to the internal angle of friction if the cementing is carried out under pressure,
$p_0 = \gamma' h$ = effective geostatic stress,
γ' = apparent density of the soil,
h = depth of insertion of the pile.

In the case of piles cemented into clayey soils, the lateral frictional strength is expressed by the equation:

$$f = kc_u$$

where:

c_u = undrained shear strength (undrained cohesion) of soil,
k = coefficient of about 1 if the cement-clay contact is perfect (cementing under pressure).

X.3.2.2 Holding power of horizontally loaded anchoring piles

It is not proposed to return here to the method of calculation for horizontally loaded piles based on the solution of the general equation:

$$EI \frac{d^4 y}{dz^4} = - p(y, z)$$

indicated for fixed platforms on piles (see Section VII.3).

We shall merely confine ourselves to a number of indications concerning the holding power of anchoring piles as stated by manufacturers.

The holding power of horizontally loaded anchoring piles depends:

– in particular on their diameter,

– only to a slight extent, on their length beyond about 10 m, this being a common approximate length of construction.

Table X.3.5 gives the following information for a number of anchoring systems produced by *AZ* and *DORIS:*

– first, the normal tension in the line,
– second, the line breaking tension.

TABLE X.3.5

	AZ anchoring piles		DORIS anchoring piles		
	16" (40 cm)	26" (65 cm)	8 5/8" (21.5 cm)	10 3/4" (26.5 cm)	14" (35 cm)
Normal tension in line (t)	100		100	200	400
Line breaking tension (t)	150	410	200	400	800

X.3.3 Holding power of gravity anchors

A gravity anchor laid on the sea bed, buried to a depth which depends on its weight, geometry and nature of the soil, forms an anchoring system for which the vertical holding power can be estimated in an obviously similar way to that of the lifting force of any other object resting on the bottom [10] [11].

X.3.3.1 Estimation of the holding power of gravity anchors

The vertical force needed to lift a gravity anchor is given by the general expression:

$$HP_{\text{vertical}} = P_a + R_s$$

where:

P_a = submerged weight of gravity anchor,
R_s = resistance of the soil to pullout of the gravity anchor.

The submerged weight can vary with the water content of the soil:

– if the water content is below the liquidity limit W_L:

$$P_a = P - P_w$$

where:

P = weight of gravity in air,
P_w = weight of water displaced by gravity anchor,

– if the water content is above the liquidity limit (as frequently occurs in soft muds):

$$P_a = P - P_{lw} - P_{ls}$$

where:

P_{lw} = weight of water displaced by gravity anchor,
P_{ls} = weight of soft soil displaced by gravity anchor.

The difficulty will very often lie in estimating the true depth to which a gravity anchor is buried and hence to estimate P_{lw} and P_{ls}.

The resistance of the soil to pullout of a gravity anchor is a function of the depth to which it is buried and the speed with which it is broken out, and can be expressed by the equation:

$$R_s = c_u A_l + p A_b$$

where:

c_u = undrained cohesion (or more generally shear strength) of the soil,
A_l = lateral surface area of the buried section of the gravity anchor,
p = suction pressure at the gravity anchor/soil interface, dependent on the permeability of the soil and the loading application speed,
A_b = surface area of the gravity anchor in contact with the soil.

In practice, the following values are accepted:

– a mean value $\bar{c}$ of the shear strength of $\frac{c_u}{2}$, where c_u is the cohesion at the base,

– a suction of $\bar{c}$,

therefore:

$$R_s = \bar{c}(A_l + A_b)$$

Experience shows that the pullout force depends on the geometry of the gravity anchor. For a given surface area, it is maximum for a square section.

X.3.3.2 Influence of the type of sollicitation on the holding power of gravity anchors

The holding power of gravity anchors depends on the direction of the loads and the type of sollicitation (static or cyclic).

In the case of static vertical stress, the holding power of a gravity anchor can be estimated by the equation given above (Section X.3.3.1).

For example, let us consider a concrete cylinder (specific weights $\approx 2\ t/m^3$), 10 m in diameter and 5 m in height. The submerged weight is $P_a \approx 400$ t. The pullout (or lifting) effort is approximately:

– 540 t in the case of a soft clayey soil with an average cohesion $\bar{c} = 1\ t/m^2$ (10 kPa), assuming that the gravity anchor is buried to a depth of 1 m,

– 400 t in the case of a sand where the suction force is negligible.

In the case of a subvertical load, the pullout force is obviously lower and depends on the friction of the gravity anchor against the soil. A gravity anchor would appear inappropriate as an anchoring system called upon to withstand considerable horizontal forces in soft clays.

In a case of cyclic loadings (heave due to wave action), it would invariably be wise to neglect the influence of the suction forces and examine the risk of liquefaction of loose soils.

BIBLIOGRAPHY

[1] OGG (R.D.). – "Anchors". *Handbook of Ocean and Underwater Engineering*, MYERS (J.J.), HOLM (C.H.). MacGraw Hill, 1969.

[2] BECK (R.W.). – "Anchor Performance Tests". *OTC*, Houston, 1-3 May 1972, Paper OTC 1537.

[3] MAARI (R.). – *Offshore Mooring Terminals*, SBM, Monaco.

[4] EIDE (O.). – "Marine Soil Mechanics. Applications to North Sea Offshore Structures". *Offshore North Sea Technology Conference*. Stavanger, 3-6 Sept., 1974.

[5] GOUVENOT (D.) and GABAIX (J.C.). – "A New Foundation Technique Using Piles Sealed by Cement Grout Under High Pressure". *OTC*, Houston, 5-8 May 1975. Paper OTC 2310.

[6] BREWER (J.H.) and SHRUM (S.J.). – "Tension Leg Offers Steady Base at Sea". *Offshore,* Nov. 1973.

[7] GAULT (J.A.) and COX (W.R.). – "Method for Predicting Geometry and Load Distribution in an Anchor Chain from a Single Point to a Buried Anchorage". *OTC,* Houston, May 6-8, 1974. Paper OTC 2062.

[8] COLE (M.W.) and BECK (R.W.). – "Small-Anchor Tests to Predict Full Scale Holding Power". *44th Fall Meeting of SPE,* Denver, 28 Sept-1 Oct. 1969. Paper SPE 2637.

[9] PUECH (A.), MEUNIER (J.) and PAILLARD (M.). – "Behavior of Anchors in Different Soil Conditions". *OTC,* Houston, 1978, Paper OTC 3204.

[10] BEMBEN (S.M.) and KUPFERMAN (M.). – "The Vertical Holding Capacity of Marine Anchor Flukes Subjected to Static and Cyclic Loading". *OTC,* Houston, 1975. Paper OTC 2185.

[11] LIU (C.L.). – "Ocean Sediment Holding Strength *against* Breakout of Partially Embedded Objects". *Proc. Civil Eng. in the Oceans II,* Miami Beach, 10-12 Dec. 1969.

CHAPTER XI

Pipelines

CONTENTS

INTRODUCTION

1. The stability of a pipeline, which is governed by the "pipeline-soil" interaction, depends on a great number of factors, which can be classified as follows [1]:

– the profile of the soil and the geology of the site,
– the properties of the soil: bearing capacity, possibilities of scour and sedimentation,
– the dimensions and weight of the pipeline, method of laying and the initial tension of the pipeline,
– the conditions of use of the pipeline: nature of the fluids, any possible variations in temperature and the vibrations of the pipeline,
– the hydrodynamic environment: current speeds, possibilities of vertical currents near the pipeline, waves, etc.

2. The question of whether to bury the pipeline or not will be determined on the basis of reconnaissance of the soils, the topography and the hydrodynamic environment (water depth, currents, waves, etc.). Depending on the particular case, the stability of the pipeline can be ensured:

– by gravity alone (after coating the tube with concrete the thickness of which may vary) and friction against the soil,
– anchoring,
– burying in a trench excavated after laying the tube, particularly near the coast.

3. The following will be examined in the present chapter:

– the problems involved in laying and the stability of pipelines,
– reconnaissance of the seabed and soils necessary before laying,
– application of the results of soil reconnaissance to calculation of the stability or burying of the pipelines.

XI.1 GEOTECHNICAL PROBLEMS INVOLVED IN THE LAYING AND STABILITY OF PIPELINES

Depending on the solution envisaged or imposed by local conditions:

– laying the pipeline on the seabed (Fig. XI.1.1a),

– or burying the pipeline after laying (Fig. IX.1.1b),

the problems of the soil mechanics differ.

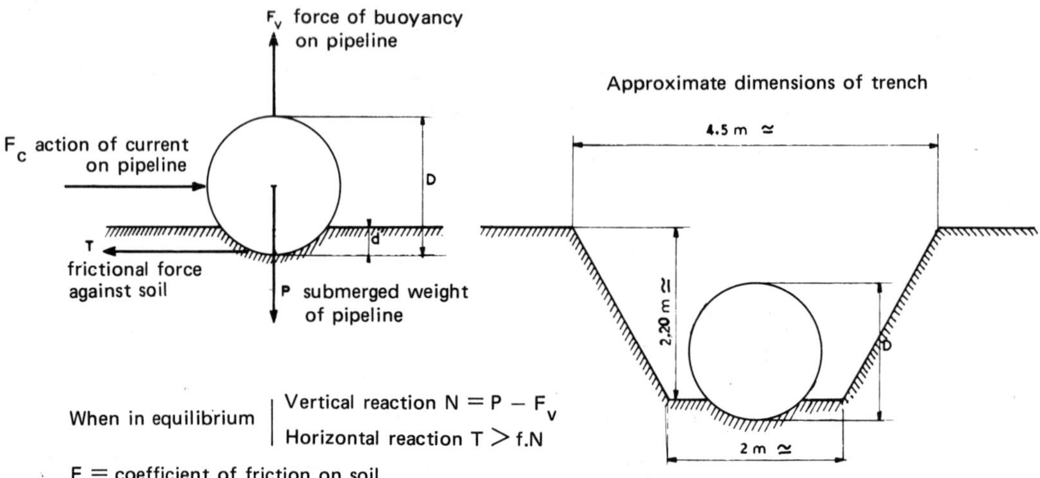

FIG. XI.1.1. – Stability of pipelines.
a) Pipeline resting on sea bottom.
b) Buried pipeline.

In the former case, the stability of the pipeline must be ensured under the action of:

– the forces of gravity,

– the lateral hydrodynamic forces (waves and current),

– the vertical bending forces caused by the span of the pipeline resulting from scour or the topography of the soil,

– the horizontal bending forces caused by curves and the initial tension in the pipeline.

In the latter case, the stability will generally be more than enough, though laying the pipeline raises problems of the shear strength of the soil:

– if the surface soil is too soft, its bearing capacity will be insufficient to enable the trenching machine to travel,

– if on the other hand soil is too highly consolidated, the trenching machine will be able to advance only very slowly.

Furthermore, the risk of liquefaction of the soils, causing the tubes to "float" must also be considered.

XI.1.1 Stability of pipelines laid on the bottom

Let us assume that the action of the hydrodynamic forces is liable to endanger the stability of the pipelines, though without analyzing the causes and extent of this action.

Verification of the vertical stability of a bottom-laid pipeline requires an examination:

– of the detailed relief of the bottom throughout the length of the route,

– of the nature of the surface formations and particularly the possible alternation of hard zones (rock or highly consolidated soils) and loose zones (where scour may occur near the pipeline),

– of the possibilities of liquefaction of the soft layers: sands, silts and soft clays.

These various characteristics will be specified by the study:

– of the detailed topography of the route,
– of the apparent nature of the soils as revealed by surface seismic prospecting,
– of the grain size distribution and platicity of the soils (Atterberg limits),
– of the risks of liquefaction of the sediments,
– of the shear strength of the soils,
– of the coefficient of friction between the soil and concrete (around the pipeline),
– of the possibilities of scour near the pipeline.

XI.1.1.1 Topography, roughness and nature of the formations

The purpose of bathymetry is to reconnoitre the profile of the possible alignment(s): valleys, hills, gradients of slopes, breaks in slopes, etc.

The morphology of the bottoms must be examined in detail:

– rock outcrops,
– boulders,
– hydraulic dunes,
– wrecks, etc.

The surface seismic prospecting will provide an initial indication of the hardness of the surface formations and in particular the degree of homogeneity that can be expected along the alignment.

XI.1.1.2 Bearing capacity and liquefaction of the soils

The bearing capacity of the soil intervenes in calculation of the vertical stability of a bottom-laid pipeline and will generally be characterized by:

– either the **shear strength** of the soil measured on cores sampled by a Kullenberg corer or by vibrocoring,

– or the value of the **limit pressure** measured with a pressuremeter.

Identification of the soils enables the possible risks of liquefaction of the sediments around the pipeline under the effect of external forces to be predicted: shocks, vibrations, wave action, etc.

XI.1.1.3 Friction of the pipeline against the soil

The **lateral** stability under hydrodynamic forces is governed by the lateral force of friction of the pipeline (coated with concrete) against the soil.

Determination of the coefficient of lateral friction $\operatorname{tg} \delta$, generally carried out on a model in the laboratory [2], could be achieved more validly by means of control samples moved in shallow water.

If the soils liquefies, the coefficient of friction tends towards zero and generally the pipeline is no longer stable.

The force of **longitudinal** friction of the tube against the soil must be estimated as exactly as possible for laying the pipeline on the ground.

The dynamic longitudinal coefficient of friction will be determined in situ by means of representative control samples.

XI.1.1.4 Scour near the pipeline

Scour near the pipeline may be harmful to both:

– the vertical stability of the pipeline, owing to the important bending loads in certain sections,

– the horizontal stability, owing to the cancelling out of the forces of friction.

The extent of scour decreases sharply as the speed of the currents decreases, i.e. as the depth increases. However, laying a structure such as a pipeline results in a local increase in the speed of the current and hence the amount of scour.

Scour is studied almost exclusively on models. However, observations (on control samples) of the erosion could well provide highly useful information.

The protection of pipelines against scour in zones which are particularly exposed is now achieved by means of sheets of plastic 4 to 5 m wide covering the pipe and secured on either side of it [3] (Fig. XI.1.2).

XI.1.2 Burying of the pipelines

Aside from problems of stability of pipelines laid on the bottom, trenching is necessary:

– if the depth of water is less than about 50 m; this will therefore be the case immediately next to coastlines,

– if rock outcrops occur over some of the path,

– if required by law (for instance trawling zones).

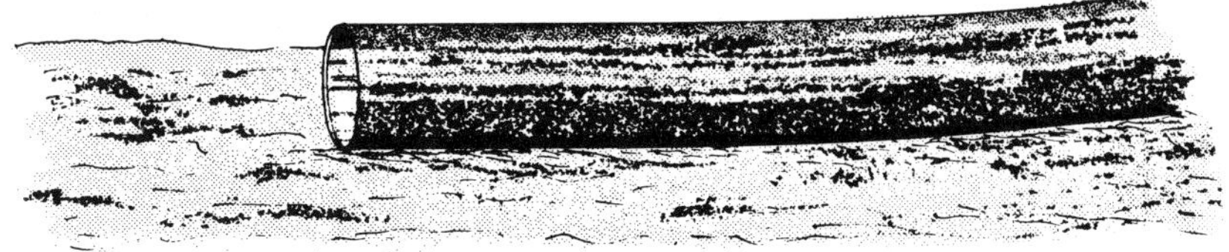
(a) No current, hence no scour

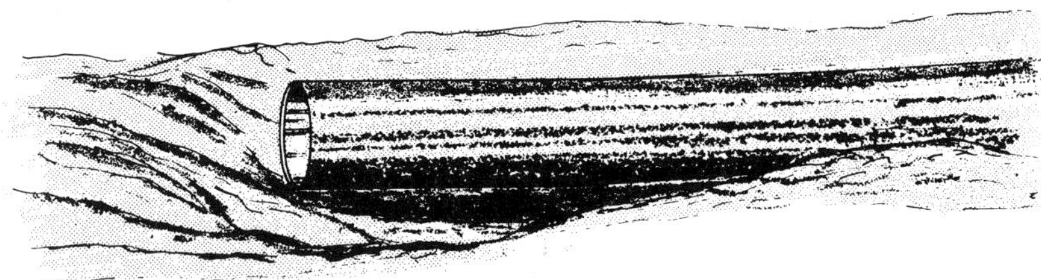
(b) Scour beneath pipe

(c) Protection against scour

FIG. XI.1.2. – Protection of pipeline against scour.
a) No current, hence no scour.
b) Scour beneath pipe.
c) Protection against scour.

Assuming burying is necessary (either completely or locally), the following should be examined:

– the possibilities of travel of the trenching machine along the route,
– the mechanical properties of surface formations,
the behaviour of the sediments which fill the trench.

XI.1.2.1 Bearing capacity of the surface soils

After the usual soil identification, the bearing capacity of the surface soils will be checked to see that it is sufficient to enable the trenching machine (crawler-mounted) to travel (see Section XI.3).

The bearing capacity will be determined:

– either by measuring the shear strength on cores,
– or by in situ measurements (penetrometer or pressuremeter).

In recent years, a great many experiments have taken place on the travel of civil engineering equipment on loose soils [4] [5].

XI.1.2.2 Depth and execution of the trench

The depth of the trench varies with that of the water and the nature of the sediments:

– from the shore down to 30-50 m of water, the depth is between 2 m in highly consolidated clays and 4 m in sands and silts,
– from 30-50 m to 100-150 m of water, the trenching depth (if any), will not generally be more than 1 to 2 m.

The shear strength, preferably measured on cores in the laboratory, will govern the choice of burying techniques employed:

– by "jetting" [6],
– by dredging (with cutter heads) [7],
– by fluidization of the sands [8],
– by ploughing [9].

XI.1.2.3 Filling of the trench

Once the tube has been laid, the trench will be filled:

– by the discharge from the burying machine,
– by natural sedimentation.

In coastal zones (depths of less than 30 to 50 m), the possibilities of scour and liquefaction of the sediments should be examined in order to counter the risks of instability, particularly as a result of currents.

In greater water depths (over 30 to 50 m), the risks of instability are less and the very need for trenching is by no means so obvious.

XI.2 RECONNAISSANCE OF THE BOTTOM AND SOILS FOR LAYING PIPELINES

As has been seen earlier (see Section XI.1), the laying and stability of pipelines on the seabed calls for the best possible knowledge of the morphology of the bottom and the nature and behaviour of the soils.

Detailed reconnaissance of the entire pipeline route must be made, covering a "corridor" about 500 to 600 m wide [10].

In addition, it would be useful to carry out a more rough and ready reconnaissance on either side of this "corridor" over a width of a few hundred metres, in order to examine the problem of anchoring the vessels and barges used for laying the pipeline.

XI.2.1 Geophysical and visual reconnaissance

XI.2.1.1 Bathymetry

The first rough route for the pipeline will be decided on the basis of bathymetric maps and the general knowledge of the morphology of the bottom.

The possible route(s) must be verified by an accurate bathymetric survey enabling detailed maps to be established featuring isobaths in 1 m steps.

The reconnaissance will cover a strip about 600 m wide centred on the theoretical route of the pipeline.

Generally, the following will also be executed:

- a central bathymetric profile along the theoretical route of the pipeline,
- two lateral profiles 100-150 m on either side of the central profile,
- two lateral profiles 300 m on either side of the central profile,
- a few transverse profiles to reset measurements and check the accuracies obtained.

In order to attain an accuracy of 1 m on the bathymetric elevations, in water depths of about 100 to 150 m, the following are necessary:

- use of precision echo sounders,
- frequent calibration in situ (influence of the currents, temperature and salinity),
- simultaneous recording of the vessel of roll, pitch and heave,
- simultaneous recording of the tides.

XI.2.1.2 Morphology (roughness)

The bathymetry has to be supplemented by a fine morphological survey, particularly in zones of considerable roughness. In addition, an attempt will be made to determine the nature of the reliefs or obstacles observed.

Two profiles will generally be recorded, set 120 m on either side of the theoretical route, by means of a side-scan sonar with an effective range of 150 m, so as to cover the entire strip of ground surveyed by the echo sounder.

The observations must make it possible:

- to identify reliefs in the range of 0.50 to 1 m,
- to distinguish changes in the lithological nature of the surface soils (back-scattered energy).

Visual observation by means of a television camera, moved along near the bottom or mounted on a submarine, for the purpose of identifying the nature of certain anomalies recorded.

This type of direct observation will however be confined to zones of considerable roughness, revealed by the previously-mentioned indirect methods (echo sounder and side-scan sonar).

The search for ferromagnetic obstacles (wrecks, anchors, cables, etc.) buried to only a slight extent (< 3 m) will be possible by means of a magnetometer with a fine resolution (one gamma) used at the same time as the side-scan sonar.

XI.2.1.3 High resolution seismics

The problems of the stability of the pipeline involve only the very upper-most layers of the seabed (a few metres). Good definition in the top layers will therefore be sought, at the cost of penetration. However, one should not forget that knowledge of the underlying geological structures improves the geologist's understanding of surface phenomena and hence makes for better interpretation.

Generally, the following will be sought:

– a penetration of about 30 m or more,
– a definition of 1 to 2 m.

These requirements can be met by various instruments in common use:

– a relatively low energy sparker giving a resolution of about 2 m and a penetration of a few tens of metres,

– a boomer (Uniboom) with a resolution of about 1.50 to 2 m and a penetration capable of reaching several tens of metres,

– sediment sounders with extremely good resolution (about 1 m), but with penetrations of from a few metres (sands) to 20-30 m (soft soils).

For instance, 3 seismic profiles will be recorded:

– the central profile,
– two lateral profiles 300 m on either side of the central profile.

Table XI.2.1 summarizes the geophysical and visual surveys of the bottom and soils required before laying pipelines.

It will be noted that the recordings advocated must be made simultaneously without interference:

– on the one hand, echo sounder and the seismic device (central profile and ± 300 m lateral profiles),

– on the other hand, the echo sounder, side-scan sonar and magnetometer (± 120 m lateral profiles).

TABLE XI.2.1

GEOPHYSICAL AND VISUAL SURVEYS OF BOTTOM AND SOILS BEFORE LAYING PIPELINES

Survey required	Techniques applicable	Survey coverage	Spacing and number of profiles	Characteristics of applicable equipment	Precision or definition
Bathymetry	Echo sounder (+ recordings: tide roll-pitch heave)	Strip 600 m wide centred on the theoretical route of the pipeline	1 central profile 2 lateral profiles to ± 120 m 2 lateral profiles to ± 300 m	Frequency 30 to 50 kHz	Precision ≈1 m
Morphology (roughness)	Side-scan sonar		2 profiles to ± 120 m from theoretical route	Frequency ≈100 kHz Range 150 m	Identification of reliefs of about 1 m and changes in facies
	Undersea television		In certain zones requiring greater detail		Identification of reliefs of about 10 cm
	Magnetometer		1 central profile or 2 lateral profiles to ± 120 m	Device with good resolving power (≈1 γ)	Identification of any metal obstacles (wrecks, anchors or cables)
Seismic prospecting (high definition)	Sparker or Bommer (Uniboom) or Sediment sounder		1 central profile 2 lateral profiles to ± 300 m	Frequency 100 to 1,000 Hz	Resolution ≈2 m
				Frequency 500 Hz to 4 kHz	Resolution 1.5 to 2 m
				Frequency 3 to 9 kHz	Resolution ≈1 m

XI.2.2 Geotechnical reconnaissance

The essential purpose of reconnaissance of the surface soils before laying a pipeline is:

– to identify the nature, grain size distribution, etc. of the soils,

– to determine their liquefaction and scour properties which would result of in loss of stability.

The extent of the zone to be surveyed is identical to that defined for geophysical reconnaissance.

The depth required will not be more than a few metres (from 3 to 5 m) and may depend on the trenching depth intended, which is generally greater the nearer the coastline.

XI.2.2.1 Surface samples and coring

The frequency of core sampling along the route of the pipeline will depend on geological criteria based on the results of the seismic survey. In general the following values are accepted:

– 1 core per kilometre of route,

– or 1 core every 2 to 3 km in zones where the seismic profiles reveal no irregularity.

The techniques applicable differ with the nature of the surface terrains:

– gravity or stationary piston (Kullenberg) type corers are used the most widely enabling penetration of 2 to 5 m or more to be reached in loose sediments (loose sand or soft clays); although disturbed, the cores are still sufficiently representative for the common measurements to be made (identification and estimation of the shear strength of the soils),

– vibrocorers (electric or electrohydraulic) are required for compact sediments (dense sands or consolidated clays),

– submerged rotary coring devices are needed for reconnaissance of rocky bottoms where knowledge of the shear strength is indispensable in order to assess the possibilities or rock clearance and then burying.

Laboratory identifications and determination of the friction of the pipeline against the bottom (soil-concrete friction) etc., require the sampling of considerable quantities of disturbed sediments, for instance by dredging (grab sampler, etc.).

XI.2.2.2 In situ measurements

The geotechnical characteristics needed for calculating the stability of pipelines are generally determined in the laboratory on samples subjected to a varying amount of disturbance.

However, particularly with sands, calculation of the bearing capacity essentially depends on the density in place, which is impossible to attain on samples which are always disturbed. In the case of sands, in situ measurements willl always be necessary, and are always desirable whatever the nature of the sediments.

Among simple, fast techniques for implementation, the following can be adopted:

– the pressuremeter released by a Kullenberg type device. It will however be observed that the correlation between the limit pressure p_l and the density in place is not always evident, which excludes satisfactory estimation of the internal angle of friction,
– the dynamic penetrometer, which can also be released by a similar device, though no truly operational model is yet available,
– the static penetrometer (light device).

Table XI.2.2 summarizes the geotechnical surveys that may possibly be made before laying pipelines, depending on the nature and consolidation of the surface formations.

XI.2.3 Laboratory tests

It is not proposed to repeat what has already been said with respect to common laboratory measurements (see Chapter I):

– identification of the soil,
– determination of the shear strength according to various methods.

We shall confine ourselves to an examination of the measurements and tests of specific nature which it is recommended to perform on the samples taken.

XI.2.3.1 Liquefaction water content of the soils

In the case of silts and clays, by extrapolating the curve obtained using the Casagrande device (for the liquidity limit), it is possible to estimate the water content corresponding to liquefaction of the soil, i.e. the condition of zero shear (Fig. XI.2.1).

It is sometimes considered that this water content is that corresponding to 0.01 blow on the Casagrande device.

The density of the liquefied soil is then determined and the risk of the pipeline floating verified.

In the case of sand, liquefaction could appear under the effect of vibrations of the pipe or cyclic loads (under wave action). It can however be assumed that the risk exists only in very loose sands with grain size distribution of less than 0.3 – 0.4 mm and of relatively low permeability.

XI.2.3.2 Coefficients of friction of a concrete-coated pipe against the soil

A pipe laid on the soil is pulled:

– either longitudinally, to measure the friction involved in laying the pipeline near the coast (landing),
– or transversaly, to measure the friction to the lateral displacement of the pipeline on the bottom, under the action of the elements.

TABLE XI.2.2

GEOTECHNICAL RECONNAISSANCE OF SOILS BEFORE LAYING PIPELINES

Operation required		Coverage of zone to be surveyed	Number of soundings	Penetration necessary (m)	Techniques applicable	Performance	Results sought
Core samples	Low consolidation of soils or sands	Strip 600 m wide centred around the theoretical route of the pipeline (The core samples and in situ measurements will be located in the light of the results of seismic prospecting)	1 coring every 1 or 2 km (dependent on geological criteria)	2 to 3	Kullenberg type corer Vibrocorer	Not usable in highly consolidated soils Applicable in consolidated soils	Identification of the soils
	Rocky bottoms		1 coring per km (dependent on geological criteria)	5 to 6	Submerged rotary corer	Penetration in very hard soils and rock	Determination of the mechanical characteristics
In situ measurements occasionally in low consolidated soils or sands			1 test per 2 or 3 km (dependent on geological criteria)	2 to 3	Pressuremeter possibly released by Kullenberg device or Penetrometer		Of advantage particularly in sands

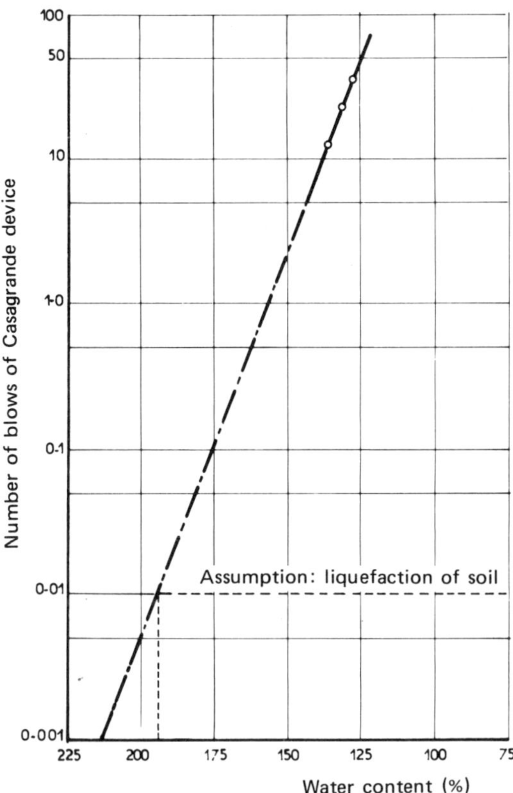

FIG. XI.2.1. – Determination of the liquefaction water content of the soil.

Figures XI.2.2 and XI.2.3. show the results of longitudinal and transversal friction tests on calcareous sand, carried out by *IFP* and *CNEXO* (1), with different tubes:

- steel pipelines,
- fibro-cement tubes,
- flexibles tubes.

Furthermore, a distinction must clearly be drawn:

– between the maximum coefficient of friction f_{lim} to be overcome to start moving the tube,

– and the dynamic coefficient of friction f_d ($< f_{\text{lim}}$) as the tube continues to move against the soil.

The values of the coefficients of friction measured generally reveal scatter from one site to the next. This scatter can be explained at least partly by the disturbance of the soil used and the operating conditions, which probably differ considerably.

Figure XI.2.4 shows a number of values of the coefficient of soil-concrete friction f_{lim} in terms of the density and nature of the sediments.

(1) *Institut Français du Pétrole* and *Centre National pour l'Exploitation des Océans.*

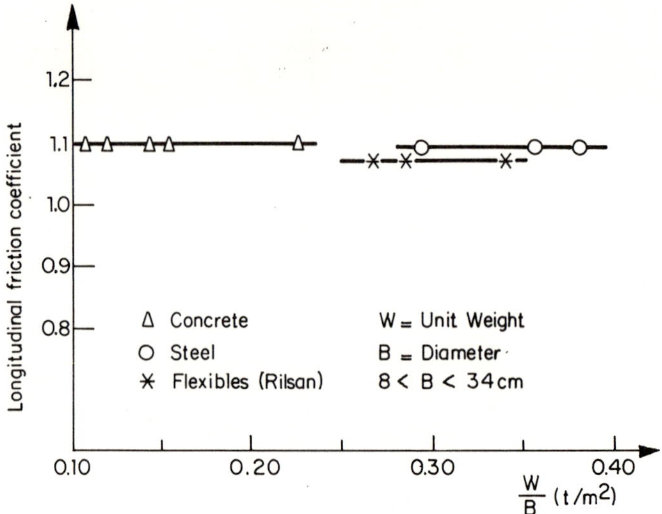

FIG. XI.2.2. – Longitudinal friction of different pipelines on calcareous sand (*IFP-CNEXO* results).

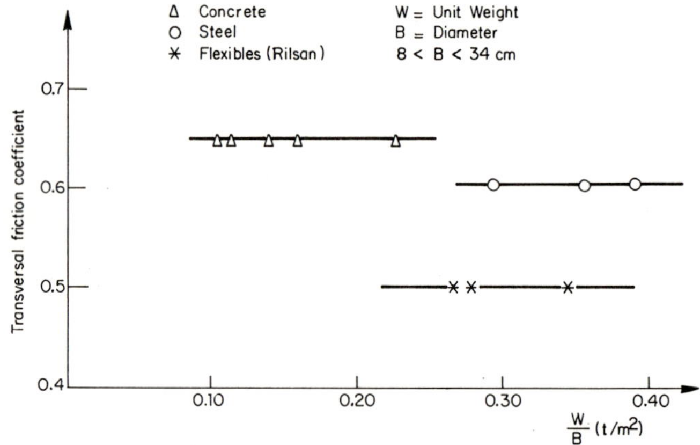

FIG. XI.2.3. – Transversal friction of different pipelines on calcareous sand (*IFP-CNEXO* results).

The values of the coefficients of friction obviously differ with the diameter of pipe used. The values given in Table XI.2.3 reveal considerable observed variations of f_{lim} and f_d with the diameter of pipes resting on a sand with a grain size distribution of 0.4-0.8 mm.

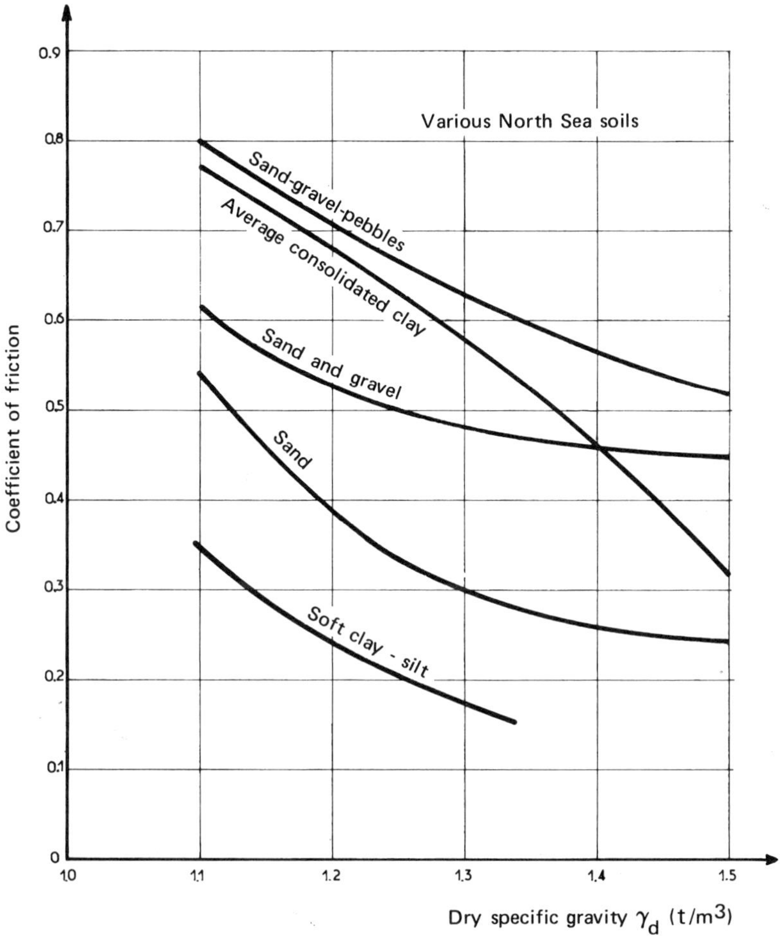

FIG. XI.2.4. – Coefficient of friction of pipelines against soil.

TABLE XI.2.3

Diameter of pipe	Coefficient of friction	
	Maximum f_{lim}	Dynamic f_d
1″(2.5 cm)	{ 1.06 1.34	0.76
9–16″(23–40 cm)	{ 0.66 0.71	0.39

Clearly, all these figures are of only relative value, depending on the characteristics of the soil and also the experimental conditions. Reality would probably be better approached by measurements made on full scale control samples.

XI.2.3.3 Rate of erosion of sediments

The rate of erosion of sediments is one of the highly complex characteristics of scour. This rate may be measured in the laboratory, thus providing at least one item of information of comparative value, depending on the nature of the surface sediments.

The observations and in situ study of actual erosion phenomena are of the greatest utility.

Table XI.2.4 summarizes the tests to be conducted in the laboratory and in situ before laying pipelines.

TABLE XI.2.4

MEASUREMENTS ON CORES AND/OR IN SITU TO BE MADE BEFORE LAYING PIPELINES

	Characteristics of soils measured	Observations
Core analyses in the laboratory	**Identification of the soils**: Nature; Atterberg limits (W_L, W_P); Grain size distribution	On clayey soils
	Water content w Density γ	Directly aboard the vessel
	Shear strength Undrained shear strength: Unconfined compression; Plane; Pocket penetrometer; "Fall cone"	Immediately aboard the vessel
	Liquefaction (on the basis of identification tests: liquidity limit)	In the laboratory
	Coefficients of friction of the pipe against the soil: transversal; longitudinal	
Pressuremeter Penetrometer (occasionally)	Limit pressure p_l Tip resistance R_p Lateral friction f	

XI.3 CALCULATION OF STABILITY AND PIPELINE BURYING PROBLEMS

From the geotechnical standpoint, a distinction should be drawn:

– between study of the stability of the pipeline laid on the bottom (Fig. XI.1.1a),
– and soil mechanics problems involved in trenching the pipeline (Fig. XI.1.1.b).

XI.3.1 Calculation of the stability of a bottom-laid pipeline

XI.3.1.1 Bearing capacity of sandy and clayey surface sediments

The bearing capacity of a foundation soil can be expressed by the following well-known general formula:

$$q = \frac{1}{2}\gamma' B N_\gamma + c_u N_c + \gamma' \cdot d \cdot N_q$$

where:

q = maximum bearing capacity,
γ' = submerged density of the soil ($\gamma' = \gamma - 1$),
B = width of foundation,
d = depth to which foundation is buried,
c_u = cohesion of soils,
N_c, N_γ, N_q = dimensionless coefficients depending on the angle of friction φ of the soil.

The pipeline can be assimilated with a mat of infinite length and a width B such that:

$$B = 2\sqrt{d(D - d)}$$

where:

D = diameter of the pipeline,
d = depth to which pipeline is buried,
B = width at soil level, where $B \leqslant D$.

In the case of a pipeline, the foundation may break up locally. Under these circumstances, Terzaghi recommends, in order to calculate the bearing capacity, that a value of cohesion c'_u and an angle of friction φ' be adopted such that [11]:

$$c'_u = \frac{2}{3} c_u$$

$$\operatorname{tg} \varphi' = \frac{2}{3} \operatorname{tg} \varphi$$

The previous formula can then be written:

$$q = \frac{1}{2}\gamma' BN'_\gamma + \frac{2}{3}c_u N'_c + \gamma' dN'_q$$

where N'_c, N'_γ and N'_q are dimensionless coefficients dependent on the internal angle of friction φ of the soil (Fig. XI.3.1).

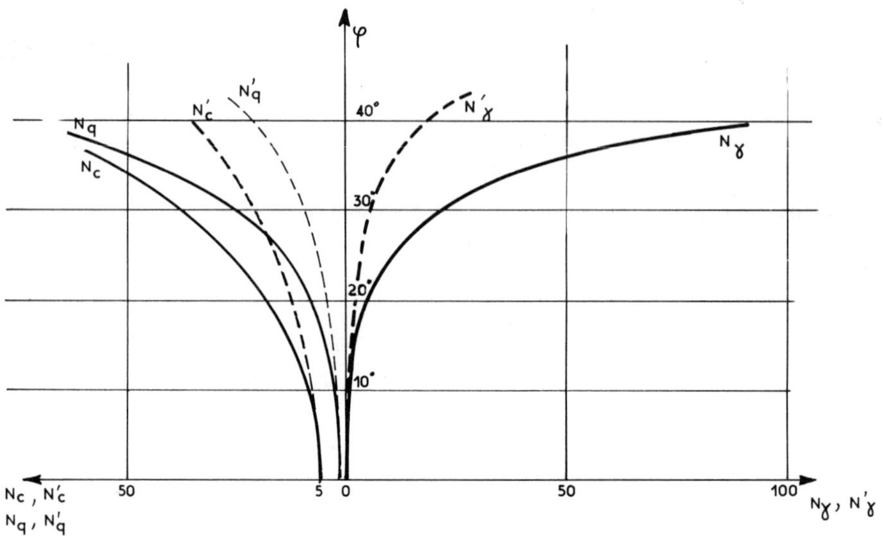

FIG. XI.3.1. – Nomogram giving the coefficients of bearing capacity in terms of the angle of friction φ (according to Terzaghi).

In the case of **sands**, where the cohesion c_u is zero, we have:

$$q = \frac{1}{2}\gamma' BN'_\gamma + \gamma' dN'_q$$

In the case of **clayey soils** (muds, clays), where $\varphi = 0$, $N'_c \approx 5$, $N'_\gamma = 0$, and $N'_q = 1$, we have:

$$q = 3{,}3\, c_u + \gamma' d$$

The linear bearing capacity of the soil beneath the pipeline can be defined by the equation:

$$Q = Bq$$

XI.3.1.2 Vertical stability of the pipeline

The vertical stability of a bottom-laid pipeline is defined by the two relationships:

$$\begin{cases} Q > N \\ N > 0 \end{cases}$$

where:

Q = linear bearing capacity of the soil,
$N = P - F_v$,
P = linear submerged weight of pipeline,
F_v = vertical force (due to the action of currents, waves, etc.) tending to lift the pipe.

Provided the requirement that $N > 0$ is satisfied, the stability of the pipeline can be characterized by the depth at which it is buried into the soil.

For example, consider a 32-in (80 cm) pipeline coated with a thickness of 12 cm of concrete, the linear submerged weight of which P is about 900 kg/m.

In sands, it can be verified that the depth d at which the pipeline is buried is invariably small and at all events less than 25 cm (Fig. XI.3.2a). The pipeline remains vertically stable in the absence of scour or liquefaction.

In muds or soft clays, the pipe is completely buried if the cohesion c_u is less than about 0,2 t/m^2 (2 kPa) (Fig. XI.3.2b). The vertical stability therefore depends on the layers underneath.

In **very soft clays** (or) generally in **liquefied soils**, the pipeline will float if the buoyancy is in balance with its weight. The force of buoyancy per unit volume of the tube is calculated in the same way as the unit weight of the liquefied soil, such that [12] [13]:

$$\gamma = \frac{G(1 + w_l)}{1 + G w_l}$$

where:

γ = unit weight of soil (density),
G = unit weight of matrix of grains (in general ≈ 2.7),
w_l = water content of soil on liquefaction (determined for instance as shown in Paragraph XI.2.3.1).

For example, if $w_l \approx 70\%$, the density of the liquefied soil $\gamma \approx 1.6$ t/m^3.

The pipeline will be prevented from rising (where $N > 0$):

– either by coating with concrete; if the fluid circulating in the pipeline is gas, the minimum equivalent density must not be less than 1.10,

– or by anchoring the pipeline; the distance between anchors must be determined in the light of the nature and profile of the soil on the one hand, and the estimation of the vertical buoyancy thrust on the other.

XI.3.1.3 Horizontal stability of the pipeline

The horizontal stability of the untrenched pipeline subject to the action of hydrodynamic forces is defined by the relationship:

$$T > F_c$$

where:

$T = fN$ = force of friction of the tube against the soil,
F_c = action of currents on the pipe.

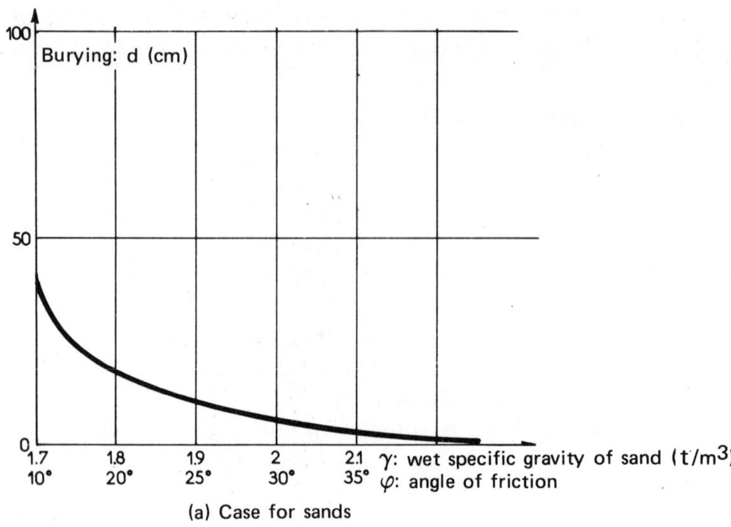

(a) Case for sands

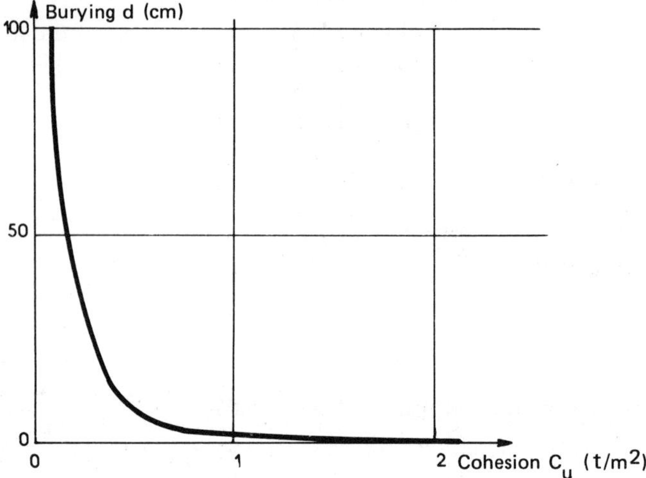

(b) For muds and soft clays

FIG. XI.3.2. – Example of calculation of the burying of a pipeline ($D = 1$ m ; weight = 910 kg/m).

The pipeline will be rendered horizontally stable by:

– the friction of the pipeline against the soil. In actual fact, determination of the coefficient f is highly imprecise (see Section XI.2) and moreover differs, depending on whether the direction is transverse or longitudinal. Generally speaking, f decreases as the compaction of the sediments increases (Fig. XI.2.2). In the case of clayey soils in particularly, friction f improves with time when there is no scour,

– the possible sedimentation as a result of currents. Ultimately, natural scour of the pipeline occurs (not allowed for in the calculation of the vertical stability),

– the anchoring of the pipeline at intervals that vary with the speed of the currents but that generally lie between 10 and 30 m in zones where the current is swift (Fig. XI.3.3) [14] [15].

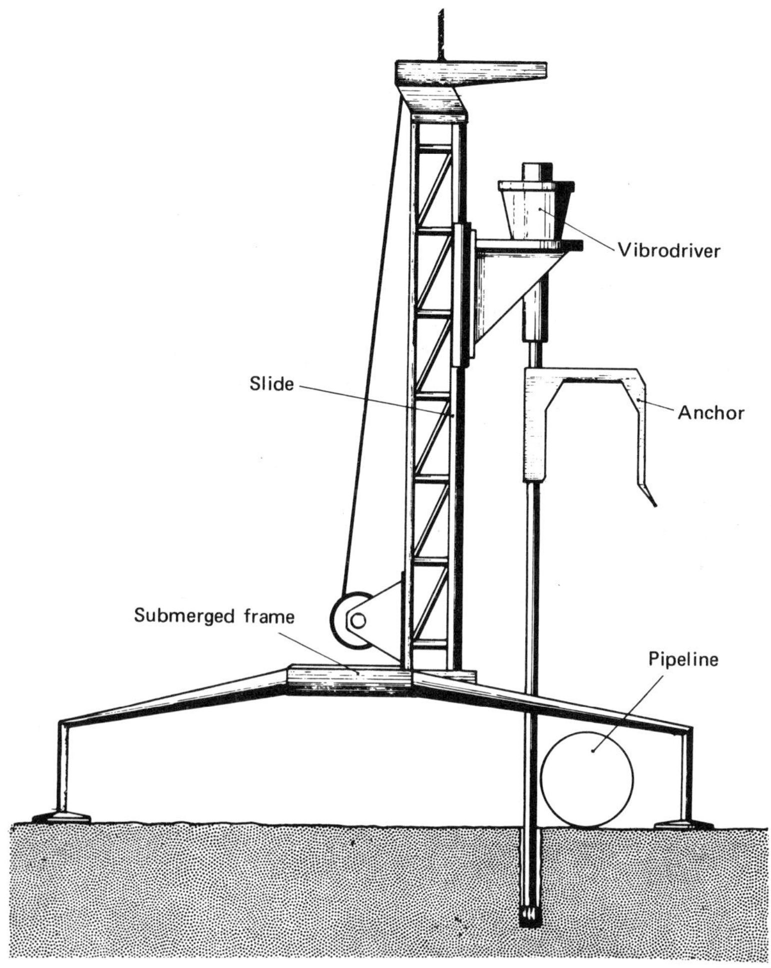

FIG. XI.3.3. – Anchorage of a pipeline *(TLM)*.

The circumstances critical to the stability of the pipeline are essentially:

– no contact with the soil as a result of the local topography. A radius of curvature less than a certain critical value R_c may cause the pipeline to break. For a 32-in (80 cm) pipe coated with 12 cm of concrete, the minimum radius of curvature that can be tolerated is about 630 m,

– liquefaction of the soil as a result of shocks or vibrations resulting either from the laying operation or from the movements of the pipeline. In can be noted that liquefaction will probably last only a short time and that, as soon as the vibrations cease, the soil will gradually recover a certain amount of shear resistance.

XI.3.2 Burying of the pipeline

The choice of the burying technique to be applied is related to the mechanical characteristics of the soil.

XI.3.2.1 Travel of the burying machine

The minimum bearing capacity of the surface soils compatible with travel of a trenching machine must be at least about 2 t/m² (20 kPa):

- the shear resistance of even loose sands generally meets this requirement,
- the cohesion of clays or muds must be above about 0.5 t/m² (5 kPa).

This requirement is not necessarily verified in view of the nature of some bottoms and possibilities of the sediments liquefying as a result of the movement of the machine.

Kinematic study of the travel of tracked machines on sandy or clayey bottoms is an extremely complex problem. We shall confine ourselves at the present juncture to a few practical conclusions.

For the same pressure on the soil, if the width of the track is increased [16]:

- the maximum possible driving torque falls,
- the settlements of the soil rise.

The coefficient of friction tg δ of the track on the soil is given by [5]:

$$\text{tg}\ \delta = \frac{T}{W}$$

where:

T = hook traction,
W = weight of machine.

In clayey soils tg δ is a function of:

$$\Delta c_u = c_u - c_{u\ \text{min}}$$

where:

c_u = cohesion of soil considered,
$c_{u\ \text{min}}$ = minimum cohesion needed to enable machine to travel.

Table XI.3.1 gives a few mean values of tg δ in terms of Δc_u for soil pressures of from 2 to 3.5 t/m² (20 to 35 kPa).

However, the coeffficient of friction tg δ of the track on the soil cannot increase beyond a maximum value:

$$(\text{tg}\ \delta)_{\text{max}} = \left(\frac{T}{W}\right)_{\text{max}} = \frac{P}{W} \cdot \frac{1}{v_t}$$

where:

P = power of engine,

v_t = relative speed of track with relation to soil (about 2 m/min for a true speed of the trenching machine of 100 m/h, allowing for a slip of about 20%).

TABLE XI.3.1

$\Delta c_u = c_u - c_{u\,min}$ (t/m²)	tg δ	
	Pressure on soils (2 t/m²)	**Pressure on soils** (3.5 t/m²)
0	0	0
0.5	0.20	0.15
1	0.35	0.25
2	0.55	0.40
> 3	tg δ_{max} : 0.65	0.50

XI.3.2.2 Choice of burying techniques

Depending on the nature and shear resistance of the soil, the trench will be excavated:

– either by jetting, particularly in the case of loose sediments, the shear resistance of which is less than 2-3 t/m³ (20-30 kPa). The pressure of the water jet will reach 175 to 200 kg/cm² (17.5 to 20 MPa) and the installed capacity on the barge 40,000 to 50,000 hp. The laying rate will be about 120-150 m/h (i.e. 3 to 4 km a day),

– or by dredging (with cutters) in the case of consolidated soils or rocks in particular. The rate of advance is much slower and will be no more than 60-75 m/h (i.e. 1.5 to 2 km/day),

– or by fluidization (liquefaction) of the sandy soils,

– or again by ploughing in the case of morainic soils.

Empirical formulae make it possible to estimate the loads to be overcome and hence the dimensions and ratings of the trenching machines to be estimated from the density and shear resistance of the soil [6].

The machine is equipped with a discharge pump which:

– disperses the sediments laterally; burial then occurs more or less rapidly depending in current action,

– or discharges them about 50 to 100 m behind in order to bury the pipeline already laid in the trench.

In coastal zones (depths less than about 30 to 50 m), the problem of filling the trench must be determined from the standpoints of consolidation, liquefaction and erosion of the sediments. Methods of calculation [17] enable the nature, grain size distribution and optimum thickness of the trench-filling materials to be determined.

BIBLIOGRAPHY

[1] REESE (L.C.) and CASBARIAN (A.O.P.) – "Pipe Soil Interaction for a Buried Offshore Pipeline" *43 rd Fall Meeging of SPE*, Houston Sept. 29-Oct. 2, 1968. Paper SPE 23-43.

[2] LYONS (C.G.) – "Soil Resistance to Lateral Sliding of Marine Pipelines" *OTC*, Houston, April 29-May 2, 1973. Paper OTC 1876.

[3] *Seditech Data Sheet.*

[4] WIENDIECK (K.W.) and FREITAG (D.R.). – The Role of Ground Crawling Vehicles in the Ocean". *Civil Eng. in the Oceans II*, Miami, Dec. 10-12, 1969.

[5] Collection of research reports edited by BIAREZ J. – Supplement to a course given at *Ecole Centrale des Arts et Manufactures*, Paris, Volume 2.

[6] REYNOLDS (J.) – "Pipe Trenching and Burial". *Joint Conference on Pipelining in the North Sea.* London, 9 april 1975.

[7] BANZOLI (V.), DI TELLA (V.), DOSSI (L.) and GAVA (P.). – "New Concept of Underwater Remote Controlled Tracked Vehicle for Deep Water Trenching Operations". *OTC*, Houston, May 1976. Paper OTC No 2587.

[8] BOOM (J.) – "Fluidization in Pipeline Burial" *SPE Spring Meeting.* Amsterdam, 8-9 April 1976. Paper SPE No 5764.

[9] BROWN (R.J.). – "Examining New Pipe Burial Methods", *Offshore*, April 1978, pp. 68-75.

[10] KUHN (H.). – "Etude du sol pour la pose des pipelines". Personal communication, April 1975.

[11] TERZAGHI (K.) and PECK (R.B.) – *Soil Mechanics in Engeneering Practice,* John Wiley and Sons, New-York.

[12] BONAR (A.J.) and GHAZZALY (O.I.). – "Research on Pipeline Flotation" *J. of the Transportation Eng., Div.*, May 1973.

[13] GHAZZALY (O.I.) and SUNG JOO LIM. – "Experimental Investigation of Pipeline Stability in Very Soft Clay", *OTC*, Houston, May 5-8, 1975. Paper OTC 2277.

[14] SHORT (T.A.) – "Pipeline Anchoring System". *Offshore,* June, 1967.

[15] *TLM Data Sheet* on pipeline anchoring system.

[16] CAQUOT (A.) et KERISEL (J.) – *Traité de mécanique des sols.* Gauthier-Villars, Paris, 1966.

[17] CHRISTIAN (J.T.), TAYLOR (P.K.), YEN (J.K.C) and ERALI (D.R.) – "Large Diameter Underwater Pipeline for Nuclear Power Plant Design against Soil Liquefaction." *OTC*, Houston, May 6-8, 1974. Paper No. OTC 2094.

ACHEVE D'IMPRIMER
EN AVRIL 1979
PAR L'IMPRIMERIE LOUIS JEAN
05002 - GAP
N° d'éditeur 444 – N° d'impression 193
Dépôt légal : 2e trimestre 1979

IMPRIME EN FRANCE